DISTANT SIGNALS

How Cable TV Changed the World of Telecommunications

BY THOMAS P. SOUTHWICK

A PRIMEDIA Intertec Publication
Overland Park, KS 66212-2216

Distant Signals: How Cable TV Changed the World of Telecommunications
© 1998 PRIMEDIA Intertec
All rights reserved

1st Printing, January, 1999

Published by
PRIMEDIA Intertec
9800 Metcalf Ave.
Overland Park, KS 66212-2216

ISBN 0-87288-702-2
Library of Congress Card Catalog Number: 98-89927

Table Of Contents

Acknowledgements VII
Introduction IX
Chapter 1 — Pioneers 1
Chapter 2 — 'A Hell Of A Business' 23
Chapter 3 — The Big Players Arrive 45
Chapter 4 — The Dark Ages 69
Chapter 5 — The Satellite Breakthrough 95
Chapter 6 — The Programmers 125
Chapter 7 — Growth 157
Chapter 8 — Deregulation 187
Chapter 9 — Golden Age 221
Chapter 10 — Regulation 257
Chapter 11 — The Dawn Of Digital 291
Chapter 12 — A Bright Future 325

For Christena

Acknowledgements

This work would not have been possible without the collections of oral histories and other material at the National Cable Television Center and Museum in Denver, Colorado, whose president, Marlowe Froke, was unfailingly generous in sharing all these resources with me. I also made use of the reports in such industry publications as *Television Digest, Broadcasting, Cable TV Business, CableVision,, Channels, Multichannel News* and *Cable World*. Reports from such consumer publications as *Business Week, Forbes, Fortune, The Wall Street Journal, The New York Times, Congressional Quarterly, Time* and *Newsweek* were also helpful. I am also grateful to nearly 100 cable industry executives who spent time sharing their recollections and their records with me. The original series of articles on the history of cable in *Cable World* was made possible by the sales efforts of Donna Briggs and her colleagues and by the editing skills of Peggy Conger and Matt Stump. Stewart Schley, a former editor of *Cable World*, was kind enough to give the manuscript a final read and make valuable changes.

On a personal level I am grateful to Peter Barton and Steve Cunningham who encouraged me to undertake this project, and to the senior management at Intertec Publishing— particularly Larry Lannon, Cameron Bishop and Debby Bounds— who made the book version happen. Paul Kagan, Larry Gerbrandt and Paul Maxwell all helped me in my career as a cable trade journalist/publisher.

I also owe a great debt to my father, Paul Southwick, who taught me how to write "as if I were telling somebody the story out loud," to my mother, Susan Heider Southwick, who instilled in me the discipline to stick with such a long undertaking and to my grandfather, Werner Heider, who passed along to me his lifelong love of history.

My wonderful children – Grant and Caitlin – were encouraging and interested in the project and patient during the times I was away for research and interviews. Most of all I am grateful to my friend, my partner, my editor, my best critic and strongest supporter, my wife Christena, who always believes any dream can come true.

Credits: Cover photography by David Cornwell; design by Caitlin Southwick; production by Publication Design Inc. Satellite picture courtesy of the National Cable Television Center and Museum.

This book was based on a series of articles that appeared in Cable World Magazine and was made possible by the following companies:

A&E Networks
Adelphia
Barco
Bresnan
Cablevision
CBS Cable
Comcast
CommScope
Cox Communications
C-Span
Discovery Communications Inc.
Communication Trends, Inc.
Disney Channel
Encore Movie Network
Food Network
Game Show Network
General Instrument
Golf Channel
Harron Communications
HBO
HGTV
Home Shopping Network
MediaOne
MTV Networks
NBC Cable
Outdoor Life Network/Speedvision
Paul Kagan Associates, Inc.
Philips Broadband Networks Inc.
Playboy TV
Portin Parker Consulting Inc.
Prevue Network
Rainbow Programming Holdings, Inc.
Renegade Productions
Request
Showtime Networks, Inc.
TCI
The Weather Channel
Times Fiber Communications, Inc.
Time Warner Cable
Trinity Broadcasting Network
Turner Networks
TVKO
USA Networks
Viewer's Choice
WCW
WWF
YCTV
ZDTV

Introduction

Television, during the second half of the 20th century, changed the world more profoundly than any invention since the printing press. In just 50 years TV revolutionized government, diplomacy, warfare, education, economics, family structures, entertainment and nearly every other aspect of life.

Yet television did not reach its full potential until the development of cable and its related business, satellite-distributed programming. Broadcast television was limited in the audience it could reach and the programming it could deliver. Its economics drove it to offer the lowest-common-denominator programming to audiences in the biggest population centers.

Broadcast television delivered nothing at all to those outside the range of its signals and very little choice for anybody with interests beyond basic sitcoms or dramas.

The advent of cable changed that. It enabled television to reach a universal audience with a wide array of programming.

The first printing presses churned out books in Latin, a language most people did not speak. Not until printers began to produce works in the vernacular languages of the masses did the technology reach its full potential.

Cable played the same role for television, allowing it to reach a much wider audience with a much richer array of offerings.

The first cable system began service in 1948. This is the story of how, in just 50 years, the industry grew from modest, home-made operations into a business delivering hundreds of channels of programming into 68 million television homes in the U.S. and millions more around the world. And it is a story of how the business changed not only television but the entire world of telecommunications from telephony to Internet access.

Although cable was first invented in Britain and a group of Canadians made a major contribution to the development of the industry, the story is a quintessentially American tale. Its cast of characters includes small-town entrepreneurs and big Wall Street players, backyard inventors and high-tech firms, Presidents and city council members, crackpots and visionaries. Many of the early entrepreneurs were from small towns such as Tuckerman, Ark.; Astoria, Ore.; or Casper, Wyo.

The systems they invented and nurtured were quickly discovered by big companies, some interested in making and selling equipment, others aiming to extend media empires into new territory. Later they were joined by a band of programmers who believed they could make green what Federal Communications Commission Chairman Newton Minow in the 1960s called the "vast wasteland" of television.

Along the way they were tossed by the winds of economic and political change. Sometimes they sailed with the wind, sometimes it nearly capsized them. They faced the opposition of some of the most powerful interests in the nation: the broadcast television networks, the phone companies and various levels of government.

At times they were their own worst enemies, taking advantage of a near-monopoly status to raise rates, ignore customer service and build themselves one of the worst public images of any business in the nation.

Always they were driven forward by the rewards that came to those who fed America's insatiable appetite for television. In a single decade (1948-1958) the number of American homes with television sets grew from less than a million to more than 42 million.

Out of all this change and turmoil, the founders of the cable business created billions of dollars in wealth. And they also gave birth to a communications system more powerful than any known previously on the planet.

Pioneers

The pale November sun had already dropped behind the Front Range of the Rocky Mountains when Bill Daniels pulled into Murphy's Restaurant in Denver for dinner. He was cramped and bored after the long drive up from Hobbs, N. M. It was 1952.

Daniels was on his way to the oil fields of Wyoming to start his own insurance business.

At five feet, seven inches, with blue eyes and dark hair, Daniels had the straight back, square jaw, and cocky air of a fighter pilot. He also had the flattened nose of a boxer unafraid to take a punch. He was both.

While at New Mexico Military Institute before World War II he won the state Golden Gloves boxing title. He would have gone further had he not broken his hand on an opponent's head during the regional championships.

He turned down an appointment to the U.S. Naval Academy to sign up as a Navy fighter, making $187.50 a month. Two weeks before he completed flight school, the Japanese attacked Pearl Harbor. Daniels saw duty as a fighter pilot in North Africa and some of the grittiest fighting in the South Pacific. In the waning days of the war, he won the Bronze Star for heroism when a kamikaze plane hit his carrier, the U.S.S. Intrepid, and Daniels risked his own life to carry wounded shipmates to safety. He was called back for service in the Korean War.

By 1952, his days in the boxing ring and fighter cockpit were behind him. Like millions of others in his generation, he had returned from war to find there was no longer a place for him in the family business and that the old home town seemed awfully small and tame after dodging flack and downing Zeros. Even a stint in the Blue Angels didn't give him the kick he was looking for. Nobody shot at the Blue Angels.

Daniels was restless, itchy, looking for something exciting, something big, something quick. He was looking for something that would ensure he would never face the fate of his father who had died five years before, broken financially and emotionally by the Depression.

Daniels found it in the bar in Murphy's Restaurant.

As he walked in, the ex-Golden Gloves champ saw a device showing a live boxing match. The show was "Wednesday Night Fights," and the device Bill Daniels was watching for the first time in his life was a television set. He was hooked: "Most people you meet today, they were raised on television. But I didn't even see television until I was 32, and I thought 'My God, what an invention this is.'"

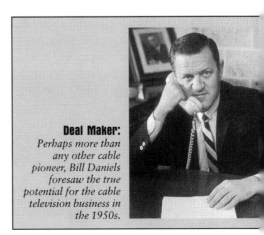

Deal Maker:
Perhaps more than any other cable pioneer, Bill Daniels foresaw the true potential for the cable television business in the 1950s.

Pioneer Brothers:
Daniels found the technical yin to his business-minded yang in two young Texas engineers, Gene and Richard Schneider.

He drove on up to Casper, Wyo., and opened up his oil insurance business. But he kept returning to Murphy's for the Wednesday night fights. And he kept thinking there must be a way to bring this magical device into Wyoming, which had no television at all. "I couldn't get it out of my mind. How do you get that great invention to a small town that didn't have any television stations?"

An article in one of the Denver newspapers caught his eye. It described a new enterprise that had been doing just what he had in mind: bringing television into areas beyond the reach of broadcast signals. It was called cable television.

He went back to Casper and brought together a group of oil men to raise the money to build a cable system for Casper. Each of them put up $5,000 and borrowed additional funds from the local bank. Daniels had $250,000 to work with.

But he had no idea how to build a cable system and needed some engineers. One of his partners recommended his two nephews, Richard and Gene Schneider of Texas, both trained engineers, both ex-GIs and both looking for something exciting, something big and something quick. Daniels called them up and talked them into getting into the cable television business.

They wouldn't be the last to start down the road to multi-million dollar fortunes by answering a phone call from the fast-talking insurance salesman from Hobbs.

The two brothers came to Casper and, with the help of a broadcast engineer named Tom Morrissey, began to construct the first cable system in the state. The Schneider brothers provided the laconic, methodical, technical yin to Daniels' voluble, shoot-from-the-hip and financial-minded yang. It was a complementary mix. Gene headed up the operations side and Richard was chief engineer.

Before anybody could watch TV in Casper, there had to be a signal. Cable systems up to that point had simply built a big tower, placed an antenna on top and then run wire to all the homes in the town. But the Denver signals couldn't reach Casper, even with a huge tower. The farthest they came was Laramie, 150 miles to the south.

So Daniels and the Schneider brothers, now operating as Community Television Systems of Wyoming Inc., went

---

*AT&T charged $7,800 a month to microwave a single broadcast channel into Casper, by far the largest recurring expense for the cable company. When additional stations launched in Denver, Casper residents wanted them as well, but AT&T wanted even more money to relay additional broadcast signals. So the cable company allowed subscribers to vote on which station's programs they wanted to watch on the single-channel cable system. A cable system employee would then sit up on the mountain at the microwave relay station and literally switch channels for the entire town. From 7:30-8 p.m. on a given night, the system might be delivering Denver's channel 2. At 8 p.m. it might switch to channel 4 and an hour later to channel 7.

to AT&T and persuaded the giant telephone company to relay Denver channel 2's signal from a tower in Laramie via microwave hops. It was the first attempt in the nation to relay television signals to remote areas via microwave. Before AT&T agreed to do it, the telco had to have a check for $125,000. Daniels plunked it down before he had a single subscriber or any concrete evidence that his business could be successful.*

By the middle of 1953, Daniels' company was stringing wire and signing up its first subscribers, at a hefty $150 installation fee and $7.50 per month for eight hours a day of single channel, black-and-white television service. That was a sizable chunk of change for most people in those days. But, as Gene Schneider later recalled, "They really didn't care what it cost. They just wanted it... We were just signing up people to subscribe like crazy."

The company held a grand opening in the Casper Armory and attracted about 12,000 people in a town of just over 20,000.

It was a smashing success. One year after launching, the system had 4,000 subscribers. It had already invested some $600,000 in plant and equipment. Most of that money, Daniels told the local paper, remained in Casper. It was an example of an instinct for public relations that was to serve him well throughout his career.

The system was also a huge financial success. "I wouldn't want to say it too loud right now," Schneider later remembered, "but it paid for everything the first year. To the oil men it was like hitting an oil well."

Bill Daniels didn't invent cable television. In fact, many years later, after making hundreds of millions of dollars in the business, he would be hard pressed to explain in any detail how cable worked. Nor did he ever really run a cable system for any length of time. He had no patience for operations and left that to others better suited to such an activity. He didn't even build the first system in the country.

But Daniels was the Billy Graham of the cable business. He was a tireless evangelist and passionate pitchman who won thousands of converts. On the strength of his sermons about the potential of cable, hundreds of individuals and dozens of corporations set aside other businesses and priorities and invested billions of dollars and people hours into a business some had never even heard of before meeting him.

In a nation of salesmen, he was one of the greatest. He was also one of the few of the early cable entrepreneurs to remain centrally involved in the business throughout its first five decades.

After building Casper, Daniels and his cohorts went on to wire Rawlins, Wyo., and Farmington, N.M. And pretty soon Daniels got to know every other cable operator in the country. His contacts and his endless boosterism led in 1956 to a stint as head of the newly formed National Cable Television Association trade organization.

As head of the trade organization, Daniels fielded countless calls from cable system buyers seeking sellers, from sellers seeking buyers and from operators seeking funding. He would always try to help and in the process found his niche in the business. In 1958 he formed Daniels & Associates with a partner, Denver lawyer Carl Williams. For decades to come Daniels would be at the center of many of the biggest cable deals.

It's easy to see why Daniels built a reputation as the "father" of the cable television business. But it's a title he felt should belong to those who built cable systems before he did. Still, he was the ultimate matchmaker, and most of the matches he made wound up in financial heaven. Among the companies and individuals Daniels helped bring to the altar were:

• TelePrompTer Corp. CEO Irving Kahn built a business bringing closed circuit telecasts of sporting events to arenas around the country. In 1959 he sent his top associate, Monroe Rifkin, out to Denver to talk with Daniels about how cable might be used to deliver closed-circuit sporting events. Rifkin urged Kahn to get into the business. In a few years Teleprompter was the nation's largest cable operator, buying most of its systems through Daniels.

• Time Warner Cable. Rifkin eventually left TelePrompTer to work for Daniels. They and a group of investors founded American Television & Communications Corp., later sold to Time Inc. By 1997 it had become the largest cable company in the world.

• Cox Communications. J. Leonard Reinsch, then head of the Atlanta-based publishing and broadcasting empire, was introduced to cable by Daniels and bought his first system from Bill, forming the basis for what would become the nation's third largest cable system operator by the early 1990s.

• Tele-Communications Inc. Texas cable operator Bob Magness purchased the system in Bozeman, Mont., from Daniels, laying the foundation of TCI, which reigned as the largest cable operator from 1981 through 1997. Magness was its chairman.

• Sammons Communications. Daniels met Texas businessman C.A. Sammons at a dinner and, after hearing Daniels' pitch, Sammons agreed to buy the cable system in Bradford, Pa., the first acquisition for what would become Sammons Communications, later one of the nation's largest cable companies. "I remember my commission was $50,000 for about six weeks' work. I thought: 'That's all right'," Daniels said later.

Daniels wasn't averse to owning cable systems himself. In fact, he entered and exited the cable operating business almost as many times as he did the institution of marriage. ("They have a bumper sticker in Denver," he once joked. "Honk if you haven't been married to Bill Daniels.")

In 1960 he sold his interest in Community Cable Systems to the Schneiders but then turned around and began buying systems again. By 1980 he had sold out again, this time to Newhouse Corp., for $100 million—115,000 subscribers at $870 per sub. Then he went out and did it again, selling all but two systems to United Artists Cable in 1988 for almost $200 million—88,000 subscribers at $2,216 per sub.

What made Bill Daniels a success was his uncanny instinct for people. He would study the biographies of people he was about to meet. He bought the first Lear Jet in Denver and used it to fly to meetings with prospective clients within hours after they had called to inquire about buying or selling a system. He remembered the names and the birthdays of children and spouses. He worked seven days a week and 24 hours a day and would have worked more had he found a way.

He loved to bet the ranch, and outside of cable he lost more often than he won. He spent $425,000 on an unsuccessful run for Governor of Colorado in 1974. He blew $300,000 backing boxer Ron Lyle and another $200,000 on an attempt to win the Indianapolis 500 car race.

His biggest disaster was the American Basketball Association's Utah Stars, which cost him $5 million and drove him into bankruptcy. It provided his biggest disappointment and his greatest moment.

"I cried like a baby the night I had to fold the club," he told *Channels* magazine in 1985. "All my life I had said I would never declare bankruptcy, but I didn't have a choice."

Five years later, he was back in Utah, armed with $750,000 he used to provide refunds to all holders of Utah Stars season tickets, none of whom had ever expected to see their money again. It was something no professional sports owner had ever done. But for Daniels it was worth every penny: "In terms of pride and ego it was the high point of my life, without question."

There are hundreds of Bill Daniels stories. John Saeman, who served for many years as Daniels' top executive, tells of the time that Walt Disney Co. held a big party for the cable industry at Disneyland to launch the Disney Channel. As the party ended, guides began to shepherd the hundreds of cable operators down Main Street to the monorail train waiting to take them to their hotels. But Daniels was off in a corner talking about a possible deal.

When the last train was ready to leave, the ushers insisted the Daniels group move. As the train began to blow its whistle, the group started to jog. Daniels, a heavy smoker, soon ran out of breath. The rest of the group reached the loading platform and shouted back to Daniels to hurry up or he would be left behind. He tried to break into another trot, but soon gave up.

Then he stopped, cupped his hands to his mouth and yelled at Saeman:

"John. . . . buy the train!"

Up or down, in cable or out, Daniels loved the grand gesture and would never allow himself to be left behind. With an Old West sense of manners and generosity, he treated his customers, his clients, his employees and just about everybody he met as if they were royalty.

He routinely tipped bellboys and waiters $100. He made millionaires out of dozens of his associates, as he called those who worked for the companies he owned.

Every time Daniels made a deal that brought in a substantial profit, he would share it with his associates, all of them, down to the parking lot attendant. In 1984, for example, he sold his system in

Relays:
Early cable engineers raised microwave towers to import TV stations over mountains to subscribers who couldn't receive off-air signals.

Anchorage, Alaska, for $11 million. He kept $5 million and shared $6 million with the rest of the Daniels associates.

(The windfall prompted one Daniels associate, when asked what she intended to do with her five-figure share, to reply "I plan to spend half on cowboys and whisky and the other half on frivolous things.").

This extraordinary generosity fostered extraordinary loyalty among the people Daniels recruited into the business and those who worked for him. It made it much easier to put up with the long hours and other demands he made (including cleaning up one's desk each night before leaving the office). And it put in place a network of former Daniels employees at most of the major cable operating companies all of which at one time or another needed the services of Daniels & Associates.

But Bill Daniels might have stayed in Casper selling insurance to oil companies had it not been for the people who invented cable television in the first place. Just who it was who strung the first wire is a matter of some dispute. The fact is that the first cable system ever built was probably in the United Kingdom where, prior to World War II, several companies hooked up multiple units in apartment buildings to a single community antenna receiving the newly launched TV service of the British Broadcasting Company.

In the United States several towns have a claim on the title of cable's birthplace. What seems most certain is that cable TV was invented independently by different people around the country at about the same time.

That, more than any final judgment about who was first, tells how hungry Americans were for television in the late 1940s and how simple cable's technology was.

What all the towns with a claim to be the birthplace of cable had in common was their location in an area where TV signals could not be received clearly with a standard, rooftop antenna. But these towns were close enough to broadcast stations so that a signal could be received by an antenna mounted on a very tall tower or placed on top of a nearby hill. And each of these towns harbored an enterprising entrepreneur with some knowledge of electronics.

Among the records at the Cable Television Center & Museum in Denver, the earliest written account of a system using a community antenna to deliver a TV signal via coaxial cable to multiple sites is an article from the front page of the weekly *Tuckerman (Ark.) Record* of November 18, 1948.

Tuckerman in 1948 was a town of about 2,000 people, ninety miles northwest of Memphis, Tenn. Its primary

Rooftop:
In 1948, Ed Parsons wired a Seattle television signal into the lobby of the Astoria Hotel in Astoria, Ore. So many people crowded in to get their first glimpse of TV that the hotel manager asked Parsons to remove it.

business was to serve the cotton farms in the surrounding area.

The town had an appliance store owned by James Y. "Jimmy" Davidson.

Davidson was the son of a Little Rock optometrist, jeweler and inventor who held several patents in the field of radio. In 1930 young Jimmy and his two sisters were left orphaned following the deaths of first their mother and, eleven months later, their father. The children were shipped off to Alabama to live with relatives they had never met. Times were hard, and the Davidson children often found themselves on the wrong end of a shaving strop if they didn't earn the money the foster parents thought they should.

At the age of 13 Jimmy ran away with, he later recalled, "all my belongings in a gunnysack and exactly 97 cents in my pocket which was left over from a dollar bill I had found along a roadside." He hopped a freight train and rode back to Arkansas where he was taken in by his mother's half brother and his wife.

He waited tables in a restaurant, pumped gas, traveled with a tent show, fixed radios and other electronic devices and did odd jobs at farms around the area. He never went back to school.

In 1938 he landed a job managing the new movie theater in Tuckerman. But Davidson had a kind of nervous energy still evident 60 years later. It was the same kind of ambition that was bred into Bill Daniels and so many other children of the Depression, grounded in the fear that if you didn't take all the work you could possibly get today, you might not have the chance tomorrow.

In his spare time, Davidson ran a radio repair business in Tuckerman, building on the love of electronics he

Arkansas Entrepreneur:
Louis French, Naomi Toler, and Jim Davidson in 1991.

had inherited from his father and honed as a teenager who would sometimes earn a little money by fixing radios.

During World War II he was recruited for the Navy's Signal Corps where he added to his communications knowledge and where he had his first up-close look at another device for which he had a passion: the airplane.

After the war he returned to Tuckerman to run his appliance store. It was successful enough to allow him to buy a small airplane, the first of many he would own.

In November of 1947 Memphis radio station WMC received from the Federal Communications Commission a license to build and operate a television station, one of the first in the South. It was big news for Davidson.

"I was very excited and eager about the project and made many trips to Memphis in my little airplane, following closely the progress on the transmitter, tower and studio construction," he later recalled. But Tuckerman was too far away from Memphis to receive the signal clearly on a normal rooftop antenna. The only way to get a decent signal was to put an antenna on top of a very tall tower.

Davidson and his sole employee, Louis French, built a 100-foot-high tower on top of a two-story building adjacent to the appliance store. In the fall of 1948 WMCT-TV began test transmissions, in part to whet the appetite of appliance dealers who would have to sell lots of TV sets if the station would have an audience.

"We received every one of these test patterns," Davidson recalled. The station's engineer would call regularly to find out about the strength and quality of the signal. The transmissions sometimes included film and slide presentations. Davidson ran a length of coaxial cable from his tower down to the appliance store and hooked up his new stock of TV sets.

One day, French recalled, he and Davidson were working on the tower and Davidson wondered aloud whether it would be possible to run a cable from the tower to the home across the street owned by Carl Toler, the operator of the local telegraph office. The two ran a cable from the tower across the street (attaching it to the wire that held up the town's only traffic light) over the roof of the theater, through the branches of a cherry tree and into the Toler home. It worked.

People were amazed. "People flocked to our store each night to watch the black-and-white pictures in our store window," Davidson recalled. Some accused me of trickery, saying 'It's impossible to throw pictures through the air.'"

Across the street, in the Toler household, the reaction was the same. "We were shocked," Carl Toler's widow, Naomi, recalled 50 years later about the idea that moving pictures could be sent almost 100 miles through the air and be received in her living room.

Children came by the house every afternoon on the way home from school to see this marvelous invention, even if the only thing on was the test pattern, Toler recalled. There in Tuckerman the school children were able to follow news of the debates about the founding of the United Nations. The Tolers agreed to pay $150 for the hookup and a fee of $3 a month for the right to stay hooked up the to the community antenna.

The price, French later recalled, was equal to what the appliance store would charge for installation and maintenance of a large rooftop antenna in a private home.

What really got the town's attention was the telecast on November 13, 1948, of a live football game between the University of Tennessee and the University of Mississippi. To show the game Davidson hooked a TV set in the local American Legion hall to his antenna.

In a front-page article, the *Tuckerman Record* reported that the town's "first television program was received here Saturday (Nov. 13) and was made visible to the public courtesy of Jimmy Davidson at Auto-Electric Company who installed a television set in the Legion Hut and invited the public to see it.

"The program was a trial broadcast of the Tennessee-Ole Miss football game from Memphis and while the reception was not clear at all times those witnessing the broadcast could get the idea of what the real thing will look like when the 'bugs' are worked out."

Davidson recalled that the event drew a standing-room-only crowd at the store, the Legion Hall and the Tolers' house. He and French were kept busy running between the three locations to keep the TV sets in tune.

When the Memphis station began full-time transmission on Jan. 1, 1949, Davidson looked for other locations to build community antennas. He completed his first full system in nearby Batesville, which had a larger population than Tuckerman. Davidson and French tested signal strength in a variety of locations near Batesville and even toyed with the idea of using a balloon to hoist the antenna up above the town.

Eventually they built a rhombic antenna, a network of antenna wires mounted on steel towers and covering several acres, to receive sufficient signal from Memphis, 114 miles away. Davidson had learned of rhombic antennas during his service in the Signal Corps, and the one in Batesville was the first ever used to receive broadcast signals for retransmission through a cable system.

Working 18-hour days and seven days a week, Davidson and French wired Batesville and began to build other systems in Arkansas, financing each from the revenue generated from the others. Davidson also founded a company, Davco, that supplied equipment and expertise for other would-be cable operators in Arkansas and other southern states, eventually expanding it nationwide.

Davco specialized in building a complete cable headend (the office where the signal was received from the antenna and then retransmitted to the system) and then flying it intact to the location where it was needed and where it could be installed in a day. It was an enterprise the combined Davidson's love of communications and love of flying.

In the 1970s, Davidson "semi retired," selling some of his businesses and turning operation of the others over to his son.

While Davidson was working to bring television to Tuckerman, a similar scenario was playing out 2,200 miles away.

Ed Parsons owned the local radio station in Astoria, Ore., a town of 10,000 at the mouth of the Columbia River. He was an engineer by training and an inventor by instinct. As a boy in Portland, he had played in his father's garage and discovered electronics, to the dismay of some of his school mates.

"Another youngster and myself in the sixth grade, because I had access to spark coils, we made things a little bit difficult for the teacher, shocking kids and so forth," Parsons said years later. "And to divert us she presented us with a book, *Marconi's Experiments in Wireless*. That changed our channels."

The boys began to experiment. The father of Parsons' friend worked for the telephone company. Using old equipment they found in their fathers' garages, the boys built a wireless set to communicate between their houses. By high school, Parsons was building radio receivers and selling them to a local radio store.

Parsons, a lanky six-footer whose unruly shock of hair turned prematurely white, worked as an engineer at Portland radio station KGW. He was also a pilot and operated a flying school. During World War II, he heard that the radio station in Astoria was about to shut down because of poor financial performance. He drove down and bought it. In 30 days he had it in the black, mostly because he instituted live news broadcasts which attracted a much bigger local audience.

In the back room of the station he ran a radio repair shop, fixing radio receivers for people in the town so they could tune in to his station. He also re-

9

paired all the other electronic equipment people brought him, including the radar, depth finders, automatic pilots and other devices used by the local fishing fleet.

At a broadcasters' convention in Chicago in 1947, Parsons and his wife, Grace, first saw television. She wanted it in her home. At that time there was no broadcast station close to Astoria. Nevertheless, his wife insisted they buy a TV set.

Parsons ordered a nine-inch model from a firm in Chicago and had it flown to Astoria. It cost him, he later recalled, more than $1,000. "I told the wife we were wasting our money with the (set), but at least I would try to get her television." He thought she might end up using the set as a table.

But Parsons' wife had faith. "She figured I was an engineer, so there was no reason why she shouldn't have television."

In the spring of 1948 radio station KRSC in Seattle announced plans to launch a television station later that year. Over the summer it began to test its signal. Parsons knew the Seattle station manager and worked with him to find the signals.

"I took an FM receiver, changed its frequency range and put a meter on it and a pair of headphones so I could listen and tune by the video and the audio. You could hear the blanking bars (used to test the video signal). And you could hear audio in the earphones.

"I put a field strength meter across the discriminator to show how much signal was being received. I built a couple of these sets and put one in my car and a second unit in my own airplane." He drove and flew around town, looking for the best place to receive the signal from Seattle.

After some testing, Parsons discovered he could receive the Seattle signal from a location atop the Astoria Hotel, across the street from his apartment. Then he set about to build a receiver.

Working out of his radio repair shop, "I designed and built a two-stage I-F amplifier and converter to channel 2 and a three-stage amplifier on channel 2 as a signal receiving unit." This converted the signal from channel 5 to channel 2 (necessary to avoid interference), and transmitted it, via coaxial cable, across the street to Parson's apartment where he hooked the cable to his TV set.

When KRSC began its first broadcast on Thanksgiving Day, 1948, the Parsons were the only ones in Astoria with a television set able to receive the broadcast signal.

Their apartment was mobbed.

"We literally lost our home. People would drive for hundreds of miles to see television. So I approached the hotel manager and suggested that it would be a simple matter to drop a cable down

Microwave:
As cable developed, operators constructed microwave towers capable of receiving several TV signals.

the elevator shaft and put a set in the lobby of his hotel. He thought that would be a wonderful idea. So we did."

But the hotel lobby got so crowded the manager asked Parsons to remove the set. Parsons then asked a local music store owner, Cliff Poole, if he would like to have a TV set in his store as a draw for customers. Poole agreed to buy a set and to pay Parsons to hook it up to the antenna atop the Astoria Hotel.

Soon the police called to tell Parsons that the crowds outside Poole's store were becoming a public nuisance. The police chief suggested Parsons string wire to the local bars.

"Every bar owner was anxious to have a television set (and) would pay for the cable," Parsons said. "I designed additional amplifiers on channel 2 and put the signal down. Then people began to put pressure to have cables in their homes."

Before long Parsons was stringing wire all over town. When the city council objected to wires hanging over the streets, Parsons devised a wireless system to transmit the signals from one amplifier to another across the streets and then hook up all the houses on a street via wire.

He charged people different prices, according to how much equipment he needed to use and how difficult the hookup was. An average hookup cost about $125. Customers provided their own power, and Parsons did not charge a monthly fee.

Soon entrepreneurs from other towns in Oregon and Washington began to show up in Astoria, asking Parsons how to build systems in their towns.

In Aberdeen, he ran into a problem with power. It was seven miles from the antenna site to the town, and he needed seven amplifiers. But there was no power available. "So I designed a system to put 220 volts AC up on the coaxial cable to power the amplifier plus the headend. And the system worked beautifully." It was one of the first uses of coaxial cable to transmit power as well as a television signal. Because the two operated on different frequencies they did not interfere with each other.

Parsons' inventions attracted the attention of *Popular Mechanics* magazine, which printed an article about him in its April 1950 issue, the first article about cable TV to appear in a nationally distributed consumer publication. The story focused on how Parsons developed his system and particularly what kinds of antennas he used. It did not talk about the possibility that cable operators could actually charge custom-

Astoria Monument:
Although cable pioneer Ed Parsons moved to Alaska soon after he brought Seattle TV signals to Astoria, Ore., on Nov. 25, 1948, he returned to the lower 48 states in 1968 to didicate a monument to his experience. Pictured on Coxcomb Hill in Astoria are (l-r): NCTA President Frederick Ford; Marcus Bartlett, VP, Cox Broadcasting, which owned the Astoria, Ore., system in 1968; Parsons; and Lew Davenport, the system's general manager. Falcon Communications owns the system today.

NCTA Founder: Martin Malarkey built his first system in Pottsville, Pa., in 1949. By 1951, he was joining other operators to organize the National Community Television Council, the forerunner to today's NCTA.

ers for the service. It also stated that Parsons' transmission system could deliver the signal only 2,000 feet. Finally, it reported that Parsons' next step would be to build a satellite broadcast station in Astoria that would receive the signal from Seattle and then rebroadcast it.

But the article did recognize that the Parsons system could be the beginning of something big. "Astoria is only one town — the beginning," the story concluded. "But what works in Astoria may work all over the Northwest and the nation, too."

But Parsons himself never recognized that his invention could be used to launch a business based on a monthly fee for service. He did sell equipment to his customers and provided consulting services to other cable operators in the northwest region. By 1953 he was so burned out he moved to Alaska. He never really made any money from cable television.

Nor did Parsons receive any widespread publicity about what he had done beyond the articles in *Popular Mechanics* and a couple in trade journals (including an article in the Aug. 13, 1949, issue of *Television Digest*, most likely the first trade journal article about cable in the U.S.). But most of the articles about Parsons were focused on the technical aspects of his system and did not really raise the possibility that Parsons' inventions might lead to a business.

Years later the city of Astoria erected a plaque near the old hotel where Parsons built his first system and declared that it was there that the first cable television system in the U.S. was constructed.

While Davidson and Parsons were building their systems in Arkansas and Oregon, the business was also taking off in Pennsylvania. The Keystone State was the Mesopotamia of the cable industry. Its hilly terrain and proximity to major television markets such as Philadelphia, New York City and Pittsburgh made it a natural for cable. And the first to develop cable in Pennsylvania were television set retailers, seeking to extend the reach of broadcast signals to more homes so they could sell more television sets.

One of these was John Walson, who ran an appliance store in Mahanoy City, Pa., and worked part time for the local electric utility. Mahanoy City was a community of 10,000 about 75 miles northwest of Philadelphia and beyond the reach of the city's broadcast television signals. When Walson began to sell TV sets in his store, he had to take customers to the top of a nearby hill to show how the sets worked. There he could get a decent signal.

Finally he decided just to string the twin lead antenna wire down to his store. He boosted the signal using "a small, top-of-the-set booster." This was in June, 1948, Walson later recalled, about four months before Parsons and Davidson hooked up their systems. But Walson said that the only records documenting his achievement were destroyed in a warehouse fire 30 years later.

(If Walson did indeed launch his system in June, 1948, he managed to keep it a pretty tight secret at the time. The launch of a cable system was regarded as big news in most towns in the U.S. But there are no newspaper accounts at all of Walson's cable system until the 1950s. Casting even more doubt on Walson's claim were the reports he himself filed for *The Television & Cable Factbook* from 1953-1966 in which he stated that his system started in 1950, according to *Factbook* editor Al Warren. It was not until 1967 that Walson began to list 1948 as the starting date in his reports to the *Factbook*).

Whenever he strung the first wire, soon thereafter his customers began asking if he could hook them up as well, and Walson began to string more and more wire around town, using the poles of his employer, Pennsylvania Power & Light Co. He found he could charge people $2 a month for the service.

It didn't take him long to discover the limits of twin lead wire, which would go out whenever it rained. "It wasn't very practical, and the phone would almost ring off the hook every time it rained," Walson later recalled. Within a couple of years he had replaced the twin lead with coaxial cable and by 1957 he had one of the largest systems in the nation with some 14,250 subscribers.

Walson called his cable company Service Electric Company. It would enter the history books again 20 years later, in a better documented event, when it became the first cable system to make an affiliation agreement with Home Box Office.

Another Pennsylvanian who discovered cable television about this time was Martin Malarkey, a radio and television appliance store operator from Pottsville, 90 miles northwest of Philadelphia. In the spring of 1949 he visited New York City and stayed at the Waldorf Astoria Hotel. In his hotel room he found a 10-inch television set that delivered a picture far better than anything he had seen before. "I noticed that the wire coming out of the back of the set was a round, black vinyl-covered wire. I had been used to seeing a flat twin lead coming out of the back of the set going up to the antenna. My curiosity was piqued."

He called down to the front desk and arranged to meet the hotel engineer who explained that he had just installed a master antenna system that linked all 500 rooms in the hotel to a single antenna on the hotel roof. The equipment was supplied by RCA Corp.

"This was just fascinating to me, and I went back to Pottsville and figured that if they could feed that many receivers from one antenna on a vertical plane, why couldn't you take an antenna and put it up on top of a mountain and run a line down and do the same thing on a horizontal plane?"

Malarkey, because he was a television set salesman, had a relationship with RCA. He called his local distributor. "They told me to go see RCA directly. So I went down to Camden, N. J., to talk to RCA. They were so in-

Service Electric:
John Walson built Service Electric Co. from his first system in Mahoney City, Pa. Decades later, he would become HBO's first affiliate.

trigued by the idea they offered to send the engineers up to Pottsville to experiment at no cost to me if I would buy the equipment to experiment with."

The engineers found they could pick up the signals from the Philadelphia broadcast stations, amplify them, and run them via cable a couple of miles down the mountain. "The pictures were absolutely gorgeous," Malarkey remembered.

Initially he wanted to build the antenna in the hopes of selling more TV sets. But that quickly changed. "Word got around and people started coming in the store offering me money and asking: 'How much do you want to hook me up to this system that you have?' I realized it was a service that could be sold."

Malarkey priced his service based on what it cost people to get decent broadcast signals. "I concluded that the average (rooftop) antenna we were installing at the time cost between $150 and $200. Consider also that those antennas were so huge, so cumbersome and so high above the rooftop that they were a continuing maintenance problem and the average person was spending $40 to $50 a year to maintain those antennas.

"Not knowing how much it was going to cost to build this system — I had nothing to guide me — it appeared to me that a $150 connection fee was a reasonable fee and the $3.75 a month was a reasonable monthly charge for the service I was providing."

After a year he had a thousand customers and had begun to wire other Pennsylvania systems.

Malarkey also decided he would need permission to cross the city streets and to put his wires on the poles owned by the local telephone company. He approached a lawyer in Pottsville whose father was on the city council and invited the attorney to join his company. He received the city's blessing.

He then paid a call on the vice president of the Pennsylvania Power & Light Co. and offered him a share in the enterprise. Pretty soon he had permission to use the utility poles. The regional vice president of the local phone company became a shareholder and Malarkey had permission to use the telephone company poles.

"So after I got all of these people in bed with me and I got all of the permissions I needed, I went out and I started to hire people who had experience with the telephone company and power company in putting their lines up."

For initial financing Malarkey used his own money and his family's. "I gambled every last penny that I had at that point in time on getting the system started."

His brothers each kicked in a few thousand dollars, but his father resisted. "I don't think he had the grasp of the potential. But he did begin to realize what I had when I gave him back his entire investment that he'd made during the first year of operations. Every year after that, as long as he lived, he got dividends that were equal each year to five times what his original investment was. He got the message."

Attorney:
A former FCC staffer, E. Stratford Smith helped cable operators in the early 1950s wade through a sea of regulatory issues. He was the National Community Television Council's first full-time paid employee.

The system was a huge success. Customers were banging at his door demanding service. Would-be cable operators from around the nation were also asking him for advice. He started to charge consulting fees. One of his first clients was Bill Daniels, who flew in from Wyoming to get some tips on how to build a system. Daniels paid Malarkey $500 for a day's worth of instruction and would later remember that he had been "glad to do it."

Malarkey also attracted the attention of Uncle Sam. The Federal Communications Commission dispatched a common carrier bureau lawyer, E. Stratford Smith, to Pottsville to find out about this new business.

"I told him as much as I could," Malarkey said, "with the hope that he would...tell me where the FCC was coming from and whether I could anticipate any problems from the FCC."

Malarkey also received a visit from a representative of the Internal Revenue Service, which wanted to slap an 8% excise tax on the revenue from subscribers. Malarkey went to court to challenge the tax and eventually won.

These visits caused Malarkey to call the other cable operators he knew and invite them to join him at the Necho Allen Hotel in Pottsville in the fall of 1951 to form the industry's first trade association. They called it the National Community Television Council, the forerunner of the NCTA. Nine operators kicked in $1,000 each to get it going, and Malarkey was elected the first president.

"The first person I called was Strat Smith," who had left the FCC and started in private practice, Malarkey recalled. Smith signed on to handle the group's legal work. Initially he offered to work for only a token fee, betting that as the business expanded and more problems arose, he would be earning larger and larger fees. It was a wise gamble. Eventually Smith was persuaded to become the organization's first full-time paid employee.

Smith suggested that the group change its name to the National Community Television Association. Later it was changed again to the National Cable Television Association, which remained for the next five decades the leading industry trade group and lobbying organization.

According to Jim Davidson, who came from Arkansas to attend the early organizational meetings of the NCTA, one of the first issues had to do with ethics. "It seems that a television dealer who also owned a community antenna system would charge higher prices to connect a subscriber if the subscriber chose to buy his television set from a competing dealer," Davidson later recalled. "It was pointed out that these quick-dollar entrepreneurs, in an effort to make a fast profit also resorted to slipshod construction. These unethical operators were, fortunately, in a minority."

Malarkey expanded his business to other towns in Pennsylvania, Virginia and Maryland. By the late 1950s he had one of the largest cable companies in the country.

Malarkey sold most of his systems in 1960. "Early on, after I graduated from college, I decided I was going to make a certain amount of money and retire. I'm essentially a very lazy individual, and I made up my mind that when I attained that goal I was going to relax and enjoy life." He was 41 years old.

But Malarkey didn't stay retired. Existing and prospective cable operators continued to call on him asking for advice. In 1963 he joined with engineer Archer Taylor to form Malarkey Taylor & Associates, which remained for many years the industry's premiere consulting group on engineering, franchising, construction, finance and operations.

Malarkey, Davidson and Walson started their cable systems initially as a way to get better pictures on the TV sets in their appliance stores. Parsons was just trying to please his wife. The first cable system that was constructed from the outset with the express purpose of selling monthly service to customers was begun in Lansford, Pa. It was also the first system to receive widespread national publicity, setting off a nationwide cable construction mania. And it was the first system to use equipment sold by a new company that would play a pivotal role in the history of the business, Jerrold Electronics.

Lansford was a town of 2,300 homes about 70 miles from Philadelphia, and just down the road from Mahanoy City, where Walson had his business.

In 1950 Bob Tarlton, a 40- year-old veteran of World War II, was working in an appliance store he and his father owned in Lansford. The store had just begun to sell TV sets.

About the only place near Lansford that could receive the local Philadelphia broadcast signals was an adjacent town called Summit Hill, to the south of Lansford and about 500 feet higher. The Tarlton radio store was located in Lansford at the edge of Summit Hill.

The Tarltons wanted to demonstrate the best possible signal in their store. So they strung a dual-lead antenna wire from the store across a couple of streets up to an antenna in Summit Hill. The Tarltons soon were getting requests from customers for similar service.

Some of them couldn't be refused.

Tarlton remembered "a friend of my father's, named David John Stevens, came to see my father. He was quite a loud individual. He was an executive secretary of the United Mine Workers and he was on (UMW President) John L. Lewis' board in Washington, D.C."

Stevens was expecting a visit from Lewis for the weekend and wanted to make sure there was a working television set in his home so the union leader could watch. But Stevens lived in a valley near a river with no decent TV signals.

The Tarltons said there was nothing they could do. But Stevens insisted they try to set up some sort of antenna. So they went down to the house and found a nearby hill where they constructed an antenna and wired up Stevens' home. The United Mine Workers president had his television.

The development opened the floodgates. People were demanding that they get the same type of service. But Tarlton knew the limits of twin-lead wire. It could not be allowed to touch anything, was highly susceptible to interference and could be very temperamental in the weather.

The Tarltons worried that if they sold

Video-on-demand:
Union boss John L. Lewis wanted to watch television, and Bob Tarlton, left, and his father found a way to get it to him.

TV sets and antennas to their customers and the wire went out in a few months it would leave a lot of angry customers. His concern was that "inside of six months to a year they'll no longer have television or it will be so poor they'll come back to us and demand their money back... that stuck in my mind as an important consideration." It made him think of using coaxial cable, which he had used extensively in the service during World War II.

Coaxial cable is a copper wire surrounded by insulation, wrapped in turn by a copper mesh. The entire structure is then encased in plastic. Because the various layers of the cable have the same axis, the package was called coaxial cable.

The wire mesh layer prevented signals from leaking out of the copper core and prevented interference from outside. Although the structure has been modified somewhat, with aluminum sheathing replacing the wire mesh, for example, the basic structure of coax has remained constant throughout the history of the cable industry.

"Coaxial cable is impervious to weather," Tarlton said. "It's impervious generally to outside interference. It's like a water pipe. What runs through is confined and delivered on the other end." Coax is also capable of carrying vast amounts of information, unlike twin lead or common twisted pair telephone wire. This would become enormously important in the 1970s and 80s when dozens of new program networks came on line.

The problem with coax was that it would impede the signal much more than a twin lead. After a mile or so the signal would begin to degrade and would need to be amplified.

Tarlton had heard about a company

Coax:
Coaxial cable consisted of an inner core of copper wire surrounded by a layer of insulation and an outer sheath of woven copper or aluminum all wrapped in plastic. It was less prone to interference than antenna wire and able to carry far more information than telephone wire.

in Philadelphia called Jerrold Electronics that was making a system to provide television service to department stores and apartment houses. He got a couple of their amplifiers and began to play around with them. He liked the amplifiers because they were smaller than those manufactured by other companies.

"(Jerrold's) piece of equipment was unique to the extent that physically the size was manageable. RCA's was a huge monstrosity. I would say something about three feet high. (Jerrold's) equipment was no more than 12 inches by 12 inches or maybe eight by ten inches high."

But the Jerrold amplifiers had been designed to boost the signals only once. Tarlton needed equipment that would amplify and reamplify the signal several times. The problem was that the Jerrold equipment, if used in sequence, would degrade the signal, particularly the sound, so that it was unusable.

"With one amplification it was sufficient... but the minute you got to a second, third or fourth or fifth amplifier the sound kept sliding way down and you had no more sound," Tarlton recalled. "So I just fiddled with it a bit. Didn't do any redesigning. I just re-tuned them a bit and flattened them out a little bit and brought the sound up a little bit, enough so we could reamplify it.

"I had maybe half a dozen, and then

I ran an experiment on it and lo and behold, it really worked."

By the spring of 1950 Tarlton figured he had a complete system that would work. He got together with all the other television set dealers in town — Rudy Dubosky, Bill McDonald and George Bright — and an attorney and state legislator named William Z. Scott.

"I wanted to have a combined effort," Tarlton said, "because I felt that the result would be a unified effort rather than everyone doing it his way."

The other dealers wanted to offer the system to customers for free, hoping to make money by selling more TV sets. But Tarlton persuaded them that it could be a business on its own and that they should charge for it from the beginning.

Each partner put in $500 and the group borrowed another $20,000 from the local bank. They decided to charge $100 for installation and a monthly fee of $3. These figures, Tarlton later said, were "picked out of the air."

The group built an 85-foot tower in Summit Hill and held the first demonstration of the new service in October 1950. Tarlton had hoped that the system would be able to attract 200 customers. The demonstration led him to believe he had underestimated the appeal of cable. One thousand people showed up in a town with only 2,300 homes.

The new company, Panther Valley Television, hired workers from the Lehigh Navigation Coal Co. to string the wire from poles they rented from the local utility for $1.50 per pole per year.

The relationship with the local phone company was critical. The manager of the company not only gave Panther permission to use the poles, but suggested how to string the coaxial cable, using a lashing machine that would attach the cable to a stainless steel wire that would prevent the cable from sagging.

One problem Panther had to solve from the outset was political. Without a decent signal from Philadelphia, most residents of Lansford who wanted to watch television did so in the bars in Summit Hill. That generated some good business for the bars.

"People traveled from all over to Summit Hill to watch television, sitting there having a beer or a drink and watching television. We sold them sets that were worth thousands of dollars," Tarlton recalled. With television signals available via cable in Lansford, the bars would see a drop in business.

"So in Summit Hill the pressure of trying to impede us came from the business community. And, of course, businesses in small communities can exert political pressure on the local governments."

At Scott's suggestion, the Panther group made the first move. They asked the Summit Hill city council for permission to build their tower and lay their wire, even though they did not have to cross any city streets. The council, fearful of the reaction from the bar owners, balked.

"To start with, they dragged their feet," Tarlton remembered. "And then they finally said: 'Yes, we will give you permission, but we want a 15% gross receipts tax.'" It was the first experience any cable operator had had with what would become known as a "franchise fee."

Panther fought the tax and the council agreed to drop the fee to 5%. Panther agreed to pay the tax but challenged it in court. The cable company eventually won in the state supreme court. But the notion that the Summit Hill council had about taxing cable was one that would occur to other city officials around the country.

The Lansford system was a huge hit. "You had to keep your door locked," Tarlton recalled. "The people were clamoring for service." Within a few months Tarlton was buying a lot of equipment, including a bunch of amplifiers from Jerrold.

Jerrold was founded in 1948 by Milton Jerrold Shapp, a 42-year-old sales rep for a group of electronics manufacturers. Shapp had come across a couple of engineers who had invented a "booster" designed to improve the signal reception of television sets. Shapp liked the device and wanted to sell it. But he couldn't find anybody interested in making it.

So Shapp decided to build and sell the product on his own. The Tarlton radio store in Lansford was one of the outlets for the boosters made by Shapp's new company.

A few months later the same engineers who had developed the booster invented a device that would allow a single antenna to serve several television sets. They planned to sell the device to appliance stores. Consumers looking to buy a television set could compare the picture quality of several models receiving the same signal from the same antenna at the same time.

As Shapp's first employee, Hank Arbeiter, later recalled: "In those days if you went to Gimbels or Sears or any of the stores that had some TV sets and you wanted to see how they worked they would have to plug in the antenna. You could only look at the sets one at a time. The need was there and this engineer came up with an amplifier that we could hook all the sets up at one time.

"One of our first installations was in Darby Mark Fine (department store) on 69th St. I think he had something like 30 or 40 sets on at one time after we got all this equipment. It was in the newspaper and everything. It was a great event.

"From that development we grew into apartment houses because it was a very similar set of conditions there. We started to get our feet wet in small apartment houses and worked our way up to fairly large ones."

The Jerrold developments generated a lot of publicity and attracted the attention of the group in Lansford. Tarlton became a big customer for the Jerrold amplifiers.

"I was buying so many I thought I would go to Philadelphia and buy them direct," he said. "Bud Green was the sales manager and chief foreman. I saw Bud, and he said, 'Oh, sure, we'll sell you amplifiers.' One day, when I was down for another batch of amplifiers, he asked me: 'What are you doing with

Pole Climbers:
The basic work of rolling trucks and climbing poles is still a fundamental activity of cable operations today.

those things?' And I told him."

When Shapp heard what was going on he went out to Lansford, met with Tarlton on Thanksgiving weekend 1950 and offered to manufacture amplifiers that would meet the specific needs of cable systems.

"His comment was, 'This has tremendous potential,'" Tarlton recalled. "He said he was going back to the factory on Monday morning and get his entire engineering staff to design equipment for this type of operation."

The Lansford system was an immediate hit with consumers, even with the $100 installation fee and the $3 a month service charge. By January 1951, 100 homes had been wired, and installs were adding 40 a week.

The event drew nationwide publicity. Stories ran in Newsweek, *The New York Times* and *The Wall Street Journal*. "Panther TV's system was not cheap," *Newsweek* noted. "Nonetheless, the company's chief difficulty is in meeting demand." The neighboring town of Coaldale, *Newsweek* added, was clamoring for service.

And *The New York Times*, in its issue of Dec. 22, 1950, declared, "The town of Lansford, which heretofore had been barred from television reception because it lies in a valley, is now receiving video programs regularly after erecting what is believed to be the country's first 'community antenna.'"

The *Times* article went on to quote Lansford mayor Evan Whilden who stated that the signals received were "just as good as you folks get up in New York." Ultimately, he suggested, the single aerial might serve 30,000 valley residents.

With the prices the *Times* reported the system was charging, readers didn't have to use a slide rule to see how much money could be made in cable.

But the most important statement in the *Times* article came from George Bright: "'There's nothing that can be patented,' he commented with a laugh. 'The system just uses standard equipment, and anybody can do it.'"

Parsons, Walson, Malarkey and Tarlton all said later that they had invented their cable systems independently. Each insisted he had never heard of the others when he began to construct his system.

Asked later about the dispute over who built the first cable system, Parsons said, "It's nothing to me if he (Walson) was first. Fine. I don't make any claim to being first because the technology was there for anybody to get it. I did it simply because of a demanding wife."

Regardless of who was first to build what could be called a true cable system, the idea spread like wildfire in the early 1950s, spurred by an unlikely ally, the federal government.

Relations between the cable industry and the government most nearly resemble the relations the ancient Greeks had with their gods. The Greek gods were enormously powerful and equally capricious. They could help a mortal one day and strike him down the next. One god could be on one side and another on the other side. They spoke in language difficult to understand, and mortals were required to pay to have the omens interpreted. The gods could be appeased through sacrifice but could just as easily be angered, sometimes for no apparent reason. The gods could never be ignored, but neither could they ever be fully understood.

Like the Greek Gods, the government

sometimes took action without knowing what the consequences would be but which nevertheless had enormous impact on the cable business. The first example of this came before the cable industry had even been born.

In 1934 the Congress, spurred by the New Deal zest for regulation, created the Federal Communications Commission and established for the first time a nationwide federal policy to govern communications. The primary responsibility of the FCC was to allocate scarce spectrum to the new medium of radio. The goal was to bring some order in the skies so that rival radio stations would not be transmitting on the same or adjacent frequencies, causing interference with each other.

World War II brought a host of advances in communications. In 1945 the FCC found the need to revisit the spectrum allocation issue to take into account all the new over-the-air communications devices: mobile radios, railroad communications devices, walkie talkies and the newly invented medium of broadcast television.

When the FCC met, there were only six television stations in the entire nation, and all were broadcasting in black and white. The commission decided to allocate enough spectrum in the Very High Frequency (VHF) band which TV used to allow for 12 channels. It adopted a rule that required that all stations with the same channel number be at least 150 miles apart and stations with adjacent numbers be 75 miles apart.

The allocation ushered in a major spectrum land rush, not the last the United States would see. By 1948, 109 stations were operational or under construction.

But the FCC found its original allo-

Ground-breaker:
Milton Shapp, seen here with his children at the ground-breaking for his Jerrold Laboratory, not only figured out how to transmit good signals over coaxial cable, he played a pivotal role in developing many early cable systems.

cation scheme wasn't working entirely as planned. There were two problems. The first was that the broadcast signals were proving to be very bouncy, ricocheting off the troposphere (that layer of the atmosphere between 300 feet and six miles above the earth) and bouncing back to earth often interfering with some station hundreds of miles away.

The second problem the FCC had not foreseen in 1945 was the advent of color. Color television signals could be transmitted in a variety of different ways, and the FCC was charged with deciding which would become the standard method of transmitting color signals.

Faced with these sticky problems, the commission decided on Sept. 30, 1948, to freeze the number of new television stations until it could resolve the thorny technical questions related to color and signal bouncing. The freeze remained in effect until 1952.

Suddenly, thousands of communities which had expected local broadcast television service were left out in the cold. Citizens in these communities were hopping mad. So were the businessmen who had planned to build TV stations in those communities. In 1950 FCC chairman Wayne Coy said that no decision in the history of his agency had caused more outrage than the broadcast station freeze.

But while the public was outraged and thousands of would-be broadcasters were frustrated, the decision provided an entirely unintended but extremely important boost for the newly hatched cable business.

In hundreds of communities around the country cable systems sprang up to slake the thirst of consumers for this new medium. By 1952, when the freeze ended, some 14,000 American homes were receiving their television signals through a coaxial cable hooked up to a community antenna.

A new species of business had been created.

Like the ancient Greeks, the early cable operators had to find a way to live and grow while dodging the thunderbolts from the gods of government and the floods, drought and pestilence sent their way by such lesser deities as the phone companies, broadcasters and theater owners.

But thrive they did. From such humble locations as Asortia, Ore.; Tuckerman, Ark.; and Lansford, Pa., sprang an industry that grew to be far bigger than any of its various parents ever imagined.

'A Hell Of A Business'

"Anybody can do it," George Bright of Panther Cable TV, the first cable system in the nation to receive widespread publicity, told *The New York Times* in December of 1950. He should have said "anybody with money."

The technology of cable was relatively simple. The equipment was readily available. Consumers in small towns throughout America were ravenous for television service. Even after the FCC lifted its ban on construction of new television stations in 1952, millions of Americans were still beyond the reach of broadcast television signals.

The only real obstacle to growth in cable was money. As would happen periodically throughout its first half century, the fledgling industry in the mid-1950s faced a serious shortage of capital. And as it would throughout its history, the industry solved the problem with ingenuity, perseverance and some old-fashioned luck.

At the center of it all, through the Eisenhower Administration, was Jerrold Electronics and its dynamo of a founder and president, Milton Shapp.

Shapp had seen his first cable system under construction in Lansford, Pa., over the Thanksgiving weekend in 1950. He never looked back. When the Lansford system launched a few months later, all the national publicity about it included the name Jerrold. Shapp began to field questions from hundreds of would-be cable operators from all around the country.

Instinctively he recognized that his company had to be more than an equipment supplier. It needed to be the catalyst to get this new industry off the ground. He moved quickly. As news of the success in Lansford reached the offices of the major Wall Street financial houses, Shapp was at their doors.

He persuaded three venture capital firms – J.H. Whitney, Fox Wells, and Goldman Sachs – to put up a total of $200,000 to finance construction of a cable system in Williamsport, Pa. No sooner had he concluded the deal than he called Bob Tarlton, who had founded Panther Cable TV, and asked him to run the Williamsport system.

Later Tarlton recalled, "Milt said, 'Bob, I've got a problem. I just sold a system to a venture capital group in New York City.' He said he needed someone who knew how to construct and adapt a system to the Williamsport area."

Tarlton's system in Lansford was up and running just fine. His son had finished high school and was headed to art school in Philadelphia. So a move to Jerrold, based in Philadelphia, would enable him to give his son a place to live while he went to school.

Cable's Great Salesman: *Milton Shapp, as head of Jerrold Electronics, helped jumpstart the cable business in the 1950's by providing cable operators with the equipment and the financing to build systems in return for an equity stake in their operations.*

And, of course, Shapp did a great sales job. "I always give Mr. Shapp credit for being a very, very fine salesman," Tarlton said. "The very fine salesman that he is must have convinced me to go with him."

So Tarlton signed on with Jerrold and headed to Williamsport. The challenges were much greater than in Lansford. It was a bigger town and required a much longer run from the antenna site to the homes: seven or eight miles, Tarlton recalled. This produced the first major equipment failure the industry was to experience, but not the last.

Most coaxial cable in those days was produced for the military in lengths of 100 feet or less. The only company willing to make it in the longer lengths needed by cable operators was a New Jersey firm called Plastoid. But making cable in such long lengths required pulling the copper core through the insulator in several segments. This was accomplished by a device called an extruder, which every so often would grab hold of the copper and pull it through the cable.

Each time the extruder grabbed the cable it left a small, hardly noticeable, dent in the copper. When these cables were installed in Williamsport, the dents were sufficient to interfere with the sound portion of the television signal.

Tarlton recalled: "We were tearing our hair out after having installed a good system, a good portion of it, and we were getting ready for a demonstration. We planned a grand opening where we'd throw the switch and, lo and behold, there we have television.

"Well, we were going to have the grand opening at the Lycoming Hotel in Williamsport. And for a week or two before we had all kind of problems. We couldn't get Channel 4. We couldn't get any decent sound. We had picture but no sound and we couldn't figure it out."

With the help of the manufacturer the problem was identified, the coax replaced and the system made to work. "Williamsport," Tarlton noted, "was a learning process because it was the first large system that any of us had any experience with."

The Williamsport system had another problem: competition. Two other local cable companies had announced plans to build, and both had started construction by the time the Jerrold system turned on. But Jerrold had made a key decision. While the other two systems concentrated on the wealthier areas of town, figuring they would be the likeliest customers for an expensive service such as cable TV, the Jerrold system wired the areas with the highest densities of homes, mostly middle-class neighborhoods.

It was a gamble, but it paid off. System manager Ray Schneider remembered the day after the first demonstration:

"We had an office on Williams Street. I drove up the next morning. I couldn't get near my office. People were lined up for a block and a half, with their $135 (installation fee) in their hot little

Door-to-door Pioneer:
Ray Schneider helped launch the Williamsport, Pa., cable system in 1952. He went on to be VP-CATV marketing at Times Wire & Cable.

hands. Some people came in and said, 'I'll pay you half now and half when you install it.' We said, 'I'm sorry we can't do that.' They dug right down and got the other half out and laid it on the counter. In one year we had twice as many customers hooked up to the Williamsport Jerrold Cable System as both our competitors combined."

It was a lesson that cable never forgot. Almost without exception, the most lucrative areas for a cable system are not the high-end homes where residents have other entertainment options and sometimes regard television as a diversion for the lower classess. Rather the blue collar and middle-class suburban areas have produced the highest penetration rates for cable systems.

Schneider also organized a professional sales force, perhaps the first for any cable system in the country. After signing up all those customers who came to the office, Schneider hired a group of furnace salesmen who had been trained to sell door-to-door on a salary-plus-commission basis. The effectiveness of the door-to-door campaign was increased because the system had built in the high density areas where a salesman could make more calls per day than in the less populated neighborhoods.

Within two years, Williamsport had become the largest cable system in the country. In three years, it was sold for $1 million, all of which was profit for the original investors.

With Williamsport under way, Shapp and Tarlton began to barnstorm the country, preaching the gospel to potential investors, possible partners and city councils. Shapp raised the money, manufactured the equipment and did the deals. Tarlton built the systems.

In each community, he took care to

Ragtime Operator:
Ben Conroy entered the cable business in Uvalde, Tex., through a Jerrold Electronics contract. An accomplished ragtime pianist, Conroy built a 340,000-subscriber MSO that he sold to Times Mirror in 1979.

visit his former colleagues, the TV set retailers, and enlist their support for the new cable system.

His tour was extensive: Clarksburg and Fairmont, W. Va.; Berlin, N.H.; Wenatchee, Wash.; and points in between. Among the towns he visited was Uvalde, Texas, where Jerrold was partnered with a young man named Benjamin J. Conroy, Jr.

Ben Conroy, like so many of the cable pioneers, was a veteran, a Naval Academy graduate who had seen service during World War II and Korea. His father, also named Ben, had had some success as a wholesaler of furniture and children's toys and moved the family to Brooklyn from Vermont when young Ben was a child.

In 1955, after combat duty in Korea, Conroy was assigned to the US Naval Academy training program for midshipmen. He grew restless shuffling paper and took a month off to decide his next career move. One day while he was lounging at the beach, a friend handed him some literature from Jerrold. Conroy brought it home to his father.

The two did some homework and went down to see Shapp in August,

1954. He offered them the standard Jerrold deal:

The Conroys would finance the system in return for a 51% ownership. Jerrold would supply engineering and operational expertise in return for a 49% share. The Conroys would agree to purchase all their equipment from Jerrold and pay Jerrold a maintenance fee of 25 cents per subscriber per month.

The operators called it a "yellow dog" contract, but it was the only game in town. (The term "yellow dog" came from politics. In the South in the first half of the 20th century, voters so consistently favored the Democrats over the Republicans that they would sometimes say they would vote for a yellow dog over any Republican. They became known as "yellow dog" Democrats.)

Conroy had done a little communications work in the Navy, but nothing that would qualify him to build a cable system. He also had no business training, though his father had acquired a pretty good working knowledge from running his wholesale company.

They spent some time thinking about it and then took the Jerrold offer. Shapp told them he had two deals in the works. One was in Bemidji, Minn., which sounded chilly to Conroy. The other was in Uvalde, Tex. The Conroys took the Uvalde deal, and Ben resigned from the Navy.

Together, the father and son agreed to put up $84,000 to secure a 51% share of a franchise in a city they had never seen to build a cable system they knew almost nothing about.

Jerrold had a regular training program for would-be operators and sent Conroy to school, first at Clarksburg, W.Va., and later in Williamsport. In January of 1955 Ben Conroy – born in Vermont, raised in Brooklyn, a Navy veteran whose main talent was operating 16-inch guns on the battleship New Jersey – first laid eyes on Uvalde Tex.

There wasn't much to see.

It was a town of about 9,000 residents, roughly one-third of whom were Mexican Americans (something Jerrold, based in Pennsylvania, hadn't taken into account). Only a very weak signal could reach Uvalde from the nearest broadcast station in San Antonio, about 90 miles away.

In Conroy's words: "At that time that part of Texas was in very severe drought conditions. The economy there was mainly farming and ranching. A lot of ranchers went broke. It was, however, a good time for the irrigating farmers. So it was a split kind of an economy. It was tough going for a lot of people."

Many of the Mexican-American families spent a good deal of the year up north, working on the farms, and had little interest in English-language television in any case.

Together with the local Jerrold contact in Uvalde, a TV repairman named Faber Spires, Conroy walked the town, measuring how long the distances were between the utility poles where the cable would be strung. The system began construction in April and hooked up its first

Cable Growth Through The 1950s

Year	Systems	Subscribers
1952	70	14,000
1953	150	30,000
1954	300	65,000
1955	400	150,000
1956	450	300,000
1957	500	350,000
1958	525	450,000
1959	560	550,000
1960	640	650,000

Source: Television Digest

subscribers in June. Conroy and Spires each made $100 a month salary.

Trained in the Navy to run things by the book, Conroy followed the Lansford model religiously. He charged an installation fee of $125 per home and $4 per month for service (plus an 8% excise tax).

It wasn't as easy as it had been in Lansford, and it wasn't as cheap as Jerrold had projected. But Conroy stuck it out, putting up with comments about the "Yankee" and even enduring a "snipe hunt" for the amusement of the local jokesters. He struck up a friendship with Uvalde's most famous citizen, former Vice President of the United States John Nance Garner.

"We'd go over to Garner's about 5 p.m. some days, and, as they would call it, 'strike a blow for liberty.' He'd break his bottle of Jack Daniels out of the ice box and (say) 'Well, let's strike a blow.' And we'd chat. He'd talk about the old days, and he'd want to know when this cable was going to get in here. He was one of my first customers. He bought two sets, and he wanted two connections so we split a connection. We had two channels so he wanted a set for each channel. We had a picture taken of him watching Dave Garroway, and I sent it up to NBC to Dave Garroway. We got a letter back from him, 'Glad to see Mr. Garner watching.'"

(Garner reemerged into the national limelight briefly in 1960 when Texas Senator Lyndon Johnson was offered the vice presidential nomination by John F. Kennedy. Johnson called Garner to ask his advice. Garner told Johnson the vice presidency wasn't worth "a bucket of warm spit." Johnson took the job anyway.)

But not everybody was ready to sign up as quickly as Garner. In a town where many residents were barely making a living, several hundred dollars for a TV set plus $125 for cable installation and another $4 a month for service was a fair chunk of change. After six months in operation, Conroy had only 250 subscribers.

Conroy found he needed to promote the product constantly. He went door to door with a yellow legal pad making notes of the addresses of any houses with rooftop antennas, figuring those already had TV sets and were the most likely subscribers. He came back and knocked on those doors to sell service himself. He worked on joint promotions with the local TV set dealers. He wrote ads out by hand and delivered them to the local newspaper.

He offered a discount on installation to anybody who would bring in a chicken to donate to charity. He offered a $10 rebate on service to any subscriber who brought in a new customer. "Anything to get people in," Conroy said. "They weren't flocking to the door by any means. We added on about 300 a year pretty evenly as we went. We didn't crack that Mexican-American market until we were able to bring in (Spanish language) Channel 41 by microwave from San Antonio in about 1960-61."

Conroy tried to think of other ways to make money. He sought the help of local motel owners, asking them to install cable, but made no sales. Finally he gained permission to put TV sets in the motel rooms and wire them himself. Then he hooked up a coin box in each room and charged two bits to watch. He split the take with the motel owner. It wasn't quite Spectravision, but it did add a few bucks to the bottom line.

With the marketing problems and other issues, the Uvalde system ran in

From Cotton Seed to Coax:
Cottenseed buyer Bob Magness was introduced to cable when he ran into two men who were building a cable system in Texas. Inspired, Magness went on to build a system in Memphis, Tex., which would be the foundation for TCI.

the red for most of the rest of the decade: "You can figure that at 300 people (which was the subscriber total after the first year) at $4 a month, that's not a lot of monthly revenue. So we were in the hole." By 1958 the system owed Jerrold $35,000 for equipment, and the senior Conroy paid a visit to Milton Shapp.

Jerrold, in the meantime, had run into some problems of its own. There were other cable equipment manufacturers now on the scene (Blonder Tongue Labs, C-COR, Philco, RCA, Spencer Kennedy Labs, Technical Appliance Corp., Transvision, and National Antenna Corp, for example). But Jerrold had roughly 80% of the business, partly because of the sales acumen of Shapp and partly because of the "yellow dog" contracts Jerrold signed with so many of its partners in cable operations, pledging them to buy only from Jerrold and giving Jerrold part-ownership of the systems. By 1957 Jerrold was facing an antitrust lawsuit filed by the U.S. Department of Justice.

When the senior Conroy walked in to Jerrold, Shapp was ready to dump his interest in the system in Uvalde. The two agreed that Shapp would sell his 49% of the system and the Conroys would pay Jerrold what they owed for equipment. "Basically we paid what we owed and we got the 49%. We ended up sole owners," the younger Conroy recalled.

By 1960 the system had dropped its installation fee to $10 with a monthly service charge of $8.95. The new deal made cable affordable to many more people, and the Hispanic population began to sign up to get the Spanish station from San Antonio. As the subscriber rolls grew, the system moved into the black. Conroy turned his attention to expansion.

He hooked up with a neighboring operator, Jack Crosby, and the two began to seek franchises in other communities. By the mid-1960s they had formed an alliance with some other operators including Gene Schneider from Casper, Wyo., creating a holding company that Schneider eventually took control of and renamed United Cable Television Co., later one of the nation's largest cable companies.

When that group broke up, Conroy and Crosby bought out their original systems in Uvalde and Del Rio, Tex., won additional franchises and built their own cable company. By 1979, when it was purchased by the Times Mirror Co., publisher of the Los Angeles Times, the company served 340,000 subscribers.

Another Texan building systems in the late 1950s was a former cotton seed buyer named Bob Magness. A native of Oklahoma, Magness had served as a rifleman with Gen. George Patton in World War II and returned home to get a degree in business from Southwestern State College.

He went to work for the Anderson-Clayton Co. buying cotton seed for use in margarine, salad oil, cattle feed and other products. One day, while calling on a ginner near Paducah, Tex., he ran into some folks whose truck had broken down and needed a ride.

He drove them back to town, stopped for a hamburger along the way and learned that they had been involved in constructing a cable system in Paducah. Later, after a friend mentioned what a good business cable was getting to be, Magness tracked down the hitchhikers and picked their brains. By 1955, he had his first system up and operating in Memphis, Tex. It was the beginning of what would become the largest cable company in the world: Tele-Communications Inc. And it would make Magness a billionaire.

Among others who built early cable systems which grew into major cable companies were John Rigas, a theater owner in Couderport, Pa, who acquired the town's cable franchise for $100 in 1953; and Alan Gerry, a TV appliance dealer, who built his first system in Liberty, N.Y., in 1955. Rigas's system was the first in what would become Adelphia Cable Co., and Gerry's was the foundation for Cablevision Industries.

While Magness, Conroy, Rigas, Gerry and dozens of other entrepreneurs were putting up their family savings and borrowing from partners or local banks to build small cable systems, Bill Daniels was beginning to hit pay dirt in his quest for big investors to back the business.

After serving as chairman of the National Cable Television Association in 1956-57, Daniels hung out his shingle, or, more accurately, sent a mimeographed sheet around to all his friends in cable announcing he was in the brokerage business.

The first big bite came from Joe Sariks, who wanted to sell his cable systems serving 7,500 subscribers in Bradford, Pa., and nearby towns in New York state. "He wanted one million bucks, in cash, for his property," Daniels recalled. "He priced it; I didn't. We had no idea what Joe's property was worth. So I went to see Joe and got some numbers together and I struggled with those, 'cause I'm not an accountant."

Daniels worked out a deal with Sariks to sell the system for $1,050,000, with the $50,000 representing Daniels' commission.

A few weeks later Daniels was the speaker at a dinner in Rapid City, S.D. Next to him sat a broadcast executive from Dallas named Charles Sammons. Sammons listened to Daniels' pitch for cable and later wrote to say he wanted to get in the business.

Daniels offered him the Bradford property. When Sammons looked at the figures he found the system was generating in excess of $50,000 a month in cash flow. He asked only one question:

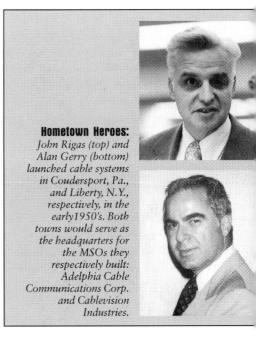

Hometown Heroes: *John Rigas (top) and Alan Gerry (bottom) launched cable systems in Couderport, Pa., and Liberty, N.Y., respectively, in the early 1950's. Both towns would serve as the headquarters for the MSOs they respectively built: Adelphia Cable Communications Corp. and Cablevision Industries.*

Sammons History:
Bill Daniels sold Charles Sammons the cable system serving Bradford, Pa., in the late 1950s. It was the cornerstone of Sammons Communications. Sammons (right) is seen here with a young Jeff Marcus, a former Sammons executive who later founded Marcus Communications.

"Will you take a check?' As Daniels watched, Sammons wrote a check for $1,050,000, a price equal to 1.75 times the system's cash flow, or $133 per subscriber.

Daniels later recalled "Sammons, having been around the business world for years, knew you usually buy a company for five to 10 times (cash flow). Today it's 10 times cash flow. But what I didn't realize, and Sammons did, was here's a hell of a business. It will pay out in 18 months and even if the business goes to hell and there's some kind of a wave that brings television over the mountains, I have minimal risk."

Sammons bought some more properties from Daniels and eventually hired him to manage them in exchange for 20% of the equity, a formula Daniels used with other major investors who didn't want to actively manage their properties.

Another early Daniels customer was Bob Magness. In 1958, Daniels sold the system in Bozeman, Mont., to Magness. Over time, Magness and the company he would later found (TCI) would become Daniels' biggest customer.

Sammons' banker was the Bank of New York. As Daniels found other would-be investors he introduced them to the bank, which quickly became the largest lender to the fledgling industry.

But bringing other banks and financial houses into the business proved to be a tough sell, Daniels found. "In the early days, in my dealings with the banks and investment bankers, the principals in the company had to personally guarantee the loans (which) leaves very little risk to the bank. The next step was they agreed that they would lend money from three to five years without personal endorsements on the loan. Then it grew to five to seven years.

"The cash flow is high, and everything is rosy today," Daniels recalled 30 years later. "But in the early days it was tough convincing them that some kind of invention wasn't going to put us out of business overnight."

There were plenty of candidates to put cable under. As the TV frenzy swept the nation, entrepreneurs, backyard electronics buffs, inventors, crackpots and broadcasters themselves considered a wide array of techniques to extend the broadcast signals. Boosters on broadcast towers and construction of mini broadcast towers (known as satellites) were the two most commonly suggested solutions.

One enterprising businessman in the Rocky Mountain region began to market a "rhombic antenna," that was 80 feet high and 150 feet wide. The antenna was intended to be placed on a mountain top where it would reflect broadcast signals down into an otherwise unserved valley. The materials, the investor told *Television Digest*, had cost only $25. The problem was how to mount such an unwieldy structure on a mountaintop. Nevertheless, he reported getting more than 5,000 inquires.

Despite the development of these alternative systems, the growth of cable continued. Many of the early entrepreneurs sold out, becoming wealthy beyond their dreams in the process.

Fueling the fires of system sales in the 1950s was the federal income tax law, which allowed operators to write off or depreciate the cost of their equipment on their taxes over a limited number of years (typically three to five). After that time, the operators were unable to get a tax break for the amount they had spent for equipment.

But when a system was sold, the new owner could begin the depreciation cycle all over again. This anomaly of the tax law, written to encourage industries to invest in new equipment, had the unintended consequence of making it almost irresistible for cable operators to sell their systems after five or six years of operation. Some of them would then take the money and buy back into the industry.

Meanwhile, the broadcast television business was booming. After the FCC lifted its freeze on new stations in 1952 it began handing out new construction permits by the bushel. The number of stations grew from 109 during the freeze to more than 300 two years later. Many of these stations were serving smaller- or medium-sized communities where cable also had gained a foothold.

For most of the '50s, broadcasters looked on cable simply as a way to extend their signals into areas that couldn't otherwise be reached. But as broadcast stations began to spring up in towns where cable also was present, the two were bound to clash.

One of the first conflicts came in Asheville, N.C. There the city council banned construction of a cable system because the prospective operator of a local UHF broadcast station said he wouldn't build as long as he had to compete with the big Charlotte station being imported into Asheville via cable.

In Memphis, Tenn., the local broadcaster informed area cable systems it regarded use of its signal as piracy.

In Fairmont, W.Va., the owner of a UHF broadcast station called for federal legislation to limit cable systems, stating that the local system would carry his station only two days a week, threatening his "very existence."

Copyright holders also were uneasy about the new medium. In Reno, Nev., the producer of the popular "Cisco Kid" TV series took the cable system to court, charging that the rights to the series had been sold exclusively to the local broadcast station and that the cable system therefore was violating the contract by importing the San Francisco station

---

Following The Money

Jerrold Electronics founder Milton Shapp laid out the economics of cable to Television Digest in the mid 1950's:

Expenses:
- Building an antenna tower and running cable to town: $2,500 to $3,500 per mile.
- Wiring the town: $3,000 per mile or $25 per home with a 135 homes/mile density.
- Fixed costs (trucks, legal, offices, test equipment): $5,000 to $10,000/system.
- Startup office costs: $2,000 per system.
- Total prelaunch construction for 2,000-home community: $80,000.

Revenue:
- Installation: $125 per home.
- Monthly fee: $3.50 to $3.75 per home.

aring the same show. Television producer Arche Mayers told the 1954 National TV Film Council that cable systems "cheat" the film industry because they paid no copyright fees on the programs they import from distant broadcast stations.

In Casper, Wyo., the applicant for a local TV station license dropped his bid, noting that he could not compete for viewers with the Denver stations that were being imported by Casper's cable system.

Many broadcasters speculated that as stations continued to grow, they would begin to put out of business the cable systems that previously had been the only way to receive television signals. But cable defied these predictions and continued to expand.

In Kalispell, Mont., the owner of a local broadcast station shut his operation down, blaming competition from the distant signals imported on the local cable system. The local cable system offered to buy him out and run the station itself.

The storm clouds were clearly brewing. In the spring of 1956 they burst when a group of television and radio stations in western states filed with the FCC seeking federal regulation of cable systems.

Cable, they argued, was making it difficult for local broadcast stations to survive in small and mid-sized communities. The stations simply could not compete for viewers or national advertising with distant signals imported by the cable system from bigger cities.

The broadcasters asked the FCC to declare that cable was a common carrier (like the phone companies) and to regulate cable's rates and services the way it regulated other common carriers.

By mid-decade, the larger stations had become uneasy about cable as well. Their concern was fueled in part by the prospect of pay television that would bring uncut movies and other fare into the homes of TV viewers, causing them to switch off the programming offered by broadcast stations.

One of the first such experiments took place in Palm Springs, Calif. International Telemeter, a subsidiary of movie studio Paramount Pictures Corp., built a cable system with the express purpose of testing pay-per-view movies. It launched the "pay as you look" service in November 1953, offering customers the chance to watch the premiere of "Forever Female" with Ginger Rogers for $1.35. A local theater owner was charging $1.15 to see the same film in the theater. The system also offered the Notre Dame-USC football game for $1.

Some 73 homes in Palm Springs were equipped with coin boxes that allowed viewing of the movie and football game. But when theater owners, alarmed at the prospect of studios selling directly to consumers, formed a group to protest fee TV, Paramount retreated, shutting down its Palm Springs pay-as-you-see operation in the fall of 1954.

The most widely publicized pay television operation in the 1950s took place in Bartlesville, Okla. A chain of theaters, Video Independent Theaters, decided that the best way to compete with television was to join it. VIT won a franchise to build the cable system in Bartlesville with the express purpose of offering a channel of uncut motion pictures.

Bartlesville was picked because it was a town well within the range of three major broadcast stations in Tulsa. VIT figured that if it could make the operation work in Bartlesville it could challenge broadcast television anywhere. It offered the 28,000 residents of Bartlesville a channel of first-run, uncut movies for a monthly fee of $9.95. At first, response was positive, but the number of subscribers quickly fell to about 300 from the initial figure of over 1,000. A cut in price to $4.95 a month helped, but the operation was forced to close after less than a year.

The project was a victim of multiple problems. Local broadcast television stations increased their offerings of movies; the local telco hiked its charges for pole rentals in Bartlesville, and subscribers expressed a preference for a system that would allow them to pick the movies they wanted and pay only to see those, similar to what they experienced in the theaters.

But company president Henry Griffing said his operation was simply ahead of its time and predicted that one day pay television would be as pervasive in American homes as vacuum cleaners or refrigerators. Jerrold announced plans to produce equipment that would allow viewers to select and pay for single movies.

All of a sudden it appeared to big broadcast stations that cable, once a nice way to add a few extra viewers in areas beyond broadcast signals, was likely to be a major competitor within the broadcasters' home turf. They began to join their smaller market colleagues in calls for regulation of cable.

The dispute spilled over into congressional hearings where lawmakers, sensitive to the power of broadcasters, took a skeptical view of cable's impact on the broadcast business. The Senate Commerce Committee hired a special counsel, Kenneth Cox, to conduct hearings on the industry.

A Broadcast Industry Friend:
While a member of the Senate Commerce Committee staff in the late 1950's, Kenneth Cox wrote a report urging widespread regulation of cable. He would go on to be an FCC commissioner from 1963 through 1970, when commission rules curtailed cable's growth.

In his report to the committee, Cox took the position that the public interest was best served by the widest possible distribution of broadcast stations. He concluded that cable in many small towns was interfering with the development of local broadcast stations by importing distant signals that competed with the local stations.

He noted that broadcast stations were regulated by the FCC and required, before they were granted licenses, to provide an array of public services. Cable systems, he pointed out, were free of such requirements. Cox urged Congress to give the FCC the authority to regulate cable. "It seems clear," the Cox report stated, "that the TV industry cannot thrive and grow, to the greatest ultimate public interest, if it continues to exist only half regulated."

Broadcasters had always wrapped themselves in the red-white-and-blue cloak of public service, claiming that their businesses deserved special status because they provided essential services to the public. Among those making this claim for broadcasters was the godfather of broadcasting, David Sarnoff, chairman of RCA and NBC. Speaking before the 1953 convention of the National Association of Radio and Television Broadcasters, Sarnoff took a shot at those who would charge viewers for the chance to watch television.

Of such operations, Sarnoff said: "It would be a negation of the philosophy upon which American broadcasting has been established – freedom to listen and in television freedom to look. It has become part of the American heritage."

Customer service problems also were beginning to plague cable systems. This was a natural outgrowth of the speed with which many cable systems were constructed. In addition, many of the homemade or early prototype versions of amplifiers, splitters and other cable equipment weren't very good. The pictures the systems delivered were consequently often filled with snow or other so-called "artifacts" that could give viewers a splitting headache after a half hour of watching.

Initially, cable customers were happy to pay to get any type of signal at all. But as time went on and some of them had a chance to compare what they were getting with what others closer to the broadcast signals were able to receive for free via rooftop antennas, they began to express dissatisfaction.

A group of subscribers in Walnut Creek and nearby towns in California took their complaints to the state and asked the California public utility commission to regulate the industry the same way it regulated telephone companies. The PUC agreed, and the case went to court.

But the most powerful opponents to cable in the 1950s were the phone companies, and the contest between the David cable operators and the Goliath telcos would continue for decades.

Cable systems needed to put their wires on the utility poles sometimes controlled by the electric companies, but more often by the telcos. The cable systems paid to do this. At first the phone and electric companies saw the rent from their poles – usually ranging for $1.50 to $3 per pole per year — as an added source of revenue which cost them nothing.

But as cable became more powerful, pervasive and lucrative, the telephone companies began to consider how they might enter the video delivery business themselves, or at least benefit from higher pole attachment fees. By the end of the decade several major telephone companies had begun to deny cable companies access to poles. Others hiked the rates they were charging, in some cases doubling them.

Several of these cases also went to court, and in the initial round of skirmishes cable systems won the right to string their wires at reasonable rates. But the battles continued, and the issue would not be finally resolved until Congress passed the Pole Attachment Act in 1978.

In 1958 the FCC, in its most important ruling of the decade for the industry, voted unanimously to reject the request of the western states broadcasters that the FCC regulate cable as a common carrier. Common carriers, the agency said, were characterized by a willingness to carry any message from anyone willing to pay to have it delivered. Telephones and telegraphs operate this way.

But cable systems, the commission said, were not common carriers because cable operators decided which signals to carry based on the needs and desires of their customers, not simply on the ability of a would-be sender to pay.

The commission was asked several

times to reconsider the issue and continued to come to the same conclusion. In April, 1959, for example, it stated that if the FCC were to regulate cable in an attempt to limit its impact on local broadcasting, the FCC would have to act as a "censor" deciding which programs a cable system could carry and which it could not.

Blocked at the FCC, the broadcasters began to lobby heavily for action in the Congress. They were able to persuade half a dozen members of Congress to introduce various versions of legislation to regulate cable and to limit its impact on broadcasting.

The issue came to a head on the floor of the Senate in dramatic fashion. Throughout 1959, the Senate Commerce Committee's communications subcommittee, chaired by Sen. John Pastore (D-R.I.), held hearings on a variety of legislative proposals designed to follow up on the Cox report.

The hearings and the legislative proposals for regulation split the cable industry into two bitterly opposed camps, one favoring federal legislation that would preempt local and state laws and the other group opposing any attempt to regulate the industry at any level.

Throughout the 1950s the industry had been forced to fight many different wars on many different fronts. Cable was battling the telephone companies, the broadcasters, copyright holders, state and local governments and the federal regulators. These battles took place in state and federal courts, city councils, the Congress, the FCC, in state legislatures, governors' mansions and state public utility commissions.

An influential faction within the cable industry, including the board of directors of the NCTA and its chief

Capital Turnabout:
John Pastore (D-R.I.) watched the cable industry back off its support for a 1960 bill he pushed that would have regulated the industry. Cable engineer Archer Taylor, who witnessed the Senate debate on the bill, said Pastore was "absolutely livid with rage" that cable changed its mind on the measure.

counsel, Stratford Smith, concluded that it would be better to have a single set of regulations on the federal level, administered by the FCC, that would supersede all local and state laws. This, they reasoned, would be preferable to having dozens, if not hundreds of sets of regulations at various levels of government, in various stages of implementation and challenge. They reasoned that the FCC, which had already expressed its unwillingness to regulate cable systems, would be a sympathetic forum in which the industry could make its case.

But other cable operators were bitterly opposed to any type of regulation at any level. Many were conservative Republican businessmen who instinctively recoiled at any type of regulation by the government.

"I can remember this," Bill Daniels said years later. "I can remember that, having fought in World War II and Ko-

Engineering Expertise:
Archer Taylor was an early cable engineer who eventually joined up with Martin Malarkey to form the inflluential Malarkey-Taylor & Associates consulting firm.

rea, and having been in the cable business such a short time and having the government tell me what I could or could not do, I deeply resented it, because I thought 'now wait a minute, goddamit, I've been out there busting my ass and getting shot at, and all I want to do is make a living.'"

When Senator Pastore's subcommittee held its hearings in 1959, Smith testified on behalf of the NCTA that the industry would be willing to support legislation to allow the FCC to regulate cable systems under the same standards that applied to broadcast television stations.

One of the cable operators who favored the legislation was Charles Clements, who had built one of the first systems in Washington state, home to Senate Commerce Committee Chairman Warren Magnuson (D-Wash.). Clements was actively involved in the legislative efforts and served on the NCTA board of directors during the late 1950s.

"I was very active early, politically, with Senator Magnuson," Clements later recalled. "I was party chairman in his district and became very well acquainted with Senator Magnuson...who was chairman of the Senate Interstate and Foreign Commerce Committee which had jurisdiction over broadcasting and communications. A lot of people thought that we ought to put cable television under some sort of control. I worked with Senator Magnuson on that."

But while Clements and others on the NCTA board and staff were working to promote legislation regulating cable, other cable industry figures were beginning to move in the opposite direction. "We had a lot of internal opposition to it in our industry," Clements recalled, "which I didn't realize, from Jerrold Electronics, who were also cable operators. A fellow in Oklahoma, Henry Griffing (chairman of VIT which had built the Bartlesville system), was very, very powerful politically."

The Commerce Committee approved a bill that would have done four things:

• Required cable systems to obtain a license from the FCC.

• Required the FCC to include limits in the cable system licenses that would protect the interests of local broadcast stations that might be threatened by the startup of a cable system in their community.

• Required cable systems to carry all local broadcast signals, without degrading their signals.

• Prohibited cable systems from importing programs that were already being aired by the local broadcast station.

For the hard liners, this was a far cry from the minimal regulation that they had envisioned when the NCTA board asked for FCC jurisdiction over cable. They particularly bristled at the "non-duplication" clause that would have prohibited them from importing into a community any program that duplicated a show being aired by the local station.

As Glenn Flinn, a cable operator from Tyler, Tex., told the Pastore committee: "I believe that a TV signal, once broadcast, is free to be received by anyone, whether an individual or a community antenna."

Backers of regulation continued to believe that the Pastore proposal, while not perfect, would still be preferable to the continued battles at state, local and federal levels. They persuaded Pastore to accept several softening amendments and to reject a proposal that would have required cable systems to obtain permission from any broadcast station before retransmitting that station's signal on the system. Pastore went along with the changes proposed by cable with the understanding that the industry would back the final version of the bill.

But even these changes weren't enough to satisfy the anti-regulation hawks in the cable industry. The big break came when Jerrold president Milton Shapp changed his mind on the bill. He joined with VIT's Henry Griffing to send an urgent telegram to all cable operators urging them to come to Washington to oppose the bill.

As Montana cable operator Archer Taylor remembered it "Henry, I think, was the one that persuaded Milt and said, 'Look, you don't want regulation. You like it the way it's been. You can do anything you want to do until they stop you.' So he and Milt Shapp turned the industry around, and they set up a grass roots campaign to go into the halls of Congress and lobby against the bill that they had been lobbying for."

Shortly before the Senate took up the bill, the NCTA board reversed its positions, declaring that it was now opposed to the measure. Griffing persuaded the two senators from his home state of Oklahoma, Mike Monroney and Robert Kerr, to lead the opposition to the bill on the floor of the Senate.

Just a few months before the Kennedy-Nixon presidential election, just days after the Soviet Union shot down a US Air Force U-2 "spy" plane and brought the world to the brink of a nuclear war, and only weeks after the Congress had finished debating one of the most important civil rights bills in history, the United States Senate devoted two full days of often herated debate to the issue of how to regulate cable television.

But sometimes the most innocuous issues can generate the most rancor. *The New York Times* reported that the debate at times degenerated into a "shouting match" between senators. And *Congressional Quarterly* magazine noted that "the Senate spent two days of heated debate" on the bill.

Backers of the legislation charged that the cable industry had broken its word on the bill. They noted that the NCTA had called for federal regulation just a year before and that several amendments backed by cable had been incorporated into the legislation.

But these changes in the bill weren't enough to placate the industry hawks. "Amendments worked out in earlier negotiations with CATV spokesmen were adopted on the floor but failed to save the measure" from cable industry opposition, *Congressional Quarterly* reported.

The halls of the Senate were filled with lobbyists on all sides of the issue. The NCTA called on all its members to participate, and senators were swamped with calls and letters from cable operators back home urging opposition to the bill.

"We've never seen anything like it in this industry," *Television Digest* reported. "At very last minute, CATV operators pushed panic button, admittedly, called in colleagues from entire

Working the Halls: Clarksburg, W. Va., cable system owner Sandford Randolph, a prime mover in forming the Cable Pioneers, was influential in lobbying his state senators to vote against a 1960 bill that would have regulated cable.

country. Some 100 responded, talked to virtually every Senator in Washington. Campaign was naive and crude by most lobbying standards, but it worked."

The debate became highly charged and very personal. Senator Kerr, in particular, launched a rhetorical barrage at the NCTA and Smith. Taylor, who was sitting in the Senate gallery with Smith listening to the debate, later recalled: "Senator Kerr of course, he was owned lock stock and barrel by interests in Oklahoma, mostly oil, but Griffing was a powerful enough one so that when Griffing said 'vote against it,' he voted against it. So (Kerr) got up and made this impassioned speech about how the lawyers were just walking all over these poor little people that are running cable television systems: 'Strat Smith is the guy; he's just terrible, he's just taking these people down the primrose path.' And here he (Smith) is sitting right along side me and he can't say a word.

"It was vicious. And he (Kerr) cried, you know. Kerr could weep at the drop of a hat; it was important to weep." Smith also remembered Pastore pointing to Smith in the gallery and telling the Senate: "That man assured me the cable industry would support this bill." Pastore, Smith recalled, added some comments about lawyers and lobbyists that were later deemed too intemperate and were stricken from the *Congressional Record*.

Behind the scenes, the opponents of the bill sensed victory and intensified their lobbying. Sandford Randolph, manager of the system in Clarksburg, W.Va., and president of the NCTA, persuaded his home state senator, Democrat Jennings Randolph (no relation) to execute a "live pair" on the Senate floor. (This procedure occurs when a senator who is absent for a vote asks a colleague on the other side of the issue to refrain from voting. The "live pair" would be executed by one senator standing up and saying "My colleague, Senator Randolph, cannot be here today. He would normally vote against the bill. I would normally vote for it. As a courtesy to him, I am withholding my vote.")

In the end, the tearful Senator Kerr and the cable industry hawks won. By the margin of a single vote, 39-38, the Senate agreed to send the Pastore bill back to committee, effectively killing it. The bill probably would not have become law, at least that year, in any case, because the House had not scheduled a vote on a similar measure.

But the exercise left a lasting legacy.

The cable industry engendered an enormous amount of bitterness among some of the most powerful Senators and among those in the industry who had earlier lobbied on behalf of the legislation.

As Clements later noted. "I had to go to Magnuson's office, and he had this bill introduced on behalf of the industry and then the industry defeated it. The two factions of the NCTA were not communicating with one another, and a deal one group would make was likely to be abrogated by the other.

"We had a lot of internal opposition

to it in our industry which I didn't realize," Clements noted. "I thought everybody in the industry wanted it. I was mistaken. So was our Washington, D.C., staff."

The NCTA's reversal of position and subsequent divisions provided fodder for its opponents. But the industry also alienated some of its strongest supporters and some powerful lawmakers who had been neutral. Senator J. William Fulbright of Arkansas, for example, had been a supporter of the cable industry and had been a prime sponsor of the bill. But after the debate, he told Arkansas cable operators, according to Taylor, "Look I'll always support you in the Senate, but don't ever come and ask me to front for you again."

And Senator Pastore, a short, hot-tempered lawmaker who had been willing to work with the cable industry was, according to Taylor, "absolutely livid with rage because, he said, 'I had no ax to grind at all, I have no cable systems in my constituencies. I supported you on this bill and now you're coming along and asking me to turn it around.'

"Boy, he was just-double crossed. He was livid with rage," Taylor remembered.

Soon after the vote it became evident that the industry's political woes were not over. The FCC, able to read the mood on Capitol Hill, recommended legislation to prevent cable systems from carrying programming duplicated on local stations and to require operators to get permission to carry signals from distant stations.

The industry dug in its heels. It voted to oppose any regulation and announced a search for a national spokesperson to carry its case. Smith, who had testified before Congress about the NCTA willingness to accept regulation, stepped down from his post as NCTA general counsel.

The old adage says history repeats itself. That's certainly true in this case. The 1960 Senate melodrama was to be reenacted almost without a script change exactly 30 years later when the NCTA board again reversed its position on a piece of regulatory legislation and was successful in killing it at the last minute. Like the 1960 event, the 1990 change of position left many Senators charging betrayal and would return to haunt the industry a few years later.

The cast of characters even overlapped. Two senators on the floor to vote on the 1960 bill (Strom Thurmond and Robert Byrd) were still around for the debate in 1990. And the fathers of both George Bush and Al Gore were both present in the Senate to cast votes in 1960. In 1990 Bush, as President, and Gore, as senator from Tennessee, were to play key roles in the debate over cable regulation.

Dodging the various bullets that local, state and national governments sent their way, cable operators continued throughout the 1950s to expand their business. By the end of the decade, according to a *Television Digest* estimate, some 650,000 U.S. homes were getting their television signals from 640 cable systems. On average, more than one new cable system had launched every week since the Lansford, Pa., system began operations in 1950. An average of more than 180 new customers signed on for cable service every single day during the 1950s.

Bill Daniels was right. It was a hell of a business.

Most systems were making money

hand over fist. One of the nation's largest systems in the mid-1950s was in Clarksburg, W.Va. In 1956, the 7,000-subscriber system reported profits of $264,000 on income of $415,840. That amounted to an astounding 64% cash flow margin. The system was owned by Fox, Wells, one of the first big Wall Street firms to heed Shapp's summons to enter the cable business.

Within a year the system was sold to NWL Corp., another New York investment company, for just under $900,000, or a multiple of just under 3.5 times cash flow.

Meanwhile, in the courts, the industry won a huge victory when the U.S. Circuit Court of Appeals struck down the IRS proposal to impose an 8% excise tax on cable system subscriber fees. The excise tax, the court ruled, was not intended to apply to cable systems which did not even exist when Congress passed the law imposing the tax on utilities.

C-COR Legacy:
Jim Palmer, chairman and CEO of C-COR Electronics (shown here in 1977), led the equipment manufacturer through much of its history. C-COR built a cable powering system in 1953, and would go on to introduce integrated circuits in amplifiers, and 320 MHz and 1 GHz amplifiers over the next five decades.

The victory was a big one for the NCTA and Smith.

The NCTA and Smith also were deeply involved in a California case in which the state supreme court struck down an attempt by the state Public Utilities Commission to regulate cable as a utility. The case was the first such one in state courts.

To combat growing customer service problems, the NCTA instituted a standards and practices committee chaired by Meadville, Pa., operator George Barco. The group was charged with establishing a set of guidelines for operation of cable systems.

But customer service problems would continue to plague operators through the remainder of the century.

Partly it was because of the technology. "At the time everything was tubes," Casper, Wyo., cable operator Gene Schneider recalled. "There was no such thing as a transistor. Those tubes used to go out overnight, the worst time of night, right in the middle of a fight or something and blow the whole bit."

Schneider (no relation to Ray) remembered keeping his fingers crossed every time there was a big sporting event. "I can recall times when the wind was moving the dish and the pictures were flopping in and out. Whenever you had a big event like a Rose Bowl Game or a World Series you always were just on pins and needles hoping, 'My gosh, I hope this thing doesn't go off.'"

The cable itself was subject to weather. Moisture would get inside the dielectric and cause snowy pictures. During cold weather, the copper core would shrink and disconnect from taps or amplifier hookups.

But other problems were the result of more than just first-generation technology. Some operators built their own equipment. (The National Cable Television Center & Museum has some amplifiers that were welded inside coffee cans.) And some operators cut corners to save on cash.

Ray Schneider, who later worked for Times Fiber selling coaxial cable, remembered one operator in Michigan who would call every so often to see if he could buy the odds and ends of cable that were left on the spools after all the standard lengths had been sold. Schneider was happy to sell him these for next to nothing. He estimated that this operator had wired his system with thousands of cheap, short pieces of cable spliced together. Each splice, of course, was a potential source of problems.

In one of the first articles on the economics of the new industry, *TV Digest* interviewed Shapp, who estimated that the cost of building a system would average out to about $5-$6 per person in the community. The costs broke down as follows:

• Antenna tower and cable run to town, approximately $2,500 to $3,500 per mile depending on the distance.

• Wiring the town, about $3,000 per mile or about $22-$25 per home in a community with 135 homes per mile.

• Other fixed costs such as trucks, legal expenses, offices, test equipment, etc., added $5,000 to $10,000 per system. Preliminary operation costs such as maintaining an office, pre-launch promotion and so forth would add $1,500 to $2,000.

This would bring pre-launch construction and other costs to about $80,000 for a community of 2,000 homes. Costs per home for installation were pegged at $18-$20 per home. Most operators, he said, were charging about $125 per installation and a monthly fee of $3.50 to $3.75. Operating costs, he said, would be "modest" with the average system employing between four and eight people.

Using those numbers, he projected that a system signing up 500 homes in the community could pay back its startup costs in less than two years. After that, the vast bulk of revenue would be profits. No wonder the number of cable systems was increasing by one a week.

And the number of channels served by these systems was increasing as well. In October, 1953, Jerrold installed its first five-channel system and told cable operators it would cost about $400 a mile to upgrade from the old three-channel service. And the introduction of color proved no problem for the robust coaxial cable. In Pennsylvania, Jim Palmer led his manufacturing company, C-COR Electronics, in developing cable powered amplifiers. Blonder Tongue Labs introduced split-band amplifiers to replace the old strip amps and provide 12 channels rather than five channels of capacity. And Spencer Kennedy Laboratories introduced the first broadband amplifiers based on a concept originated

Early Customer Service:
Decades before the NCTA implemented customer service standards, George Barco, a Meadville, Pa., cable system operator was working on guidelines that would help operators serve customers better.

by a British engineer for EMI. This amplifier provided the platform to deliver even greater channel capacity.

In Arkansas, cable operator Jim Davidson founded a distribution company that would assemble a complete headend at Davco's plant and then fly it to the system where it could be installed in a single day.

With such a growing and lucrative business at stake, it was not long before juicy franchises were the subject of warring parties. City councils were asked to choose between competing bids for the right to wire city streets for cable. This process degenerated rapidly, drawing out the worst instincts of local politicians and the worst instincts of cable operators.

One of the first big contests was in Dubuque, Iowa, where a group backed by Jerrold competed against a local operation proposing to use equipment manufactured by Spencer Kennedy Labs. Two local professors hired as consultants by the city council recommended Jerrold, but the council voted for the better-connected local group instead.

Jerrold then took its case to the people, garnering 400 petitions to place the issue on an election ballot and then winning a special election that generated a bigger voter turnout in Dubuque than any election except the 1952 presidential contest. The effort paid off. By the end of its first year in operation, Jerrold had more than 2,200 subscribers.

Without question the 1950s was the Jerrold era in cable television. The company had grown from a side business started by Milton Shapp to a major corporation with annual revenues of $8 million by 1959. Shapp had brought the first serious money into the business and built the first really big systems, securing most of their business for Jerrold with exclusive supplier contracts and keeping part-ownership, as he did with the Uvalde, Tex., system.

Jerrold also entered into arrangements with favored companies that supplied equipment Jerrold did not make itself. Times Wire & Cable of Wallingford, Conn., for example, had the inside track to provide most of the coaxial cable in Jerrold-owned and operated systems, giving it the lion's share of the coax market in the 1950s.

Shapp was a tireless and outspoken booster of cable, making speeches around the country touting the business and predicting tremendous growth. Jerrold supplied much of the equipment for the Bartlesville pay TV system, and by 1956 the company let it be known it was considering buying first-run films directly from Hollywood producers for showing on pay-per-view basis on cable systems.

All this was bound to draw the attention of Jerrold's competitors and eventually the U.S. Department of Justice. This was particularly true because of the atmosphere surrounding television in the late 1950s.

TV had exploded into the national consciousness in a way that affected almost every aspect of American life. In less than 10 years the number of TV households grew from less than 100,000 to more than 40 million. Viewers were glued to "I Love Lucy" and "Show of Shows."

And TV demonstrated its political power as well, destroying the career of anti-Communist witch hunter Joseph McCarthy and bringing such events as the national political conventions into almost every living room in the nation.

The power of television, increasingly

concentrated in the hands of the broadcast networks, threatened the stability of the political process, and politicians scrambled to gain a hold on the medium. This movement was accelerated with the payola scandals of the late 1950s when a shocked nation learned that the most popular quiz show on the air had been fixed.

Congress jumped quickly into the fray, investigating the payola scandals and launching an investigation of television itself.

In such a highly charged atmosphere, Jerrold stood out as an almost irresistible target. Competitors and cable operators alike had complained steadily about Jerrold's tactics. Montana operator Archer Taylor decided not to use Jerrold equipment. When Taylor's partner ran into the local Jerrold representative in a Montana barroom one day, the rep shouted across the room that Jerrold would "run you right out of business." Taylor's partner wrote letters recounting the incident to the President, the FCC and members of Congress. He wasn't the only one to complain of heavy-handed tactics by Jerrold.

In 1957, the U.S. Department of Justice filed suit charging Jerrold with restraint of trade and other violations of the Sherman and Clayton antitrust acts. The Department's suit cited Jerrold's practice of insisting on exclusive contracts with cable companies, its ownership of cable systems and its alleged threats to build competing cable systems in communities where the cable operator refused to do business with Jerrold.

Jerrold strongly denied the charges, and Shapp noted that his company competed with much bigger entities such as RCA and Philco for the cable equipment business.

The case dragged on for more than two years, consuming considerable resources for Jerrold. During that time the company began to abandon some of the practices that had landed it in such hot water. It eliminated its exclusive contracts and began to shed ownership in cable systems, selling nine of them to H&B American, a Los Angeles-based construction and transportation company. Once H&B got a taste of the cash flow cable could generate, it jettisoned its other businesses and quickly catapulted itself into the position of the nation's largest cable operator.

When a decision in the Jerrold antitrust case finally arrived, in 1961, it proved to be mixed. The court restrained Jerrold from buying any more cable systems, but did not expressly order it to divest those it already owned. It also prohibited the exclusive contracts which Jerrold said had already been discontinued. But the court did not go further in punishing Jerrold, as the Justice Department had requested.

Nevertheless, the lawsuit effectively marked the end of an era in which Jerrold so completely dominated so many aspects of the cable business — operations, manufacturing, franchising, finance, lobbying. Shortly after the suit was filed, Jerrold acquired the Harmon Kardon consumer stereo equipment company, expanding outside cable for the first time. It then sold the rest of its cable systems. Finally, in 1961, Shapp sold his one-third interest in the company to oil and entertainment executive Jack Wrather for $4.4 million.

Shapp remained as Jerrold's president but shortly exited the company to concentrate on politics. He was elected governor of Pennsylvania in 1970 and was a candidate for the Democratic nomination for president in 1972.

Jerrold (the name was changed in the 1990s to the GI Communications Division of General Instrument Corp. and again in 1997 to NextLevel Systems) would continue to be a leading supplier of equipment to the business and a key developer of new technology for cable for the rest of the century. It would even be able to get back in the cable operating business briefly in the mid-1960s.

But it would never again play the central role in so many aspects of the industry as it did in the 1950s. The baton of leadership in the industry was about to pass to others.

At the head of the list in 1960 was a rotund 43-year-old from New Jersey, the son of Russian Jewish immigrants, the winner of a baton-twirling scholarship to the University of Alabama, the nephew and namesake of composer Irving Berlin, the holder of a patent on a new device that allowed actors and politicians to read scripts without moving their eyes from a camera, a leading promoter of heavyweight boxing events, and a consummate showman himself.

His name was Irving Berlin Kahn and his company was TelePrompTer.

Canada

While the Americans were getting their cable industry up and running, the industry was also getting under way in Canada. The first and most ambitious system was built in the 1950s in Montreal by Rediffusion Ltd., a British company that had constructed small master antenna and single-channel cable systems in the United Kingdom following World War II.

According to Rediffusion chief engineer K.J. Easton, in his book "Thirty Years in Cable TV," by September 1952, Rediffusion had run 200 miles of cable past 58,000 homes in Montreal.

But the system quickly ran into problems. It was constructed to carry only two channels.

The idea was that one would be the fledgling Canadian Broadcast Corp. channel and the other a channel of French-language films for the big French-Canadian population. But CBC didn't like the idea of competing with a film channel, in any language, and forced Rediffusion to take it off.

Second, it was possible for nearly all the homes in Montreal to receive the CBC channels and U.S. channels from across the border without having to pay for cable. Finally, unlike the British, Canadians preferred to purchase their own TV sets rather than rent them, so the combined TV rental-cable service fee that worked for Rediffusion in the U.K. failed across the Atlantic.

But while Rediffusion ran into problems with its big Montreal system, entrepreneurs began to spring up in small towns all across the country, importing broadcast signals into communities that could not get off-air signals.

Easton credits two backyard electronics buffs from London, Ontario, Ed Jarmain and Harry Anderson, with building the first system in Canada in 1951 after they had paid a visit to Martin Malarkey in Pottsville, Pa.

By 1957 the National Community Antenna Association of Canada had been formed. By 1959 *TV Digest* reported that some 200 systems in Canada were serving more than 135,000 homes.

The Big Players Arrive

Irving Kahn lived large.

At five and a half feet and roughly 250 pounds, he cut a wide swath when he moved across the floor of a cable convention or into the best seats at a four-star restaurant or championship boxing match.

Kahn was one of those improbable "only in America" stories. He was born in 1917, the son of Russian Jewish immigrants and nephew of the great composer Irving Berlin for whom he was named. Kahn grew up in New Jersey. While at Boy Scout camp one summer he learned to twirl a baton. The only high school in his area that had a pipe, drum and bugle corps and could use a baton twirler was St. Benedict's Prep, a Catholic school.

Kahn later recalled: "When I got to the point where I was really good, I went down to St. Benedict's Prep in Newark and talked to a guy. I said 'You don't have anybody who can really twirl. How about it?' And he said 'Sure, be glad to have you.' They paid me five dollars a night for every competition. I got really good at it. I won the Eastern State Championship." That opened up the possibility of a scholarship to the University of Alabama.

In 1938 the Alabama football team went to the Rose Bowl. Twirling the baton at the head of the Crimson Tide's Million Dollar Marching Band was the Russian Jewish kid from New Jersey who had won the state championship on behalf of St. Benedict's Prep.

After college Kahn did some PR work for dance bands in New York City, trading gossip for plugs with columnists such as Walter Winchell. He worked briefly for 20th Century Fox Studios before serving in the Army Air Force in World War II, primarily as a public relations man.

After the war he went back to Fox. One day an actor walked into Kahn's office. As Kahn later recalled, "He's got a butcher roll of paper with handwritten lines on it, like cue cards, but on a roll." It was the first TelePrompTer.

Fox didn't want to take on the new business, so Kahn quit in 1950 and formed his own company, TelePrompTer Corp. He sold TelePrompTer service to everybody, from the broadcast networks and movie studios to David Rockefeller and David Sarnoff. He even went to the Democratic and Republican National Conventions in 1952, personally working with Dwight Eisenhower and Harry Truman to teach them how to use the machines.

Renting out TelePrompTers helped Kahn pay his $1,000 per month tab at New York's swanky Stork Club, the fee for his chauffeur-driven limousine and other basic costs of living, but it didn't

Cable Showman: *Irving Kahn cut a larger than life path during his career in the cable industry. Kahn developed the TelePrompTer, and founded an MSO by the same name. He also championed the cause of pay television and later fiber optics.*

make the shareholders of the company rich. Other than the patent, the company had no real assets. Nor was it making much money. In 1959 TelePrompTer Corp. posted a loss of $175,000 on revenue of $3.5 million.

So Kahn looked around for other businesses to enter. His colleague, Hub Schlafley, had a line on a new type of large-screen television set, and Kahn decided it could be used to televise closed circuit events, such as big boxing matches. Kahn signed a deal with heavyweight champion Floyd Patterson and his manager, Cus D'Amato, giving TelePrompTer the rights to televise the Patterson heavyweight fights. It also led to Kahn's first run-in with the government. The Attorney General of the state of New York charged TelePrompTer with attempting to monopolize heavyweight boxing and threatened to break up the company. The case was later settled, but Kahn was already looking for other businesses.

Just how he met Bill Daniels is a matter of some confusion. Each claims he called the other. But the best memory may be that of Monroe Rifkin, then TelePompTer's executive vice president and chief financial officer.

While Kahn was fond of lunching with TV stars and entertainment industry moguls at five-star New York restaurants, Rifkin was more of a pastrami-on-rye-at-his-desk kind of guy.

Born in New York, Rifkin had received a degree in finance from New York University, worked for Touche Ross for two years and then been hired as controller of TelePrompTer.

He remembered that one day he was eating lunch at his desk "and this dapper little guy comes up to the receptionist and says 'I want to speak to whoever is in charge.'" It was Bill Daniels. He said he wanted to "educate" TelePrompTer about the opportunities in the cable business.

Since Kahn was out to lunch, Daniels spent some time in Rifkin's office telling him about cable. Midway through the conversation Rifkin asked, "Okay, so what can these systems be bought for?' Daniels said, 'Three to three and a half times cash flow.' And I said, 'Okay, educate me some more.'"

In August of 1959, Rifkin, who had never been west of Chicago, flew to Denver to see some of the cable systems Daniels had for sale. "I remember I was met at the airport by this gorgeous, statuesque woman with two huge dogs. It was Bill's wife-of-the-moment, Eileen."

Daniels took Rifkin to Wyoming to see the Casper and Rawlins systems and down to Silver City, N.M., to visit a system owned by Bruce Merrill.

Rifkin remembered one system manager who operated out of a dingy office behind a gas station. Rifkin asked him why he didn't have better offices. He said he didn't want everybody in town to know how much money the system was making.

"I flew back to New York City and

Kahn's Sidekick:
Hub Schlafley teamed with Irving Kahn to develop large-screen TV sets for the closed-circuit boxing business.

told Irving we ought to sell everything else and buy cable systems. It was a license to get rich."

TelePrompTer did just that. In December of 1959 it announced plans to buy systems in Silver City and Farmington, N.M., and Rawlins, Wyo., for $747,000. The next month it announced a public offering of stock to raise $2 million to buy another 10 systems, a microwave relay service and a radio station.

Kahn jumped into cable headlong. He made a big splash.

By June of 1960 he was the main attraction at the NCTA convention in Miami Beach where he treated attendees to a live, big-screen, closed-circuit showing of the Floyd Patterson-Ingemar Johansson heavyweight title fight. The event also was delivered to some 25,000 subscribers in 13 cable systems nationwide. Those who watched it were asked to pay a fee of $2 on the honor system.

This kind of event, Kahn told the convention two days later, would be the industry's future. He urged cable operators to sign up to buy a device Kahn had invented called Key TV. The system allowed viewers to order events and movies by pushing buttons that sent a signal from the TV to a recorder on the utility pole outside the house. Servicemen would then collect the data from the recorder and respond to the consumer requests.

Kahn vowed to outbid theaters for first-run rights to movies. He said he would buy the rights to the World Series and charge cable subscribers $2 to watch the Fall Classic. He defended the scheme as good for baseball on the theory that the funds could be used to help struggling minor league teams. Kahn offered to renovate Carnegie Hall in return for the pay TV rights to its concerts. And he predicted that one day cable would be the dominant medium throughout the country, even in its largest cities, even in New York.

Meanwhile, Rifkin was slogging through dusty towns of the Great Plains and Rocky Mountains, buying up small systems. He set up an operating division of TelePrompTer to run the systems, and hired a group of folks to keep track of them, headed by Ray Schneider, who had managed the pioneering system in Williamsport, Pa.

Rifkin and Kahn had a new twist in cable system purchases. Because TelePrompTer was perpetually short of cash, they offered to buy many of the systems using TelePrompTer stock. This was a good deal for many sellers. It enabled the seller to avoid paying huge chunks of the sale proceeds in taxes, to participate in the future growth of TelePrompTer and to convert his stock to cash as needed.

But sometimes Rifkin came across a would-be seller who didn't want stock. Such was the case with a small system in Washington state, just north of the Columbia River.

At the time, the head of TelePrompTer's closed circuit business was a young man named Robert Rosencrans. Rosencrans had come to Rifkin one day and said that if the company ever came across a system for sale which TelePrompTer couldn't buy, he would be interested. Rifkin passed the Washington state deal on to Rosencrans who put together a group of backers to buy the system, laying the foundation for Columbia Cablevision.

Rosencrans was one of several major cable executives who got their start at TelePrompTer. Others who worked for the company in those days included Jack Gault, who later ran the New York

City system for Time Warner, and Leonard Tow, a former business school professor who started Century Communications.

From the time they first met, Bill Daniels had tried to hire Rifkin. And, as Rifkin found out, "Bill doesn't give up." The chief obstacle was Rifkin's wife, who didn't want to leave the East Cost with their three children.

Daniels set about to woo another man's wife. He brought the Rifkins out to Denver for a weekend in the spring of 1963 and arranged a cocktail party at the Denver Country Club so they could meet other couples with young children living in the area.

"He showed us the best of Colorado," Rifkin recalled. "It wasn't all bad." Nor was the promise of an equity interest in the business.

Rifkin left TelePrompTer and came to Denver to head Daniels' cable operating company, "and basically to do everything else." He never went back east.

One of Rifkin's first tasks was to fly to Los Angeles to meet a Canadian businessman named Jack Kent Cooke who had been studying the entertainment business and had concluded he wanted to get into cable. Cooke had made his fortune working with Canadian media mogul Roy (later Lord) Thomson. After leaving Thomson, Cooke decided to become a U.S. citizen. He didn't have the patience to wait the normal five years required of most immigrants so he arranged for a bill to be passed in Congress granting him instant citizenship. It was typical of Cooke's style.

Cooke plunged headlong into American culture. He became a sports nut, acquiring an empire that eventually included ownership of the Los Angeles Lakers of the National Basketball Association and the Washington Redskins of the National Football League. When he first moved to the U.S. he rented a home in Pebble Beach, Calif., with multiple television sets. Cooke was intrigued that the sets had such fine pictures even so far away from the broadcast stations. He learned that this was possible because of cable television. Cooke looked into the business and liked what he saw.

The cable deal Rifkin put together for Cooke involved the purchase of four systems serving 16,000 subscribers in La-

Teleprompter Gave Them Their Start

Monroe Rifkin was Irving Kahn's CFO at TelePrompTer and went on to launch American Television and Communications in 1968.

Leonard Tow (left) and Jack Gault (below) both worked at TelePrompTer. Tow would go on to start his owned company, Century Communications. Gault would rise through the ranks at Time Warner Cable.

guna Beach and Barstow, Calif.; Graham, Tex., and Keene, N.H. The purchase price was $4.6 million. It was one of the largest cable deals ever. Cooke called his company American Cablevision.

"I was amazed at his intelligence," Rifkin said of Cooke, "Particularly his ability to discuss tax, legal and accounting matters. He had the best professional talent he could find. Nothing got away from him." Among those he hired to help him do cable deals were a young attorney named Jay Ricks and Bill Bresnan, the chief engineer of a cable system in Rochester, Minn.

Cooke could be tough to work for. As Bresnan, who later became chief executive of Cooke's cable business, recalled, "Jack insisted on total loyalty and total honesty. But the most important thing for him was the process. Process was more important than results. If you did something the right way and failed, he would back you. If you did something the wrong way, even if you succeeded, he could be difficult."

One day Cooke was at a Los Angles Lakers game. The Lakers were ahead by one point and had the ball with less than a minute to play. They were running out the clock. With just a few seconds left Laker superstar Jerry West took a shot from the corner and sank it, putting the team ahead by three. They ended up winning by three.

But Cooke was livid. He marched down to the locker room and waded through the players — most of them a foot and a half taller than the diminutive Cooke. When he came to West he looked up, pointed his finger as close to the All Star guard's face as he could reach and read him the riot act for taking a shot that, had it missed and the other team scored, could have cost the

Cooke's Process: *Jack Kent Cooke entered and exited the cable business several times, and was an executive for whom "the process" was more important than the end result.*

Lakers the game. "You dumb son of a bitch," Cooke bellowed at the future Hall of Famer, "learn how to play this game right."

Cooke applied the same standards to his employees. In 1973 his cable company, then the largest in the country, was attempting to cross the one million subscriber mark by the end of the year and had put on a big marketing effort to get there. The public relations director at headquarters had devised a slogan, "A million or more by '74," which was stamped on every piece of mail that went out of headquarters.

Cooke got one of the envelopes and called to find out who had been the source of the slogan. He was delighted. "That's the kind of enterprise we need," he said, and called the employee into Cooke's office for a pat on the back. As the session ended Cooke, almost as an aside, asked: "How do you print the slogan on the envelopes?" The employee answered that it was done with a rubber stamp.

"Oh," said Cooke, "and how much did it cost for that rubber stamp?"

"I'm not sure," said the employee. "My secretary bought it."

Cooke's eyes narrowed. "Young

man," he said, "that is not the right answer. The right answer is 'Mr. Cooke, I will find out.'"

So the employee made a quick call to his secretary and discovered that the rubber stamp cost $29.

By day's end Cooke had him fired for paying too much for a rubber stamp and, most of all, for failing to pay close enough attention to expenses.

But for those who met his standards, Cooke could be a great boss. Bresnan, who Cooke called "Willie," was one. Like Cooke, Bresnan had raised himself out of near poverty through sheer hard work. When Bresnan was five, his father, a small-town postmaster, died, leaving four young children and their mother who took in sewing to put meals on the table. Bresnan attended parochial school in the days when the nuns were quick to take a ruler to the back of the hand of any young man showing signs of straying from the straight and narrow. Bresnan learned to work hard and to live by the rules, even if the rules sometimes made no sense to him.

To earn pocket money Bresnan raised chickens and fixed radios. After the chicken coop burned down, turning his inventory to toast, he went into radio repair full time.

He entered a technical school in Mankato, Minn., and became a salesman for Northwest Electronics, a cable distributor. He chose to take a 6% commission and pay his own expenses rather than take a 4% commission that included a per diem from the company. To save money the six-percenters would stay at cheap hotels and drive beat-up cars, but their hunger to make it put them far ahead in the bottom line.

One day Bresnan heard about a new cable system being built in Mankato and arranged to become the distributor for Jerrold Electronics so he could supply hardware to the system. Soon he expanded the business to serve the new system in Rochester which, in 1958, hired him as deputy chief engineer. (Soon after, the chief engineer skipped town, and Bresnan moved into that position.)

In March, 1965, Cooke bought the Rochester system and called up the chief engineer. "I hear you're a pretty good engineer," Cooke told the petrified Bresnan. "Why don't you come out tomorrow so we can talk." So Bresnan flew to Beverly Hills where he was met by Cooke's chauffeur-driven Bentley and taken to the best room at the Beverly Hilton Hotel. Cooke took him to dinner at the Escoffier Room where they were entertained by Cooke's personal string quartet. Cooke asked him what he wanted to drink and without waiting for an answer ordered him a Scotch (the first Bresnan had ever had). Cooke asked him what he wanted to eat and

Passing The Gavel:
Bill Bresnan (left) handed over the NCTA chairmanship gavel to Amos Hostetter (right) in 1973. Bresnan started as a cable engineer in Rochester, Minn., then worked for Jack Kent Cooke before forming his own MSO in 1984. Hostetter founded Continental Cablevision with Irv Grousbeck in Tiffin, Ohio, in the 1960's.

without waiting for an answer ordered him the Dover sole, which Cooke had flown in fresh from England.

Although he remembered being scared by Cooke's intensity, by the end of the evening Bresnan had agreed to move to California to become chief engineer for all of the Cooke Cable systems. Within a year he was Cooke's executive vice president, paying as close attention to Cooke's rules as he had to the rules the nuns had enforced back in grade school. He worked for Cooke for over a decade. Thirty years later he would say the man had treated him very well and taught him more than he had learned from anyone else.

Within six months Cooke had gobbled up systems serving more than 50,000 subscribers.

Cooke wasn't the only entrepreneur to become interested in cable during the early 1960s. Others who got their starts during the Kennedy and Johnson years included:

• Ralph Roberts, a Philadelphia businessman, who formed a group of investors to buy a system in Tupelo, Miss., laying the basis for what would become Comcast Corp. Roberts bought his system from Jerrold. To run it he hired the Jerrold executive who had handled the sale, Dan Aaron.

• Amos B. Hostetter, Jr., a 27-year old investment banker recently out of Harvard Business School who ditched the East Coast with a partner named Irv Grousbeck. After pouring over maps of broadcast signals in the U.S. the two in 1964 settled on a town in Ohio as the best site for a new cable company. They won the franchise in Tiffin, Ohio, on behalf of their newly formed company, Continental Cablevision.

• Former NBC vice president Alfred

Growing Comcast:
Ralph Roberts (seated, left) started Comcast Corp. by buying the Tupelo, Miss., cable system from Jerrold Communications. Roberts hired Daniel Aaron (standing, left) from Jerrold to run the system. Roberts, Aaron and Julian Brodsky (standing, right) formed the cornerstone of Comcast. Seated at right is W.W. Keen Butcher, of Butcher & Sherrerd, which underwrote a 430,000 share stock offering for Comcast in 1972, which raised $2.7 million.

Stern in 1962 formed a company, Televents, to purchase 18 systems serving 43,500 subscribers from C.A. Sammons. Carl Williams, who headed the Daniels & Associates operating unit, was named president of the new venture. The purchase price of $10 million was twice as large as any cable deal in history.

• Charles Dolan, president of a New York company — Sterling Information Services — began to wire New York City hotels to deliver a channel carrying information about city events. He also offered New York Mayor Robert Wagner the chance to use the system to address the many conventions that met in New York. Wagner found the system a great time saver. It was an early example of telecommuting and an early indication of Dolan's keen instinct for the nuances of local politics.

Broadcaster Inroads:
Former NBC VP Alfred Stern formed Televents in 1962 to purchase 18 systems from C.A. Sammons.

Phone companies also continued to flirt with cable. Southern Bell announced it would no longer make its poles available to cable operators but would instead offer to build entire cable systems and then lease them back to operators who would handle the programming, marketing, billing and other operations. But the price the telco planned to charge for this turnkey operation was far in excess of what cable operators were used to paying when they built their own plant.

Texas operator Ben Conroy, heading up an NCTA committee on pole attachments, went over the heads of the Bell South folks, persuading AT&T to adopt a nationwide policy of renting poles to cable operators at a reasonable price. But Ma Bell also said she would step up her efforts to market turnkey services at lower prices. The skirmish ended well for the cable operators but reminded them once again of their vulnerability to the utilities that owned the poles on which cable was strung.

The biggest news for the cable industry in the early 1960s was the flood of broadcasters into the business. Among the notable broadcaster entries into cable were:

• CBS, became the first network to own a cable system when it bought the system in Vancouver, B.C., in November 1963.

• Westinghouse Broadcasting Co. purchased four cable systems and a microwave operation in Georgia for $1.1 million in 1964.

• Cox Broadcasting Co. bought its first two systems in 1962. They were Lewiston, Pa. (4,500 subs for $660,000) and Aberdeen, Wash., (9,500 subs for $1.5 million). The latter was bought from Seattle broadcaster J. Elroy McCaw. Cox, under the leadership of J. Leonard Reinsch, then embarked on a rapid expansion plan in the cable industry.

• Lucille Buford, owner of Tyler, Tex., station KLTV, bought a 50% interest in the cable system in Lufkin, Tex.

• San Jose, Calif., broadcaster Allen Gilliland, announced a joint venture with Jerrold to seek franchises in Monterey and Salinas.

• Newhouse Broadcasting in 1964 applied for the cable franchise in its headquarters town of Syracuse, N.Y., and purchased the system in nearby Rome, N.Y.

• General Electric Broadcasting formed a cable unit, headed by Douglas Dittrick, to apply for the franchise in its headquarters city, Schenectady, N.Y., and other communities.

• Harriscope, headed by broadcaster Burt Harris, purchased systems in Palm Springs, Calif., and Flagstaff, Ariz., from H&B. (When Harris walked in with his $250,000 check as a down payment on the Palm Springs system, H&B officials sheepishly told him that they had sold the system the night before to Jack Kent Cooke, who inked the agreement with H&B president David Bright on a napkin at a Los Angeles restaurant. But Cooke, it turned out, didn't act in the time frame outlined by the napkin deal and Harris got the system anyway.)

With this influx of new investors, the price of cable systems during the first half of the 1960s rose from about 3.5 times cash flow to about 7.5 times cash flow, Daniels estimated. He was in a position to know. His firm handled 80% of the deals during that period.

The business continued to be highly profitable. A study commissioned by the Federal Communications Commission and released in 1964 found the average cable system enjoyed an operating profit margin of 57%, far ahead of most other businesses, including broadcasting, which was pegged at about 30% margins. Costs for a typical 3,600-subscriber cable system were estimated as follows in terms of a percentage of gross revenues:

Salaries, 13%; pole rentals, 4%; advertising, 3%; maintenance, 2%; franchise fees, 4%; microwave costs, 8%, leaving a profit before interest, depreciation or taxes of 57 cents on every dollar of revenue taken in.

The study estimated that the average system spent $3,500 to $4,000 per mile to build its plant and took in $60 per subscriber per year. Three percent of systems offered one or two channels, 85% offered three to eight channels and 11% offered more than eight.

The NCTA estimated that there were some 1,600 cable systems in the U.S. at the start of 1965, an increase of 200 in just six months. Applications for permission to build systems were pending in another 1,000 communities, the trade group said.

An article in *Newsweek* noted that since the industry had started only three cable companies had failed. This was a failure rate of one twentieth of one percent, the magazine said, one tenth of the average failure rate for U.S. businesses.

Fueling the growth of cable in the early 1960s was a huge leap forward in cable technology.

Since its birth, the cable industry had been plagued by technological problems. Much of the equipment used to build the first systems was either made by the operators themselves or jerry-rigged from equipment manufactured for other purposes. Cable-oriented equipment manufacturers had trouble with first-generation equipment. They also were fiercely competitive and eager to feed a hungry market, leading them to put equipment into the field at times before it had been adequately tested. Even the best equipment didn't always work very well and didn't last very long.

One of the biggest sources of trouble was the coaxial cable itself. Many of the early systems – including Davidson's and Tarlton's – used military surplus

Cox Leader:
J. Leonard Reinsch led Cox Broadcasting's foray into the cable business in 1962. Reinsch is show here with fellow Georgian, President Jimmy Carter and his daughter, Amy.

cable, built originally for carriage of audio signals. The copper core was loosely encased in its dielectric and would move around. This was not a problem for carriage of audio, but created real difficulties when video transmissions were added.

The sheath, made of woven copper mesh, was susceptible to leakage of two kinds. Moisture would seep in around the taps and other connections, creating havoc with pictures down the line. Cable operators found it difficult to locate and repair these types of faults. The cable also was prone to signal leakage. As more towns became home to both broadcast and cable systems, the broadcast signal would sometimes leak into the cable, creating a dual signal that would produce ghosting on the customer's screen.

There were a number of attempts to fix this. Gene Schnieder, then head of GenCoE., remembered that in the late 1950s a new type of cable, called strip braid, came along "and everybody thought it was just great. It had low loss characteristics, and you could space the amplifiers wide apart, and a lot of other good things could be said about it.

"But after being up in the air a year or so this cable — probably due to corrosive effects on the braid— started to have suck-outs, and losses increased tremendously."

Schnieder and his colleagues found that "if you would take something like a baseball bat and drive along with a bucket truck down the alley and start beating on that cable every few feet it would come back in nice shape. So, actually it became sort of standard equipment on all our service vehicles that we would carry baseball bats." Eventually all that cable had to be replaced.

In the late 1950s Phelps Dodge, a large electronics company, began to produce coaxial cable with sheathing made of solid aluminum and a dielectric (the buffer between the core and the sheathing) made of foam rather than solid polyethlylene. This cable was far less prone to moisture damage or to signal leakage than the old coaxial. Soon other manufacturers, including Times Wire, the biggest supplier of coax, were offering similar products. The first cable system built entirely with this new type of cable was George Barco's system in Meadville, Pa.

(Sometimes clever marketing was more important than clever technology. Ray Schneider, manager of the Williamsport, Pa., system and later an executive with Times Wire, remembered how Times Wire president Larry DeGeorge handled the challenge of Phelps Dodge, a much larger company.

DeGeorge started by telling operators that the new Phelps Dodge cable,

Operator/Supplier:
Bruce Merrill was both cable operator and equipment supplier through Ameco, which built transistorized all-band equipment.

which was marketed as "semi-rigid," would crack in the winter. Then, Schneider recalled, DeGeorge "comes out with the same product, but he called it 'semi-flexible.' His wasn't 'semi-rigid'; his was 'semi-flexible'! Just that change from rigid to flexible put Times Wire number one in the industry.")

Another major headache for cable operators centered around the vacuum tubes at the heart of every amplifier and most of the other electronic elements of a cable system. Tubes were fragile and likely to go out if the temperature dropped or if the tubes were jarred or struck. Even in the best conditions they didn't last very long. And they were expensive. Montana operator Archer Taylor recalled that the amplifiers he used in an early system required twelve 6AK 5 tubes in each amplifier. Each tube cost $3.50, "which was big money for tubes."

"We found ourselves changing 12 tubes on each amplifier every three months," Taylor recalled. "That wasn't any good. That wasn't going to work."

In the late 1950s the electronics industry was revolutionized by the development of silicon-based transistors that could do the work of a tube but were much smaller, cheaper and more reliable. A researcher at Westbury Electronics in Mt. Vernon, N.Y., Dr. Hank Abajian, was the first to begin to use transistors to build amplifiers for the cable industry.

Ameco, an equipment manufacturer owned by Arizona cable operator Bruce Merrill, pioneered the development of transistorized all-band equipment. By the mid-1960s all the new systems were using equipment with transistors (called "solid state") and older systems were replacing their cumbersome tube-based electronics with solid-state components.

S-A Engineers:
Alex Best (left) and Tom Smith (right) were among the young engineers at Scientific-Atlanta who developed quadrate channellers and transistorized solid state headend signal processors. Best later became SVP-engineering at Cox Cable.

Another weak link in the cable chain was the tap, used to connect the trunk wire running along the utility poles, to the drop cable that ran into the home and the TV set. Most systems used pressure taps, which essentially cut a hole in the trunk cable so the core of the trunk would be in contact with the core of the drop cable.

Spencer Kennedy Labs made a directional tap, which required cutting the trunk cable in two and then hooking the tap to the two ends. The tap contained a splitter that would feed the drop cable. This device took more time to install than the pressure tap, but was far less prone to leakage once installed. And a single tap of this type could feed more than one home. In the 1960s Charles Clements, a Washington State cable operator, and a group of his colleagues approached the other cable manufacturers — Jerrold, Ameco, Entron and others — and persuaded them to get into the business of producing cost-effective directional taps. The development came at a crucial time, just when color television was becoming popular.

Prior to the advent of color, modest

signal leakage was not that big a deal. "People didn't care at that time," Clements said. "That was really before color and it really didn't make much of a difference if you had a few ghosts out there." But for color, he noted, pressure taps were "terrible devices."

Throughout the 1950s a small company based in Atlanta had been manufacturing antenna measurement equipment primarily for the government. Founded by a group of engineers from Georgia Technological Institute, the company, Scientific Atlanta, branched into manufacturing antennas, some of which it sold to the cable industry.

In 1965 one of the S-A engineers, Tom Smith, began to work on a problem many cable systems were having with reception of off-air signals. Broadcast signals operating on the same frequency would interfere with each other. A cable system equidistant between two broadcast stations operating on the same frequency could not pick up either one.

Smith, a native of Drew, Miss., had worked in a radio and TV repair shop and knew about this problem cable systems were having. At S-A he developed a new type of antenna, the quadrate channeller, which was able to distinguish between signals on the same frequency coming from two different directions, but pick up only one. He installed the first such antenna on the cable system in Columbus, Miss., owned by Polly Dunn. It worked like a charm, and soon S-A was selling the new antennas like hot cakes. It was not the last time S-A would make a splash in the business.

Two years later S-A set another standard, introducing the first transistorized headend signal processor. Jerrold had dominated this business for most of the decade with its Channel Commander line. But when S-A hired a young engineer from RCA named Alex Best, it set him to work almost immediately to design a solid state version.

"I remember it was my first day on the job," Best recalled. "My boss came in and tossed an instruction manual on my desk. 'I want you to design one of these out of transistors,' he said, 'and I want to show it at the NCTA show next year.'

"I said 'Fine, but what is it?'"

By the spring of 1967 Best was ready to demonstrate the first solid-state headend signal processor, SA's model 6100. The device worked but was so sensitive it was difficult to install and had to be constantly retuned. "It was temperamental and susceptible to interference," Best recalled. In particular it would get out of tune, pick up adjacent frequencies and mix them with the correct signal.

After a few years on the road visiting hundreds of cable systems to install and then retune the headend equipment, Best designed a device that helped immensely: the surface acoustic wave or SAW filter. This eliminated many of the problems of the solid state headend signal processor and sent S-A one more step down the cable system from the antenna, where it had started, on toward the consumer's home.

There were more marvels to come. In February 1962, GT&E demonstrated a technology that, 30 years later, would revolutionize the cable industry once again: the first use of fiber optics to transmit television signals. The system translated traditional television and microwave signals into light waves that were then fed through the fiber optic cable and translated back to traditional analog signals at the receiving end. GT&E said the fiber could carry 160 different television signals compared to traditional transmission modes (primarily coaxial cable) that could carry only 10 signals over long distances. The company estimated it would be five to ten years before the device would be cost-effective and reliable enough to roll out.

Cable systems also began slowly to introduce new services to supplement their broadcast signal lineup. Some systems pointed a camera at the headend on a clock, a barometer and a thermometer to offer primitive time/weather information to their local audience over unused channels. More than one system trained a camera on a fish tank or a street in the community to offer viewers something different to watch. Some provided background music from tapes at the headend to accompany these early efforts at local origination programming.

In 1964 Jerrold introduced another service, allowing customers to hook up their FM radio sets to cable and receive much clearer FM signals off the community antenna. Systems also took the first, tentative steps toward offering local advertising, even if it was just a printed placard propped up next to the fish tank or the clock on a local channel.

All these new technologies and services— both those that could be used right away and those still on the drawing boards—presented an ever more optimistic picture of cable's future. And that continued to attract the interest of investors. It attracted the attention of competitors and the government as well.

In November, 1960, John F. Kennedy was elected President of the United States by pledging to "get this country moving again." It didn't take him long to get the FCC moving, although in a direction the cable industry didn't much like.

Trailblazer Dunn:
Polly Dunn was one of cable's early female pioneers, helping to build the Columbus, Miss., system in 1953 with her husband.

By 1962 Kennedy had appointed three of the seven members on the commission — Chairman Newton Minow (who won lasting fame when he termed television a "vast wasteland"), Tennessee lawyer and JFK campaign supporter E. William Henry; and Kenneth Cox, the attorney who had written a report calling for regulation of cable when he worked for the U.S. Senate Commerce Committee in the late 1950s.

The reconstituted FCC was clearly more regulation-oriented than it had been in the sleepy, *laissez faire* Eisenhower years. Cox openly called for regulation of cable systems and argued his point that the television industry could not exist while broadcast was regulated and cable was not.

But cable held a trump card. One of its strongest supporters, Rep. Oren Harris (D-Ark.), was chairman of the House Commerce Committee. No legislation regulating cable could pass without his approval, and the FCC needed a new law if it wanted to regulate the industry. Or so everybody thought.

The only leverage the commission had over cable systems was its authority to grant microwave licenses. An increasing number of cable systems were using microwave technology to import broadcast signals from cities hundreds of miles away. Frequently this meant that cable subscribers in those communities had the choice of watching a given program on the imported stations or on the local station. This practice infuriated the local stations which had paid syndicators for the "exclusive" local broadcast rights to the shows.

By the middle of the decade some 77 microwave systems were carrying broadcast signals across thousands of miles. One of the most ambitious plans was proposed by Cox Cable, which asked the FCC for permission to bring the signals of New York City independent stations and WGN — the big Chicago independent — into the Cox cable systems in Ohio and to any other systems along the way that wanted to pick up the signals.

Looking at a map of the areas served by microwave in the middle 1960s it was possible to imagine that one day every broadcast station in the U.S. could be available in every town in the U.S. This thought also occurred to the broadcasters, who stepped up their efforts to regulate cable.

Until 1962 the FCC had granted microwave licenses based on spectrum allocation. The only reason the commission would deny a license was if the signal would interfere with some other communications device. But in 1962 the commission decided it could deny a microwave license if the grant was likely to cause economic harm to a local broadcast station. It picked a case in Wyoming, denying the microwave license application for Carter Mountain Transmission Co., which had applied for permission to import television signals from Denver into Riverton, Wyo. The FCC declared that this would threaten the local Riverton broadcaster, KWRB.

Carter Mountain went to court, but lost when a U.S. Appeals Court ruled that the FCC could regulate the distant signals. The Supreme Court declined to hear an appeal.

The issue of imported signals competing with local broadcast stations became front page news briefly in 1964 when a cable system in Austin, Tex., asked the FCC for permission to import distant signals. The local television broadcaster opposed the request, and the FCC voted to protect the station.

The case drew attention because the station was then owned by one Lyndon Baines Johnson, address: 1600 Pennsylvania Ave., Washington, D.C. The FCC vote was 6-0 to protect the local station.

For most of the 1950s and early 1960s broadcasters remained split about cable. The bigger stations really didn't mind the extension of their signals into smaller, surrounding communities. And as more broadcast companies (Cox, GE, CBS, Newhouse, Sammons, Westinghouse) entered the business, the debates within the National Association of Broadcasters over cable were more intense and heated than the debates

New York Operator:
Ever the visionary, Charles Dolan applied for a New York City cable franchise in 1964, touching off a major franchising war.

Pay TV Loses a Vote

Pay television in the 1960s continued to generate as much enthusiasm and as much opposition as it had in the previous decade. The failure of the Bartlesville, Okla., pay TV experiment in the late 1950s did not quash dreams that one day programming could be delivered to the home so that viewers could pay for what they wanted to watch. RKO General, already a major cable owner, teamed with Zenith Electronics to deliver an over-the-air pay service in Hartford, Conn., and several other companies announced similar plans, both through wired and wireless means.

The most ambitious and widely followed pay TV experiment of the decade was conducted by a California-based company called Subscription TV, headed by a former president of the National Broadcasting Co., Sylvester "Pat" Weaver, and backed by R.H. Donnelley Corp.

Weaver planned to wire communities in California, starting with Los Angeles and San Francisco, and to deliver sporting contests, films, plays and other fare that viewers would pay for on an individual basis. The marquee sporting attraction was to be the Los Angeles Dodgers whose games would be offered exclusively on pay TV.

Weaver hired the phone company to build a two way-plant delivering three channels of programming. He raised and spent some $20 million to launch the system which began operations in July, 1964.

But Weaver ran into a major roadblock. The theater owners in California were petrified that this new distribution system would put them out of business. Together with a group of California broadcasters and others who felt threatened by pay television, they launched a "Citizens Committee for Free TV" and gathered over a million signatures to put a referendum on the November, 1964, ballot to outlaw pay television in California.

STV was caught by surprise. Weaver mounted a counter effort, but the backers of Proposition 15 (as the anti-pay TV amendment was called) were effective in persuading the public that if pay TV were allowed to go forward it would soon win rights to the World Series and other events that had been available on free television and would then charge for them. It wouldn't be long, they said, before only the wealthy would be able to watch the World Series. Weaver laid off almost 150 employees while he devoted all his resources to fighting the referendum. But his efforts were too little and too late. When the ballots were counted Californians had voted for the proposition by a margin of almost two to one.

The vote had a devastating effect on pay TV all across the country. Theater owners announced they would mount a similar campaign in any state where pay TV threatened to take off, and several companies with such plans announced they would postpone or abandon them altogether.

It would be another decade before a successful pay TV venture was able to get off the ground. And the person who made it happen would be the same guy who stirred up such a hornets' nest when he applied to wire New York City: Chuck Dolan.

NAB held with the NCTA.

For several years the NAB attempted to negotiate an agreement with the NCTA over cable regulation which both sides presumed would then go to Congress and become law. But each time the two sides neared agreement, one faction or another at the NAB would scotch the deal.

Then, almost overnight, everything changed.

Broadcasters panicked when a group of cable operators, led by Sterling's Charles Dolan, applied for permission to bring cable to the nation's biggest cities.

Cable had always been a creature of small-town America. It had grown up where broadcast signals were weak or non-existent. Later it found it could prosper where there were only one or two local stations by importing distant broadcast signals into the community.

But in the mid-1960s Dolan figured he could make money even in the urban areas where there were many broadcast signals available over-the-air. In some big urban areas, including Manhattan, reception of off-air signals was problematic.

Large buildings blocked the signals, and interference from an increasing array of other communications devices made it difficult for many viewers to get a clear over-the-air broadcast signal. Rooftop antennas provided a solution, but an unsightly one. Master antennas were practical for some apartment buildings, but not all. This left plenty of fertile ground for cable, particularly if it could offer not only clear local broadcast signals, but also stations from outside the market.

Cable had been creeping closer to the major urban areas for years. In San Diego, for example, by 1965 some 3,500 subscribers were receiving cable service that offered the broadcast stations from Los Angeles.

But in September, 1964, the door flung open with a resounding bang when Sterling, with the financial backing of Seattle broadcaster and cable operator J. Elroy McCaw, applied for a franchise to provide cable service to New York City.

The audacious application by Sterling electrified the U.S. television business. Five other companies — including TelePrompTer and RKO General — quickly filed competing applications to serve all or parts of New York. As the city began the process of deciding which companies would be granted permission to build, other urban areas also became targets of cable companies.

Triangle Publications filed for a franchise in Philadelphia, and the city began hearings on the issue in February 1965. In Los Angeles, Subscription TV, hoping to recover after its defeat at the polls, announced it would use its wires for conventional CATV.

But the most ambitious plans were those of Cox which formed a series of alliances with local newspapers to bid for franchises. In Cleveland it hooked up with the *Plain Dealer*, in Pittsburgh with the *Post Gazette* and in Toledo with the *Blade*.

All this activity did not go unnoticed by the big-city broadcast stations. They hadn't much minded when cable came to the small towns. And they hadn't much noticed when some of their fellow broadcasters entered the business. After all, none of that would really threaten their highly lucrative businesses which had become the major advertising medium in most of the nation's largest urban areas.

But when Sterling filed to bring cable to New York it hit the broadcast giant in the toe with a two by four. The howling could be heard from coast to coast.

Within weeks the American Broadcasting Co. (which, like all the networks, also owned big-city broadcast stations) asked the FCC to impose regulations on cable. It was the first formal policy statement about cable a broadcast network had ever made to the FCC. And it wasn't timid.

ABC had been looking closely at cable for several years. In the mid-1960s its senior vice president of advertising and public relations, a young executive named Ed Bleier, had produced a strategic study of cable. The study examined the Los Angeles-San Diego market and concluded that as more stations became available to viewers, audiences would fracture. While the three major networks enjoyed a 95% share of the market nationwide, in San Diego where cable systems were importing stations from Los Angeles, the network share was closer to 80%.

"San Diego was an eye opener for me," Bleier recalled three decades later. "It was clear that audiences, given more choices, would spread out."

Bleier recommended that ABC become a programmer for a wide array of distribution technologies. But his recommendation was ignored, and ABC instead moved to quash cable, starting with a petition at the FCC. (Bleier soon joined Warner Communications, which was to become a major cable operator and programmer.)

ABC's filing at the FCC urged the agency to establish clear geographic areas in which broadcast signals could be received and to prohibit the export of signals beyond those areas. Only by regulating both broadcast and cable, ABC said, could the FCC "provide for coordinated development of free televi-

System Promotion:
Ever the showman, Bill Daniels used cars and women to help promote the launch of cable service in Colorado Springs in the 1960's.

sion and cable." (It was common practice for broadcasters to refer to over-the-air as "television" and cable as something else.)

For the next decade ABC would remain the most rabidly anti-cable of the broadcast networks. (At least this was true in its public pronouncements. At the same time it was bashing cable in public, ABC had hired Martin Malarkey to develop a strategy for the network to enter the business. The plan was never implemented although ABC at one point came close to buying the Televents cable operations run by Carl Williams.)

ABC's opposition to the cable industry would be remembered by cable operators and would later cause friction in the relationship between the industry and ESPN, which ABC purchased in the 1980s.

Then, in early 1965, a group called The Association of Maximum Service Telecasters, comprised of about 150 big-market stations, formally asked the FCC to bar cable systems from carrying TV signals beyond the Grade B contour (or about 80 miles from the station's signal source). The group also asked the commission to require that all cable systems carry all local TV stations and that cable companies be banned from originating programming.

The smaller stations, which had always viewed cable as a threat, continued their drumbeat. Some used their own stations to promote their anti-cable message. One station in Springfield, Mass., owned by Bill Putnam, was reprimanded by the FCC for carrying anti-cable advertisements without giving cable operators a chance to reply.

These ads stressed the message that broadcast television was "free" television whereas cable forced viewers to pay to watch. The notion that broadcast television was "free" took hold. Cable was never able to get across the point that the cost of broadcast television was embedded in the cost of nearly every consumer product. A purchaser of a bar of Ivory Soap was paying for television advertising whether or not he or she watched TV.

With cable, only those who wanted the service paid for it. But the broadcasters ran rings around the cable industry when it came to public relations, and the notion that broadcast was free was deeply embedded in the public consciousness.

Broadcasters also peddled with great success the notion that their business was somehow in the "public interest" because they offered local news and coverage of local events. This message also stuck even though many of the local stations offered very little in the way of original or locally oriented programming. And the broadcasters, while touting their public service, continually resisted efforts by the FCC to impose additional public service requirements on the stations.

The biggest weapon in the broadcasters' arsenal was unstated: their ability to influence the political process. Every politician in America was well aware that Richard Nixon had lost the 1960 presidential election because of his poor performance in the televised debates. Every member of the Senate and the House understood that broadcast television was the most effective means of communicating with their constituents and that an appearance on the local TV news show could send a politician's approval rating careening in one direction on the other.

Broadcasters rarely, if ever, employed an overt threat of favorable or unfavorable coverage in lobbying politicians. They didn't have to. The power of television was understood. And cable, with no local origination programming to speak of and no presence at all in the major population centers of the nation, could simply not compete with broadcasters on this playing field.

Nor were the cable operators very savvy in their lobbying efforts in the 1960s. While the NAB was headed by affable and well-known former Florida Governor LeRoy Collins, later to be an

Lansford Contractor:
While an engineer at Henkels & McCoy, Jerry Conn help construct the poles for Robert Tarlton's Lansford, Pa., cable system in 1950. Conn went on to work for TVC Supply before establishing Jerry Conn Associates in 1964.

official in the LBJ cabinet, the NCTA was leaderless for almost a year and a half leading up to the spring of 1965. When it finally completed a lengthy search for a new president, filing the post with former FCC commissioner Frederick Ford, the political tide had clearly turned against the industry.

Finally, the scars of the 1960 legislative battle were long in healing. Politicians have long memories. By mid decade there were few Capitol Hill lawmakers eager to rush to the defense of cable and risk getting caught in the same change of mind that had left so many industry supporters stranded and embarrassed when the cable industry abruptly reversed its position on a cable regulation bill debated by the Senate in 1960.

By the spring of 1965 the broadcasters' drumbeat of anti-cable rhetoric began to make an impression in Washington and in the countryside.

To consumers and increasingly to the FCC it began to appear that the cable operators were doing something that, if not illegal, was certainly unethical, even a little dirty.

As Montana cable operator Archer Taylor put it later: "I think the commission just felt we were bad people. They thought we were cheap and dirty people and that we were not professional. We weren't doing things right, we weren't to be trusted."

That image had been created, Taylor believed, largely by the broadcasters. He recalled getting a letter from a broadcast station manager in Billings, Mont., one day that opened "I can't really refer to you people as thieves, because thieves work stealthily at night and you guys work right out in the open."

The idea that cable was harmful to broadcasters was promoted by the coverage in *Broadcasting* magazine, then the dominant trade publication in the television industry and a major influence in Washington, where it was based and where its publisher, Sol Taishoff, would routinely dine with members of the FCC and key members of Congress.

Although the magazine in general practiced objective journalism, never distorting the facts, it certainly approached things from a pro-broadcasting point of view. One headline in 1964, for example, read "CATV: A Big Problem That's Getting Bigger."

NAB weighed in with a $25,000 study of the impact of cable on broadcast television which found, not surprisingly considering the sponsor, that cable was a bad influence. The study found that where a local broadcast station was forced to compete with signals imported by a cable system it would cost the station anywhere from $9,400 to $20,000 or more a year in lost revenue.

Cable operators fought back. In a speech in Washington in the fall of 1964, Jerrold president Milton Shapp pointed out that of the 12 small-market TV stations that had testified at the 1958 congressional hearings about the threats of cable, all but one were still in operation and clearly profitable. Shapp might have

Ford Switch: *Former FCC commissioner Frederick Ford served as NCTA president from 1965 to 1969, a tumultuous time for cable as new FCC rules severely hampered the industry.*

Emergency Alert:
Hungry to expand their revenue base, cable operators offered emergency alert services in the 1960's, designed to warn subscribers of emergency situations.

saved his breath in the face of the gale of broadcaster rhetoric.

The broadcasters' efforts paid off in April of 1965 when the FCC issued a set of rules to govern cable. Reversing its previous position which had held it could regulate only systems that used microwaves, the commission said it had the right to regulate all cable systems. It said such a policy was necessary to ensure that free, over-the-air TV would be available to as many Americans as possible. Cable, it said, was only a supplementary service to broadcast television.

In particular the commission issued regulations to:

• Forbid cable systems from carrying programs on a distant signal which duplicated the programs on a local broadcast station within 15 days before or after the local station carried the show.

• Require cable systems to carry all local stations, including those whose Grade B signals reached the cable system.

• Freeze the granting of microwave licenses for systems that were planning to serve major urban areas.

The agency also asked for comments on the following questions:

• Should there be limits on how far a microwave signal could carry a broadcast station;

• Should the FCC prohibit the practice of leapfrogging in which a cable system imported a signal from a distant city without offering a signal from a station closer to the cable system;

• Should the FCC ban crossownership of cable systems and broadcast stations in a single community;

• Should broadcasters be allowed to build translator stations to extend their signals into areas beyond the range of their main transmitter.

The impact of the FCC action was immediate. The National League of Cities urged its members to suspend granting any further cable licenses until the impact of the rules became clear.

Broadcasters glowed.

But if the broadcasters, telephone companies, movie theater owners and antenna manufacturers thought the FCC could stop the growth of cable they were dreaming. They hadn't met Chuck Dolan and Irving Kahn.

By December, 1965, the two had won the rights to build an experimental cable system in Manhattan, with Sterling getting the southern half of the island and TelePrompTer everything north of 86th street on the east side and 79th street on the west. Another cable outfit, CATV Enterprises, headed by producer Ted Gralick, won the franchise for the enclave of Riverdale.

The franchise awards came after a bruising political battle and strong opposition from New York Telephone, CBS and Universal Pictures.

Dolan and Kahn, in particular, proved to be masters of the local political scene. Each assembled a stellar cast of New York political heavyweights to press its case. Sterling's team was led by Richard Flynn, son of New York State Democratic National Committeeman Edward Flynn. TelePrompTer's team was headed by New York County Democratic Committee counsel Justin Feldman and attorney and friend-of-the-Mayor William Shea, who would one day lend his name to the home of the New York Mets baseball team.

The two battled down to the wire. Sterling, which had first proposed an installation fee of $60, cut the price to $37.50 only to watch TelePrompTer offer a fee of $19.95.

New York Telephone asked the city's Board of Estimate, which made the franchise decisions, to refrain from granting any awards, contending that construction of a cable system would damage the city's communications systems unless it was built by the phone company and leased to cable operators.

CBS and Universal Pictures both threatened lawsuits charging that the programs they produced and broadcast should not be resold by cable companies without permission from the copyright holders. Each implied that the city, if it granted a franchise, could be liable for copyright infringement.

But none of this deterred Dolan or Kahn. In the end the city granted three experimental franchises, one each to Sterling and TelePrompTer and one to CATV Enterprises.

There were a few caveats. As a condition of accepting the franchise the city insisted that the cable operators:

• Charge only $19.50 for installation, $5 a month for service.

• Provide free service to city hospitals and police stations, discounted service to other city agencies and charitable and religious organizations.

• Take out a $2 million insurance policy to protect the city against any copyright judgments.

• Earn no more than a 7% return on investment.

• Agree to reduce fees if the city determined that profits were too high.

• Agree not to introduce pay TV.

• Agree to carry all local TV stations.

• Agree not to import any distant signals.

Once again it seemed as if the networks, telcos, politicians and producers had succeeded in throwing the cable operators into the financial briar patch. But Dolan and Kahn weren't deterred. Each was willing to bet the ranch that his system world work. And each would lose the ranch in the process before being proven right.

Dolan promised his system, using some of the wire Sterling had already strung for its hotel guide service, would be up and running within four months. Kahn said his would be right behind. And, they told their investors, they would make money doing it.

The city had estimated that there were about 500,000 homes in Manhattan that could not get adequate television service. If Kahn and Dolan could get only 50,000 of those homes each to sign up each would realize $3 million a year in revenue. The total revenue for TelePrompTer in 1964 had been just over $3.5 million.

But the real key to the chances for Dolan and Kahn was the federal tax code. By depreciating the cost of their equipment, the two could avoid having to report any profits for years. This would enable them to generate significant free cash flow without having to risk violating the city's ceiling on return on investment.

The Trade Press

One landmark in the development of any community comes when it grows large enough to support its own newspaper. For the cable television community that point came in 1964 with the launch of *TV Communications*, a monthly magazine published by two brothers: Bob and Stan Searle, of Oklahoma City. The Searles moved the publication to Denver shortly thereafter.

Prior to *TVC*, as it came to be called, the primary source of news in the industry had been two Washington-based publications: a weekly newsletter called *Television Digest* and a weekly ad-supported magazine, *Broadcasting*.

Television Digest covered the entire range of the television business, from production of TV sets to the latest programming ratings. It was one of the first publications in the country to begin to cover cable, almost from its inception in the late 1940s.

Started in 1945 by by veteran Washington reporter Martin Codel, it was purchased in 1958 by publishing tycoon Walter Annenberg. The publication was taken over by its Washington editor, Al Warren, in 1961 after Annenberg decided to abandon it because, as he said in a letter to subscribers, "there is not sufficient interest in such broad general coverage of television in the newsletter from enough subscribers to justify continuance."

Annenberg didn't make very many mistakes in his enormously profitable publishing career. This was one.

Under Warren's direction *Television Digest* covered the cable business with remarkable thoroughness, accuracy and lack of bias. Warren also published an annual directory of the television business, called the *Television Factbook*, which remained through the end of the 20th century the most thorough and widely used annual directory in the business. It listed, among other things, every cable system in the country with key personnel, equipment used, franchise information and prices charged.

Broadcasting also paid attention to cable, primarily from the point of view of the broadcasters and with special emphasis on what was happening in Washington where its coverage of the legislative and regulatory process was thorough and accurate. *Broadcasting* was founded in 1931 by Sol Taishoff who remained its publisher and a key figure in the industry until he died in 1982.

The magazine, later renamed *Broadcasting & Cable*, was acquired in 1997 by the Cahners Publishing Division of Reed Elsevier, Inc., which also owned *Variety*, *Multichannel News*, and *CableVision*.

In 1974 two former *TVC* employees — sales manager Bob Titsch and editor Paul Maxwell — launched *CableVision* magazine, a biweekly, putting together the first issue in the basement of Titsch's home. Within a couple of years it had surpassed *TVC* and become the industry bible, largely because of the selling abilities of Titsch, the irreverent attitude of Maxwell and the boom in cable around the time the magazine launched.

Maxwell believed that trade publishing should be fun, he later said, and *CableVision* showed it. At the major trade shows the magazine's cover was a foldout cartoon depicting all the major figures in the business, usually in some kind of theme, such as a Wild West setting. Invariably,

Maxwell would portray FCC chairman Dick Wiley with two faces.

Among those working for Titsch in the late 1970s were Washington bureau chief Brian Lamb and associate publisher Paul FitzPatrick both of whom would go on to head cable television networks. Titsch employee Paul Levine left to found *Communications Technology* magazine.

Maxwell split with Titsch in 1979, returning to the business a year later in partnership with Capital Cities Inc., to launch *Multichannel News*, a weekly newspaper. This was in a time when news was breaking regularly but very little attention was paid by the mainstream media to cable operators.

In 1988 Thomas P. Southwick, who had been editor of *Multichannel News*, teamed with Paul Kagan to launch *Cable World*, a weekly magazine modeled after *Time* and *Newsweek*. *Cable World* was purchased by Cowles Business Media in 1994 and by Intertec Publishing in 1998.

The editors and publishers of the various trades have filled a role in their industry beyond what their titles might imply. Taishoff was at times a more formidable power in the television business than the members of the FCC, with whom he regularly dined at a restaurant across the street from the *Broadcasting* offices in Washington. Irving Kahn said he received his education about cable from Al Warren. The Searles bought and managed cable systems and played a key role in founding the Cable Pioneers group. Maxwell was a founder of the Walter Kaitz Foundation. And Titsch allowed Brian Lamb to remain on the *CableVision* payroll while laying the groundwork for C-SPAN.

All served as informal advisors, sounding boards, and rumor relays for key industry players. Most important, they provided a voice for the industry they served. They put into print the ideas, feelings and attitudes of the business. And they served as a conduit for information, especially important in a business where competition between cable systems was minimal and operators were hungry for information.

Fourth Estate:
CableVision Magazine celebrated each NCTA convention with a cover caricaturing industry figures.

As *Business Week* pointed out in an article on the N1ew York franchises "any legal ceiling on return is irrelevant since no system would show a profit until depreciation ran out." That would not occur for almost a decade, and a resale of the systems in the meantime would trigger a renewed depreciation schedule.

But the biggest card in the cable operators hand was something else entirely: color.

Color television in the 1960s had captured the imagination of the nation as completely as TV itself had in the previous decade. Millions of new sets were being sold each year. And the problems of big-city reception that plagued black and white sets were compounded when broadcast stations began to transmit in color.

As Irving Kahn put it to a *New York Times* reporter: "You ever see a ghost in black and white? You should see one in color." Color, Dolan and Kahn were betting, would bring in customers by the droves.

And always in the back of their minds was the possibility of pay television, even if they had pledged to the city they would not introduce it into New York. Kahn, after all, had pioneered the use of cable systems to offer prize fights and talked openly about using cable to carry first-run movies.

As *New York Times* columnist Jack Gould wrote in December, 1964, with 1.2 million homes already wired for cable and millions more on the way, "Dreams of a meaningful box office already in existence begin to take form despite the promises of some community TV operators that such a course never existed in their minds."

It would not be long before Dolan transformed that dream of the home as a box office into something bigger than he or anybody else imagined in that winter of 1965-66 when Sterling began to build the first cable TV system in the world's biggest city.

The Dark Ages

For the cable television industry, the Dark Ages began on a mild, rainy spring day in Washington, D.C., in the glistening white marble structure that houses the Supreme Court of the United States.

For the first 20 years of its history, the industry had enjoyed enormous growth, largely free from regulation by the government. In 1960 the U.S. Senate, by a single vote, had rejected legislation that would have allowed the Federal Communications Commission to regulate cable. Cable operators had dodged or defeated efforts to classify their businesses as common carriers at the federal level or to regulate them as public utilities in the states. City officials, anxious to bring television to their constituents, welcomed cable with open arms, imposing few limits on rates or services.

But in the mid-1960s the tide began to shift. The FCC, with new, activist members appointed by Presidents Kennedy and Johnson, limited the importation of distant broadcast signals by cable systems, saying the practice would damage the economic interests of local broadcast stations.

The new FCC rules were challenged by Southwestern Cable Co., which imported signals from Los Angles stations and delivered them via cable to customers in San Diego. The system won in the Appeals Court, which ruled that the FCC had no specific authority to regulate cable under the 1934 Communications Act.

But when the Supreme Court ruled on the case in June 1968, it took a different view. In a sweeping decision, the court ruled by a 7-0 vote that the FCC had virtually unlimited authority to issue rules governing all forms of communication by wire or over the air.

The court also upheld the FCC's contention that it needed to regulate cable to ensure the survival and growth of broadcast television. "The Commission has reasonably found," the court said, "that the achievement of each of these purposes (to protect broadcasting) is placed in jeopardy by the unregulated explosive growth of CATV."

The ruling was a blockbuster, but it was not the only decision that impacted cable that week. The second decision evened the industry's record at the Supreme Court.

Ever since 1948, when Ed Parsons asked for permission to pick up the signal of a Seattle broadcast station and retransmit it via cable, the issue of a system's right to pick signals out of the air for free and retransmit them to paying subscribers had been open to question. Program producers and broadcast stations had repeatedly taken the position that programming could not be

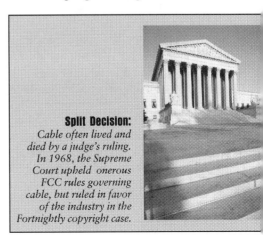

Split Decision: *Cable often lived and died by a judge's ruling. In 1968, the Supreme Court upheld onerous FCC rules governing cable, but ruled in favor of the industry in the Fortnightly copyright case.*

"stolen" by cable operators and then resold to consumers without some compensation of the copyright holders. Cable operators maintained that once a broadcast signal was sent out over the airwaves it was free to anybody who wanted to pick it up for any purpose. Cable, they maintained, was simply an effort by a group of citizens to get together to build a bigger and better antenna.

The issue inevitably wound up in court.

In 1960 United Artists Television Inc., a movie distributor and program producer, sued two cable systems in West Virginia owned by Fortnightly Corp. United Artist charged that Fortnightly was illegally picking up broadcast signals that contained UA-produced programming and then selling those signals to cable subscribers. UA asked the court to forbid the practice without the permission of the copyright holder.

It seemed to UA, and to many cable attorneys as well, that the studio had a strong case. Copyright was firmly embedded in the US system of laws. The notion that cable was simply a passive way to pick up signals that anybody could get with a big enough antenna was undermined by the increasing tendency of cable systems to use microwave links to import stations from hundreds of miles away. Cable attorneys knew their reasoning was shaky, but as NCTA counsel, Stratford Smith, said later, "it was the best thing we could think of at the time."

The industry arguments proved inadequate in the District Court, which ruled unanimously for UA, and in the U.S. Court of Appeals, which backed the lower court by a margin of 14-0.

The NCTA almost decided not to take the case any further and to work something out with UA and other copyright holders. But NCTA president Frederick Ford and special counsel Smith persuaded the board to take a shot. (Fortnightly Corp.'s systems, meanwhile, had been purchased by Jack Kent Cooke, who had agreed to assume liability for any judgment against the systems.)

By the time Supreme Court decided the case, in June of 1968, the cable industry had received two-last minute breaks:

• The ruling on the Southwestern case, which had come just prior to Fortnightly;

• The courtroom demeanor of the attorney for United Artists, Louis Nizer.

Nizer was as famous in his day as F. Lee Bailey or Johnny Cochran would be later on. He was a brilliant courtroom attorney and won the UA case handily in the district and appeals courts.

But when the day came to argue before the Supreme Court, Nizer adopted an attitude of disdain for the cable operators. His demeanor appeared to some of those in the courtroom to be a snub of the Justices themselves who had agreed to hear the case in which the outcome, Nizer implied, should have been obvious. He gave the impression to some who heard his arguments that the court was wasting his and everybody else's time by even hearing a case which was so clear-cut.

The justices didn't see it as a clear-cut issue. As they questioned Nizer it became apparent that several of them were put off by his cavalier, almost arrogant, attitude.

On top of that, the Southwestern and Fortnightly cases were argued before the court on the same day, and the Southwestern ruling came just two days before Fortnightly. In the Southwestern case the court had stated that the FCC had almost unlimited authority to regulate cable television. The reasoning in Southwestern spilled over to the copyright case as the court ruled that it was up to the Congress and the FCC to make clear the issue of cable copyright. On a 5-1 vote it overturned the lower court and ruled against United Artists.

It was a huge, upset victory for cable. The two attorneys who had argued the industry's case — Robert Barnard and Smith — had come down to the court chambers on three successive Mondays waiting for a decision. When it finally came, the two moved quickly out of the court chambers and struggled to keep their balance as they sprinted down the polished marble floors to phone industry leaders with the good news.

Had the cable industry lost Fortnightly it would have been a crushing, perhaps fatal blow. Systems would have been liable for enormous programming costs and perhaps for penalties for past copyright infringement as well. But the decision, as much as it avoided the worst, did nothing to improve cable's status. It merely upheld the status quo and sent the copyright holders scurrying to the Congress and the FCC for relief.

Southwestern also seemed at first glance to be a decision upholding the status quo. After all, it merely affirmed a power to regulate cable systems that the FCC had first asserted in 1966. But when the ruling came out, its sweeping language spurred the FCC to issue a whole flurry of new regulations governing the industry.

Within a week the Commission had issued rules that effectively froze the additional importing of distant signals into big cities, without any chance for waivers. The Commission also required systems to carry all local signals without degradation and prohibited cable systems from carrying advertising on locally originated programming.

The FCC followed up on that ruling with a series of decisions that imposed even more stringent restrictions on what cable systems could do.

By December it had issued a set of proposed rules to ban the importing of any distant signals at all in the top 100 markets and to ban systems in other markets from importing any distant signals other than those closest to the system.

Many in the cable industry viewed the FCC as the captive of broadcasting interests. Certainly the broadcasters were relentless and effective lobbyists. And certainly the rulings of the FCC were overtly designed to help the economic viability of broadcast stations.

But the FCC rulings were not the product of broadcasters buying off commission members. A majority of the Commissioners and many of the staff members genuinely believed that what they were doing was in the best interest of the public.

The two principle champions of regulation — Commissioner Kenneth Cox and general counsel Henry Geller — genuinely believed that the federal government had a responsibility to the public to make the world a better place, a pervasive feeling in Washington in the days of the New Frontier and Great Society. The FCC, they believed, had a responsibility to ensure more diversity in programming and universal access to television service, particularly local news and public affairs. They believed that the best way to do this was to encourage the growth of more broadcast stations.

Moreover they believed it was possible for a small federal agency to regulate a huge and rapidly changing industry in a way that would accomplish the government's lofty goals of programming diversity and universal access.

The actions of the FCC in the late 1960s and early 1970s had a devastating impact on the cable business, which in the end proved much more able to provide diversity of programming, including local programming, than the broadcasters. The broadcasters also found that their status as the government-ordained primary source of television was not an unmitigated benefit. The government imposed a series of rules — equal time, the fairness doctrine, limits on network programming and requirements for children's TV — on the broadcasters who howled in protest. At the end of the Johnson Administration, the Congress, over the vigorous objections of the broadcasters, enacted a law creating the Corporation for Public Broadcasting to fund a network of public stations that would compete for viewers with the commercial broadcasters.

The FCC rulings were not the only bad news the cable industry received in the late 1960s and early 1970s.

Even more disastrous for the industry than the FCC and Supreme Court rulings was the general state of the economy. The cable industry had been built with cheap money. For much of the first two decades of its existence interest rates had remained well below 5%. But in the late 1960s that began to change as the first hot winds of inflation began to stir in an economy already overheated by the war in Vietnam and the social programs of the Great Society.

The prime interest rate, the fee banks charged their biggest customers, soared from 4.5% in January 1965 to 8.5% in mid-1969. Many cable operators had bank loans that were tied to the prime rate. For them the change meant that the cost of money, far and away the biggest expense for any new system, effectively doubled in four years.

As the cost of money rose, so did the cost of equipment. The most expensive ingredient in coaxial cable, copper, began to become scarce in the late 1960s. Its price rose to 42 cents a pound in 1968 from 29 cents a pound at the start of the decade. Suppliers of coaxial cable were forced to follow suit. Times Wire increased its prices for drop cable by 6% in 1968 alone, and other suppliers hiked prices as well.

Technology, which had been a friendly Dr. Jekyll for 20 years — making it possible to build more reliable systems and deliver more services at a lower cost — turned into a Mr. Hyde in the late 1960s, particularly when coupled with the evils of the franchising process.

In 1967 Alan Gilliland, a California broadcaster, built a system for San Jose that had two cables, effectively doubling the channel capacity which heretofore had been limited to 12. Customers were

Making The Case:
Ads like this from ATC's Lynchburg, Va., system, urging TV viewers to subscribe to cable, were common in the 1960s.

given an A/B switch that allowed them to move from one cable to another to view a different set of channels.

In 1969, the first two-way cable system was built by Continental Telephone in Reston, Va. Local origination programming began to spread as cable systems televised high school football games, city council meetings, local parades and other events. Bill Daniels made news in 1968 when he became the first cable operator to purchase equipment capable of producing and delivering local programming in color.

These new technologies had a downside, particularly in the political arena. As the franchising process heated up, competing cable companies vied to offer bigger and better systems with even more local programming. They did this despite the fact that there was really no way to pay for these improvements at the time.

City councils became far more savvy about what they could demand from their local cable systems. They all began to read the trade press and the press releases from the cable companies themselves. Consultants began to go from town to town offering to help city councils with the franchising process. In 1970 the Cable Television Information Center was started with money from the Ford Foundation to help cities get better deals from their local cable companies.

Spurred on by their consultants, cities all across the country began to demand state-of-the-art systems for their communities.

They insisted on increased channel capacity even though there was little programming to put on these additional channels. They demanded more local origination programming, particularly channels devoted to covering city coun-

Local Content:
Cable operators began offering local originated programming in the late 1960s, including weather reports like this seen on ATC systems.

cil meetings, school board sessions and local news. The publication of the best-selling book, *The Wired Nation*, by Ralph Lee Smith outlined the great benefits cable would bring the country, and city officials everywhere were eager to be the first to bring the new services to their constituents.

One early dual-cable system, outside Philadelphia, offered subscribers four educational stations, a channel showing the AP news ticker, a channel showing the weather and the time, and a channel each for the color and black-and-white test patterns. Even in the sleepiest American towns it was difficult to imagine anyone paying extra money for the chance to watch test patterns.

And operators found it tough to get permission to raise rates. Rate increases hadn't mattered much to cable companies until the late 1960s. Prior to that time they had been adding so many new subscribers they didn't need more money from their existing customers. But as penetration levels approached the maximum in many towns and costs continued to rise, rate hikes became more important.

"I don't think we had any rate increases to speak of until maybe 1972 or 1973," recalled Bill Arnold, then chief financial officer for GenCoE. "I guess maybe it wasn't until the early '70s people said 'Well, you can kind of see from where we are now to the end of the houses that we can serve, and costs are creeping up. We now began to talk about needing rate increases.'"

But most franchise agreements gave city councils the power to approve or reject rate increases, and most local politicians were not eager to approve rate hikes for the cable customers, who were also voters. At the same time, they found that the franchise fees were a ready source of income for the cities without the need to raise taxes.

In community after community, officials used the franchising process to hold down the cost of cable for consumers and line the coffers of the local government. City officials then could brag about keeping more money in voters' pockets without having to cut taxes and about increasing city revenues without increasing taxes. They could have it both ways against the middle. The only one squeezed was the cable company.

A study released by the FCC in 1970 found that cities were charging cable companies an average franchise fee between 7% and 9% of gross revenues. In newer franchises the figure was 8% to 11%. In Albuquerque, GenCoE agreed to pay $25,000 for the franchise, $10,000 a year in fees until the system began to operate and up to 30% of gross revenues in franchise fees thereafter depending on the success of the system.

Then, to add to the industry's woes, the Supreme Court in 1970 ruled that cable companies could be subject to regulation by state public utility commissions. The decision upheld a Nevada statute placing cable under regulation by the PUC, paving the way for other states, such as Connecticut and Hawaii, to institute state regulations. The court rejected the argument that because the federal government was regulating cable, states had been preempted.

The combination of FCC rules, more state regulation, higher equipment costs, greater demands by the cities and increased interest rates was a near-fatal cocktail for cable in the late '60s and '70s.

Some systems folded. In Cleveland, shortly after the FCC rules were announced, the system owned by Cox and the Cleveland *Plain Dealer* announced it was shutting down, having attracted only about 1,000 subscribers after 18 months in operation and losing $20,000 a month.

The system had tried just about everything, recalled its director of marketing, Greg Liptak, who took the job just out of graduate school. The company developed a series of local programming efforts, covering high school sports, and even inaugurated a local newscast six days a week. Although the efforts won the plaudits of local civic leaders, they didn't attract enough customers to stem the red ink.

In nearby Akron, Ohio, cable operator TeleVision Communications Co. forged ahead with construction of a dual-cable, two-way system. But after 18 months and only 17,000 subscribers, it was forced to sell out in the spring of 1972 to a new player in the business, one with deeper pockets. The buyer was Kinney Services, owner of Warner Bros. Studios and soon to be renamed Warner Communications.

"We worked for 10 years to develop this company," TVC president Alfred

Stern told *Forbes* magazine, "and I could see our leadership going down the drain."

Warner also bought the systems owned by Continental Telephone, including the two-way system in Reston, Va. With Akron and Reston, Warner became an instant leader in interactive cable systems.

All across the country construction ground to a halt. *TV Digest*, which had reported the launch of more than 250 new systems in 1968, found only 90 startups in 1969. Manufacturers were devastated. Jerrold, still the biggest and most influential equipment maker and supplier, and newly acquired by General Instrument Corp., announced layoffs of 550 people, nearly a third of its workforce. Vikoa reported a net loss in 1970 of more than $8 million. Ameco reported revenues for its fiscal year 1971 of $1.5 million, less than half the revenues of the year before. Spencer Kennedy Labs was purchased by Scientific-Atlanta, Inc.

Entron, another major manufacturer, was acquired by Spedcor, a telephone equipment manufacturer which wanted Entron's cable systems and the tax losses it could take from shutting down the equipment business. As Entron president Ed Whitney later remembered: "The main thing is that the FCC put a halt on our development and we (didn't) have the financial resources. We didn't have the capability of carrying on and going on long term to try to recover. We just simply had to stop business. They knocked us dead."

The amount of money the cable companies needed to borrow in order to expand from their rural roots into the urban and suburban communities was staggering. Amos Hostetter, then executive vice president of Continental

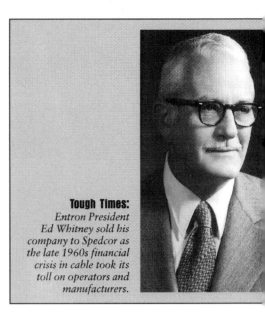

Tough Times:
Entron President Ed Whitney sold his company to Spedcor as the late 1960s financial crisis in cable took its toll on operators and manufacturers.

Cablevison, told the 1972 NCTA convention that the total amount of capital loaned in the U.S. for all construction purposes was about $35 billion a year. Of that, about half was used by the utilities — phone and power companies. Outside of that group no single industry had ever borrowed more than $1 billion in a single year — not steel, not automobiles, not railroads, not mining. Yet cable would need to raise at least $1 billion a year every year for the coming decade if it were to succeed in wiring the nation's most populous areas.

To raise more capital, some cable companies had begun, in the late 1960s, to turn to a new arena, the public markets. The first publicly traded cable companies entered the market through the back door. H&B American had started life on the New York Stock Exchange as a transportation company. It underwent a metamorphosis and emerged as a cable company when it purchased the systems owned by Jerrold Electronics in 1961.

TelePrompTer also started out as something different, a company that rented out TelePrompTer machines and produced closed-circuit events. When its chairman, Irving Kahn, discovered cable, TelePrompTer shed its other businesses. The company then found it could trade its stock for cable systems. By taking stock rather than cash, the seller could avoid paying huge capital gains taxes on the proceeds, participate in the growth of the company and sell stock to raise cash as needed.

The trail blazed by TelePrompTer and H&B was followed by a group of other MSOs, beginning in 1968. The first cable company to go public expressly as a cable operator was Cypress Communications, owned by cable operators Leon and Randy Tucker, the Cole family of Cleveland and by Hornblower & Weeks, the big stock firm. It was followed quickly by seven others. By the end of the decade 10 cable companies were publicly traded.

One of them was American Television & Communications Corp., formed by Daniels & Associates partner Monroe Rifkin.

Rifkin had been busy since joining Daniels in 1964. He had brokered dozens of systems for hundreds of millions of dollars, dealing with a wide variety of buyers and sellers. And as he structured the buying consortia, he found that the partners often would have different goals. Some would be looking for a relatively rapid turnover while others had a longer time frame. Sometimes the initial goals of the partners would change. One might find a sudden need for liquidity and seek to cash out while the others wanted to stay for the long haul. Or the death of a principal might create a sudden need for cash to settle an estate. At times the different needs of the different partners could create stress within a cable company.

One lender to some of the entities Rifkin created was Naragannsett Capital Corp., whose chairman Royal Little was a friend of George Peabody Gardner, senior manager of Paine Webber, Jackson & Curtis, the big stockbroker. Gardner first suggested to Rifkin the idea of going public.

Rifkin rolled 16 different cable companies with 69,000 subscribers in 14 states into a single company. He assigned to each partner a number of ATC shares in proportion to the value of the cable properties contributed to the new company. Then, in the spring of 1969, the company went public under the name American Television & Communications Corp.

Rifkin hired Douglas Dittrick away from GE Cablevision to become vice president of finance. Another early hire at ATC was June Travis who started out as a secretary, rose to senior vice president, became president of Rifkin & Associates cable company and still later executive vice president of the NCTA.

The ATC deal had terrific advantages for those who participated. ATC was able to sell sufficient stock to the public to raise badly needed construction capital and to retire a total of $2.1 million in

Public Markets:
Monty Rifkin, chairman of Rifkin and Associates, left Daniels & Associates to form American Television & Communications Corp., in 1968.

debt to Memorial Drive Trust and Boston Capital Corp., two lenders anxious to redeem their loans. Going public allowed the investors to sell portions of their ATC shares for cash, without having to get out of the business entirely. (Bill Daniels, ATC's largest individual shareholder, was one of the partners who needed some cash from time to time but didn't want to sell out completely.) And the offering enabled ATC to raise additional funds for working capital, upgrades and acquisitions. It consolidated the operations of the systems and gave the company additional clout for negotiating with suppliers and lenders.

The systems ATC managed were strong and getting stronger. Revenue for the company's systems totaled $3.5 million in 1967, up from $1 million in 1965. Operating income (profits before depreciation, amortization, and interest expenses) had grown from $175,000 in 1965 to more than $1.5 million in 1967. Things were looking even better in 1968.

But ATC, like all the other cable companies which went public in the late 1960s, had a problem. Its essence was contained in two sentences of the original prospectus: "ATC has paid no dividends and does not presently contemplate paying any dividends in the foreseeable future. It intends to use all available funds for expansion of existing CATV systems and the acquisition of additional systems."

This, of course, was standard operating procedure for cable companies, many of which went out of their way not to report taxable profits. In 1965, after depreciation, amortization and interest, ATC posted losses of $673,000; by 1968, its losses were more than $900,000.

ATC attempted to explain this to potential investors: "Operation of CATV systems involves a high ratio of fixed costs resulting in substantial losses in the early years when revenues are relatively low. The company's losses increased in 1965 and 1966 largely because of depreciation, amortization and interest charges resulting from the expansion of plant facilities through acquisitions and construction."

It also noted that "revenues increased in 1967 and 1968 through substantial growth without substantial additions to plant facilities.

"Starting with 1966, revenues have exceeded all expenses other than depreciation and amortization."

In other words, ATC was generating a huge amount of free cash flow that could be used to build and expand. But by writing off the cost of plant and interest it could report a loss to the IRS and avoid having to pay taxes.

All in all it looked like a pretty sweet deal. But not to Wall Street.

The major investment and brokerage houses on Wall Street had a decades-old, ironclad method of measuring the success of a company. It was based on earnings and dividends. Cable companies had neither.

The major indicator of the value of a stock was its price/earnings ratio, a figure derived by dividing the price of a share of stock by its earnings per share. No such calculations could be made for the cable companies because they had no earnings or had earnings so small they yielded enormous p/e ratios.

The attitude of the major brokers was made clear in 1969 when E.F. Hutton assigned one of its young analysts to study the new cable industry. He came back with a glowing report about a business that was growing quickly, that had a monopoly on delivery of a commodity every American wanted, and that could take advantage of some pretty terrific tax dodges.

Then his boss asked him: "Where are the earnings?" Told that the cable companies didn't report earnings, preferring to plow cash flow back into the business, the senior analyst replied in the lofty tone: "E.F. Hutton does not do reports on companies with brackets on the bottom line." (Losses on a financial report were typically indicated by enclosing them in brackets.)

The young analyst figured that made no sense. So he quit and started his own firm to follow the cable industry and look for investment opportunities. His name was Paul Kagan.

Kagan, born in 1937, grew up in what he now calls "the golden era of New York City." As a kid he was a big-time baseball fan, able to rattle off the batting averages of Joe DiMaggio, Mickey Mantle, Jackie Robinson and the other players on the hometown Yankees, Dodgers and Giants.

After graduating from Hunter College, Kagan got a job as a sportswriter at the daily paper in Binghamton, N.Y. There he covered the minor league baseball team. On slow news days he learned to make up stories based on the statistics. He might tell readers that if the team made two more double plays in July they would be third best in the league or if the second baseman continued to hit doubles at the current pace he would be on track to break the team record for two-baggers by an infielder.

In short, Kagan learned to play with numbers, to look at statistics in unconventional ways and to find stories in numbers that others regarded as boring or routine. He also began to do some freelance writing for business publications. He covered trade shows— the Toy Fair and the Point-of-Purchase convention — and learned to dig up hard news amid the hawkers and promoters. His big break came in 1967 when he went to work for *Barron's*, the weekly financial publication. From there he went on to E.F. Hutton.

Armed with all the research he had done while at Hutton and which Hutton had spurned, he printed his first newsletter in November 1969, spending "every last dollar I had" to send out newsletters to 2,500 people. After six weeks 100 subscribers had signed up at $100 each.

With that as his base, Kagan set out to change Wall Street's views on how to value cable companies. In the January 1970 issue he proposed something he called "Total Market Value" as a way to determine the relative value of the cable companies traded in the public market. The formula was as follows:

Take the current stock price, multiply by the number of fully diluted shares outstanding, factor in the long-term debt minus working capital and divide by the

50 Cents A Channel:
Comtronics Cable TV, a Tele-Communications Inc. subsidiary, was selling 12 channels of cable service for $5.95 in the 1960s.

number of subscribers served by the company. Each company could then be assigned a total market value per subscriber, which could be used to compare it to other cable companies, both public and private.

The cable industry ate it up. It provided every cable operator in the country a quick way to calculate his company's net worth. Somebody running a mid-sized system with 2,000 subscribers could look at Kagan's chart, find that the average public company was valued at about $600 per subscriber and go home that night to announce to the family that they were millionaires.

Brokers and some financial officers at cable companies had been using similar systems for years. But Kagan was the first to make these numbers public, at least to those willing to pay for his newsletters.

Kagan followed this up with a blizzard of new charts and statistics, analyses of mergers and offerings, articles on regulation, operations and technology. Above all, he hammered away at the Wall Street way of looking at the industry.

"Wall Street makes a mistake every time it looks at cable earnings based on current accounting," he wrote in the issue of Jan. 15, 1970. In October he told readers, "CATV companies can be evaluated best in terms of cash flow and not traditional taxable profits. This is simply because CATV companies net 50 cents on a dollar of revenues before depreciation, amortization and interest payments. But they cannot show 'Wall Street type earnings' because they are still infants and need all their cash for growth."

And in 1972 he was still trumpeting, "It is wrong to buy or sell a CATV stock on a price-to-earnings ratio basis. It is nonsense to talk about 20, 50 or even 100 times earnings when those earnings are for bookkeeping purposes or for unsophisticated investors and brokers who need numbers because they don't understand concepts."

Kagan's analysis was right, and eventually would be proven so. But his timing was horrible. When he launched his first newsletter cable was about to enter the most difficult period in its history. Like the industry he covered, Kagan himself "died three times in the next five years," as he later recalled.

The business was whacked by multiple hammers in the period from 1968-1974: rising interest rates, more regulation, Supreme Court rulings, lack of understanding by Wall Street. And in the most public blow, the head of the nation's largest cable operating company was indicted, convicted and sentenced to jail for bribing a city official in Johnstown, Pa., in connection with the cable franchise.

Irving Kahn, the great showman, huckster, entrepreneur, visionary and financial wheeler-dealer had a darker side. It wasn't that he was a crook, the kind of guy who stole Social Security checks or robbed gas stations. But he did have an impatience with the petty details of life that sometimes led him to cut corners or ignore the rules that apply to other people. It was the same syndrome that sometimes leads powerful congressmen who deal with matters of war and peace and billion-dollar budgets to ignore "No Parking" signs or to use money from the stamp fund to pay for greens fees.

When he got involved in the world of professional boxing, for example, Kahn found he could make more money selling TV reruns of a fight that went six or more rounds than for one that ended early. He passed the word on to some of the fighters that if they kept things going their paycheck would be a little larger.

"I got very friendly with (Heavyweight champion Floyd) Patterson through (manager) Cus D'Amato," Kahn later recalled. "I said 'Look, I'm not telling you to prolong the fight. But if there is some chance that you feel sure that you can go past six rounds you're worth an awful lot of money.' In looking back on some of those early films I think he could have polished them off a little quicker in some cases."

It wasn't exactly fixing the fights, but it wasn't something he learned along with baton twirling back at Boy Scout camp either.

The same mentality was at work when it came to a renewal of the fran-

Going Into A Tough Stretch

The mid-1960s and early 1970s were a tough time for cable operators financially, operationally and in terms of their image. TelePrompTer CEO Irving Kahn's federal conviction for bribing city officials in Johnstown, Pa., in 1971, was just the start of an era that would shake up his company and many other cable operatorions. Here¹s a look at some of the forces that hammered the cable industry in those years:

Stymied In The Courts:
The Supreme Court gave the FCC unlimited authority to regulate cable in the Southwestern decision in 1968. The FCC proceeded to freeze distant signal importation, ban ad sales on local cable programming and mandate signal quality.

Hamstrung At The Banks:
Cheap money to build systems disappeaared in the late '60s as the prime rate climbed to 8.5% in 1969.

Slammed on Construction Costs:
Copper — the key ingredient in coaxial cable — rose to 42 cents a pound in 1968, from 29 cents a pound in 1960. Only 90 new systems were built in 1969, down from 250 the year before.

Regulated – Again:
In 1970, the Supreme Court ruled that cable companies could be regulated by state public utility commissions.

Vendors Weakened:
As construction slowed, Jerrold was forced to lay off 550 workers in 1969. Vikoa reported a 1970 loss of $8 million. Ameco's revenues were halved in 1971.

Dumped on Wall Street:
TelePrompTer's financial troubles, discovered after Jack Kent Cooke's takeover of the company, caused cable stocks to lose luster on Wall Street. By 1973, analyst Paul Kagan recalls, "There was no market at all."

Strapped Financially:
Stock woes made it hard to raise money, and some MSOs found themselves in deep financial waters. TCI, for example, was paying interest on $134 million in loans in 1974, a year the company generated just $18 million in revenue.

chise in Johnstown, Pa. TelePrompTer owned the Johnstown cable system when the state of Pennsylvania passed a law requiring that all cities have formal franchises for their systems.

Most communities simply gave the franchise to the existing operator, making official a situation that had existed for many years.

But in Johnstown the mayor threatened to put the system up for bid. The process would require a tremendous amount of work by TelePrompTer just to keep what it already had. It looked like a needless headache to Kahn who was working on much bigger issues.

"I don't know about your experience with going from New York to Johnstown," he later recalled. "Weather being what it is, eight out of 10 times the flights in the winter don't land in Johnstown. For me to take a day off, with two lawyers and someone who knew the technical end of cable to go down and argue with Johnstown, we figured it would cost $3,000 to $4,000 a day. It wasn't a great town to go into with a private plane because they didn't have great instrumentation. Finally, at one of the meetings either the mayor or someone suggested that if I were to contribute $5,000 apiece to (the mayor's and two city councilmen's) campaigns we would get the franchise."

Kahn's attorney, Pennsylvania cable operator George Barco, advised against it. But it looked like an easy shortcut to Kahn. "I was sitting there thinking we were going to have to make 10 more trips back to this place. That was my mistake. In hindsight I should have given them the money but I should have brought the FBI in and we would have gotten (the franchise) anyway."

Kahn didn't do this deal very well. In the first place, by traveling from New York to Pennsylvania he violated federal, not just state, laws. Then, he gave the mayor not cash, but a check, a corporate check, which later caught the attention of an IRS agent looking into an entirely unrelated issue. But it didn't seem like such a big deal to Kahn, even when he was hauled before a federal grand jury.

"I didn't attach much importance to this. They wanted me to fly up (to testify). I was in the Bahamas with my wife and kids for Christmas on the boat, and they said come on up here and we'll just go over this. I said 'What have I got lawyers for? Don't bother me.' I just didn't take it seriously."

He should have. And he should have been more careful when he did testify before the grand jury. "They started to ask me about my company. Well, I'm naturally enthusiastic, particularly about something I believe in. They asked me a whole bunch of questions. When I came out they let us know they had me on 69 or 70 counts of perjury."

Then two of the Johnstown officials agreed to testify against Kahn in return for lesser sentences for themselves. It was a short trial. Kahn was sentenced to three five-year terms in the federal penitentiary. He got out in 20 months, the minimum.

Prison failed to dampen Kahn's ebullient spirits. While in jail he took a correspondence course in cable technology from Penn State University and became a certified cable technician. He applied for franchises in New Jersey (using the prison warden as a witness to some of the paperwork), built microwave systems to serve them and launched an extensive local origination service to televise high school football games and other events of local interest. All this he did from his cell in the Allenwood federal penitentiary.

When Kahn left prison the cable industry didn't quite know how to handle him. His conviction had been front-page news across the country and had given cable its biggest public relations black eye ever.

Kahn's conviction "cast a pall over the entire cable industry," *Business Week* reported, not just because TelePrompTer was the biggest cable company in the country, but because "Kahn has been cable TV's chief visionary and evangelist."

Just months before going to prison, he had spoken to a group of executives from major financial houses predicting that TelePrompTer would launch its own communications satellite to beam a new network of cable-exclusive programming to systems around the country, revolutionizing the industry and creating vast new wealth for those who had been wise enough to invest in TelePrompTer. His predictions may have seemed like science fiction to the conservative investment executives at the time. After he went to jail he had even less credibility.

Finally, a year after getting out of prison, Kahn was invited to speak at the Texas Cable Television Association annual convention. It was a big risk for the industry to give him a forum. Industry leaders were a little nervous about what Kahn would say in his first public appearance after getting out of jail.

When he stood up before the convention there was silence. Kahn looked around the room, cleared his throat and began:

"Now... as I was saying before I was interrupted."

The line, suggested by Kahn's long time colleague Linda Brodsky, brought down the house and opened the door for him back into the business, although never again at the level he had played on before going to jail.

His company, in the meantime, was in shambles. Just before he became a guest of the federal government, Kahn had completed an acquisition of Jack Kent Cooke's cable company, which by that time had merged with H&B American to become one of the biggest in the nation. The TelePrompTer deal required Cooke to vote his stock as Kahn dictated, precluding a takeover of the company without Kahn's approval.

But when Kahn went to jail Cooke declared the agreement null and void. He engineered a takeover of TelePrompTer. What he found when he got to corporate headquarters wasn't pretty.

"The company was virtually bankrupt," recalled Bill Bresnan, who was named chief operating officer of the company by Cooke. "It had a $40 million line of credit with Chase Manhattan Bank and had borrowed more than $50 million. It was building systems in Manhattan, Los Angeles, St. Petersburg (Fla.), Seattle. It was spending money very fast.

"The financials were a real rat's nest," he said. TelePrompTer had listed 65,000 subscribers for its Manhattan system. But when Bresnan finished looking at the books he found that there were thousands of disconnects, free installations and others listed as paying customers. The real number, he later concluded, was only 34,000.

By the time they had finished going over everything the company was forced to take a $30-million write-off. The stock had dropped 18% on Kahn' indictment. It kept plunging as the trial continued and Kahn was convicted, Cooke launched a takeover attempt, and the new management found one skeleton after another in the various TelePrompTer closets. "We were virtually bankrupt, but we were just too stubborn to admit it," Bresnan said later.

By 1973 Bresnan was ready to take drastic measures. He cut capital spending in half, laid off 20% of the TelePrompTer employees (many of them people Bresnan had himself hired and brought up through the ranks) froze the pay for all other workers and rolled back salaries for top management. Even the annual report, normally a four-color glossy brochure touting the future of cable, was printed on recycled newsprint in black and white to save a few dollars, and to demonstrate to shareholders that the company was looking everywhere for savings.

TelePrompTer stock, which had once placed a value of $1,287 on every TelePrompTer subscriber, plunged to where it was worth only $250 a subscriber. Investors bailed out by the bucketful. Kagan remained as one of the few believers, stating in his newsletter of September, 1973, that the "Wholesale dumping of TelePrompTer stock was a 'bullish sign for the faithful'."

The faithful were a dwindling band, however. TelePrompTer's woes, while more dramatic and publicized than those of other cable operators were by no means unique.

None of the other cable companies had endured a public trial and conviction of their CEO, but they were all susceptible to the rise in interest rates, demands from the cities, and the FCC rules that had slammed TelePrompTer. And the TelePrompTer mess had convinced many analysts that the cable companies

Sports Maven:
Charles Dolan parlayed his love of sports into successful cable system and programming franchises. Dolan is pictured here (second from left) with Cablevision Systems Corp.'s John Tatta, Al Rosen and Eugene McHale, president and EVP of the New York Yankees, respectively. Dolan's relationship with the Yankees dates back to the early 1970s.

were shell games, that their talk of valuation on the basis of cash flow, not earnings, was just a sham to cover up phony accounting practices.

By 1973 the market in cable stocks had ground nearly to a halt. There were few sellers at the low prices and even fewer buyers at any price. "This was far from just a bad market," Kagan lamented. "It was no market at all."

A few other analysts, including Mario Gabelli and Dennis McAlpine, followed the business. But the group of cable analysts in the late 1960s and early 1970s couldn't have fielded a baseball team.

Every public company suffered from the drop in stock prices, but none more than Tele-Communications Inc., the multiple system operator that had been started by Bob Magness.

In the early 1970s, Magness had gone on a buying spree, gobbling up a bunch of small systems and MSOs. He had financed the purchases with bank debt, which he planned to repay with proceeds from additional sales of stock after the new acquisitions had been completed and the assets could be listed on TCI's balance sheet. Such short-term bank loans were known as bridge loans.

When the stock crashed, he couldn't repay the bridge loans and was faced with financial ruin. By 1974 he was paying interest on $132 million of debt while taking in only $18 million a year in revenue. It was an impossible situation.

Back east Chuck Dolan was also feeling the financial pinch as he tried to build his cable system in New York City. A soft-spoken, slight man with a sly, almost elfin grin, Dolan had been born in Cleveland, Ohio, in 1926. His father was an inventor, the man who had developed the locking steering wheel for automobiles and an early version of the automatic transmission. Dolan remembers discovering a trunk in his attic "full of patents."

But as a child Dolan was fascinated not by machines, but by pictures, particularly movies. He bought a home projector and showed films to his friends in the basement of his home. He was a Boy Scout and used to take photos of the Scout meetings and sell them to the local papers for $2 each. When he was 15 he persuaded the Cleveland Press to buy a weekly column he wrote called "Scouting Today."

After serving in the Air Force and attending John Carroll University, Dolan began a business of compiling a weekly sports reel. He would arrange to have film of all the big college games flown to him every weekend, spend Sunday night putting together a script and highlights film to go with it and then send the prints out to stations around the country via airplane. "Some of the stations paid us as much as $15 per week," Dolan recalled. In time the Dolans (his wife, Helen helped) had 30 clients. But they had no profits. "We went broke," he later remembered.

Dolan's only competitor was a company in New York City called Telenews. He called them up and offered to transfer his clients to them if they would give him a job. They agreed, and he packed up the car and family and drove to New York.

The Dolans rented a one-bedroom apartment, and he later remembered that the walls of the kitchen could barely be seen because of all the film strips taped up during the editing process.

The company that actually syndicated the films was called Sterling Television, and in 1954 Dolan purchased an equity interest in it. He developed the business into one that produced programming, including news, for targeted audiences. It was an experience that would serve him well in the cable business.

One of the businesses he started delivered a video guide to New York City to TV sets in the rooms of Manhattan hotels. To deliver the guide, Sterling constructed a system of coaxial cables leading into the hotels.

Teleguide, Dolan later said, "turned out to be a lousy business." But while wiring the hotels Dolan became interested in cable TV. He had asked for and received a franchise from the city of New York to lay the wires needed to deliver Teleguide. In 1964 Sterling asked for permission to change the franchise so it could deliver full cable service to Manhattan.

There should have been no problem in getting the franchise. "We were the fair-haired boys" down there at the Board of Estimate that granted franchises, Dolan remembered. The franchising division, accustomed to dealing with bus lines and billboards, hadn't even realized it had authority over the underground ducts Dolan wanted to use until he showed up to ask permission to use them.

But the story leaked and the next day the *New York Tribune* ran a front-page article. Twelve other applications were filed within a few weeks, and the city, besieged, put off granting any cable franchises for a year.

Eventually the city granted three franchises: one to TelePrompTer for the upper part of the Manhattan, one to Sterling for the lower part and one to CATV Enterprises for an enclave in Riverdale.

"Financing it was a bear," Dolan would later recall in a characteristic understatement. The cost of wiring the city was $100,000 per mile. "We borrowed money from my uncles, got a $300,000 loan from Bankers Trust, and sold the motion picture syndication company for $900,000 to Cowles Media Co. It was enough to run the company for two months."

Within a few months the price got too high for Dolan's original partners, Seattle broadcaster Elroy McCaw and jet manufacturer Bill Lear, and they dropped out. Dolan then took on a new partner, Time-Life Pictures, a subsidiary of magazine publishing giant Time Inc. and operator of a group of cable systems and broadcast stations.

Time agreed to loan Dolan money he needed in return for "convertible subordinated debentures," which Dolan later recalled as "one of the great terms of my life."

Negotiating a final contract with the city was tough, Dolan remembered, particularly because of a savvy city employee named Sheila Mahony, whom Dolan later hired to head up the franchising division of his cable company. The city locked in the fee at $6 a month with an installation charge that could go no higher than $19.

When Dolan began to wire he found it a tough go. Apartment owners were reluctant to allow the cable company into their units. New York City was under rent control. Landlords were allowed to raise rents only when tenants moved out and new ones moved in. Landlords, therefore, were motivated to force tenants out as quickly as possible. Some cut off the heat or refused to fix the elevators. Allowing a new service such as cable TV, they figured, would be an inducement for tenants to stay.

Once in an apartment building, Manhattan Cable found it was vulnerable to theft of service as tenants could easily jimmy open the cable box and string a wire to their apartment without the company knowing unless it ran frequent and costly inspections.

The system also was prone to interference. The strong broadcast signals sent from the Empire State Building in-

Wiring The Big Apple:
Cable construction by Manhattan Cable TV in New York City began in earnest in the 1960s.

terfered with the signals being sent through coaxial cable, causing ghosting for some subscribers. And an outage in a single amplifier could cut off service entirely for thousands of customers in the densely packed city. From time to time a disgruntled employee would make sure that an amplifier did go out, usually in the middle of some hugely popular program.

Kagan, one of Dolan's customers, wrote of the system, "This is a potpourri of just about everything that can go wrong with a cable system."

But Dolan plugged ahead. He solved the interference problem by using dual heterodyne, set-top converters, invented by two engineers, George Brownstein and Ronald Mandell, in 1965. The signal would be sent on a frequency that wouldn't be prone to interference from the signals being broadcast over the air. The cable signal would then be converted to channel 3 by a box on top of the customer's TV set. The viewer would change channels using the converter, not the TV set which would be permanently set on channel 3. These set-top converters also would allow for much larger channel capacity because the barrier to increased channels had been the 12-channel tuner in the TV sets. The converters were expensive, about $30 each, adding still more to the cost of wiring Manhattan.

Dolan began by wiring the tony apartments on Park Avenue. (His first paying customer was the daughter of Henry Ford). There the system fairly quickly got to a penetration level in the upper 20% range. And there it got stuck, well short of the level needed to turn a profit.

TelePrompTer, meanwhile, was wiring the northern portion of Manhattan and having the same trouble. But it had developed one major technological breakthrough that allowed it to dramatically reduce costs: short-haul, AML microwave which allowed it to transmit all of its programming via microwave to various mini headends around the city. (Previous microwave technology had allowed for long distance transmission of a single channel.)

But TelePrompTer was stuck as well.

The key, Dolan thought, was the programming. It was not enough to offer customers a better signal. That would get some of them to subscribe. But to get beyond that number and into the range where profits might be in sight, the system would have to offer something that people couldn't get anywhere else. "We perceived that the solution was more and better programming," he said.

"Dolan was the first guy (in the cable business) to think in terms of what the customer wanted," said Robert Rosencrans, president of Columbia Cablevision and a sometime rival of Dolan for franchises in the suburbs of New York. (Thirty years later at the headquarters of Dolan's cable company in Woodbury, N.Y., the best parking place is reserved not for the chairman

Cable's Statesman: *Continental Cablevision's Amos Hostetter rapidly emerged as an industry leader, making cable's case before the FCC and Congress.*

or the employee of the month, but for customers seeking service.)Unlike most of his colleagues, Dolan was neither a businessman nor an engineer by training. His first love affair had been with motion pictures, and he had never quite gotten over it. He also had been involved in syndicating sports programming. This background played a key role in the coming developments.

Dolan's problem was that the city franchise prohibited him from running any original entertainment programming other than what was already on the broadcast stations, a clause that had been inserted at the insistence of the theater owners.

But Dolan wasn't about to let that stop him. He went down to the city and asked if it would be all right if the system showed movies that were not entertainment.

"So we had a big meeting to demonstrate what we meant," Dolan remembered. "We ran a long film by a producer from India to show what we meant by a movie that was not entertainment."

Dolan couldn't tell if New York franchise bureau chief Morris Tarshis actually fell asleep during the showing, but Tarshis did agree at the end that this kind of film was definitely not "entertainment." He agreed to recommend that the Board of Estimate allow the cable systems to show such non-entertainment films. He even left it to Dolan to select them, stating the city could not act as a censor.

"The theater owners went out of their minds," Dolan remembered. They formed a committee called the "Campaign to Save Free TV." They passed out petitions at theaters against pay TV. They picketed City Hall.

But the Board of Estimate, recognizing that the cable systems needed something to help them stay afloat, unanimously passed a resolution allowing the systems to show non-entertainment motion pictures. It wasn't long before the definition of what was non-entertainment began to stretch.

The concept of cable-originated programming received a further lift from an unexpected quarter: The Federal Communications Commission, now under the direction of Nixon-appointed chairman Dean Burch. Among a blizzard of new cable rules it issued in 1970 the Commission required all cable systems in the country to offer original programming and specifically authorized pay television programming.

The requirement for original programming was seen by most operators as enormously costly. Continental Cablevision's Hostetter, emerging as a leader in the industry, estimated it would cost about $50,000 per system, an amount that could not be supported, particularly in smaller markets where advertising revenue was problematic at best. The requirement for local programming was eventually struck down by the courts.

But for Dolan and other big-city operators looking for ways to boost subscribership, the FCC action was a welcome development. (Another regulation adopted by the FCC in 1971 prohibited the ownership of cable systems by broadcast stations in their home markets and by broadcast networks anywhere in the U.S. The result was to force CBS, one of the nations largest cable operators, to spin off its cable division into a separate company, Viacom International, headed by Ralph Baruch.)

The federal go-ahead for cable programming spurred a host of new initiatives in that arena. CBS, Dick Clark, *Reader's Digest* magazine and others jumped in with packages of taped programming they offered to send to cable systems for a fee.

In New York there were plenty of local sources of entertainment, and Dolan set out to tap them. In 1970 he struck a deal with Madison Square Garden to show Knicks and Rangers home games and other Garden events that were blacked out on Manhattan broadcast stations. He agreed to pay $24,000 for the package and to guarantee the gate (pay for any seats that were not sold). He picked a great year to start as the Knicks were just entering the era of Bill Bradley and Dave Debusschere that would eventually lead them to an NBA championship.

TelePrompTer also developed original programming for its Manhattan system, telecasting local college football games and community meetings.

But it still wasn't enough.

By the summer of 1971 Dolan was worn to a frazzle. He packed up the family (by then the Dolans had six children)

From The Garden:
Chuck Dolan signed a deal with Madison Square Garden in 1970 that brought New York Knicks and Rangers games to Manhattan Cable TV subscribers.

and boarded the cruise ship Queen Elizabeth II for a vacation in France. But the problems back home continued to weigh on him. It wasn't long after leaving New York harbor that he began to craft a plan for a new kind of programming service he dubbed the Green Channel.

"I can still remember," he later recalled, "sitting in the upper bunk with a portable typewriter propped up on my lap," drafting a memo proposing the Green Channel. He mailed the memo from Le Havre.

Dolan's plan was ambitious. "In the long run," he wrote, "we may think of ourselves as the Macy's (department store) of television, shopping everywhere for programs that some public, large or small, will buy. If we are successful in meeting these retail program needs of the region we are attempting to serve in 1972-1973, we will later use whatever efficient transmission systems become available, from microwave to satellite, to sell television programs worldwide to any public that signals its specific demands to us."

After returning from vacation Dolan recalled "I was so eager to hear what everybody (at his partner Time Inc.) thought" of the memo.

"Nobody had read it."

So he set up a meeting with the top brass at Time Inc., including president James Shepley and chairman Andrew Heiskell. Dolan presented his plan, which called for spending $300,000 to launch a sports and movie channel that would be shown to subscribers in Manhattan and syndicated to other cable systems around the country.

Heiskell, Dolan remembered, listened but gave no indication of how he felt. After a while the Time Inc. chairman scribbled something on a piece of paper, tossed it to Shepley and left the room.

Shepley opened the note. It said, "Go ahead."

Dolan moved into the Time Inc. building and began to put together a crew. Time Inc. assigned Tony Thompson, from the Time-Life Books division, which relied heavily on direct sales, to the project. The unit was placed under the supervision of Richard Munro, another hot-shot young executive who handled all the non-publishing ventures of Time Inc.

Dolan also brought on board a soft-spoken young attorney with dark eyes, dark hair and bushy mustache: Gerald Levin. A former divinity student at Haverford College (known for having one of the smartest group of students in the nation) he had worked for a New York Law firm on several deals in the Middle East. Thompson first met Levin at LaGuardia Airport when they were flying to Chicago for the 1972 National Cable Television Association Convention. Dolan introduced them.

Thompson recalled thinking that Levin had no experience in anything related to what they were doing and would be a certain liability. Then, as they rode out to Chicago, Thompson gradually became aware that this was a true genius. And he was a movie freak.

"He knew every movie that was ever made," Thompson remembered. "And he loved them all. He could talk with genuine passion about some little film nobody had every heard of."

Moreover, Levin had such a quick mind, Thompson realized, that he could take on any project – distribution, affiliate relations, programming – and run rings around the experts in a matter of days.

First Run: Universal Pictures "Sometimes a Great Notion," staring Paul Newman, was the first movie carried by HBO on its premiere night, Nov. 8, 1972.-

The group needed a name for the service. Each put in a few suggestions. Dolan didn't like any of them. Thompson suggested Home Box Office. "We had sports and movies. What did they have in common? Tickets and box offices. But we were offering this together with the convenience of doing it in the home."

Nobody, including Thompson, much liked the name. Levin worried it would be known as Home B.O. But nobody could think of anything better. Dolan remembered, "I said 'It really isn't very good. We'll change it later'."

Dolan set out for Hollywood to find some programming. Levin worked on finding cable systems to carry the service and some method of delivering it to them. Thompson began to develop some research. "They (the Time Inc. officials) told me 'We're definitely not going to do this, but go down there and check it out,'" Thompson later recalled.

Dolan had a tough time with the studios. "They were concerned about the reactions of the theater owners," he recalled. "Twentieth Century Fox told us 'This is no business. You're a middleman and you will be squeezed out'."

Finally they struck pay dirt with

Universal Pictures, which agreed to make a selection of films available that included "Sometimes a Great Notion," based on the Ken Kesey novel and starring Paul Newman.

Thompson was doing a bit better with the consumers. The first forays into the street produced buy rates of 40%, an astounding return for a veteran of direct mail where returns of 5% were considered excellent.

Levin set to work on the distribution angles. He found it was not possible to distribute the service in New York City right away because the studios and sports teams would not sell the rights to show their product in major urban areas. He was turned down by Storer, Times Mirror and Readers Digest systems (the last run by Peter Frame who would soon be working for HBO).

Levin found that there was a microwave link, operated by Eastern Microwave, into central Pennsylvania that could be used to deliver the HBO signal to systems in that area. Among the Keystone State cable operators contacted by Levin was John Walson of Service Electric Co., then the largest cable operator in Pennsylvania and the 12th largest in the country with almost 100,000 subscribers.

Engineers Organize

As more and more cable systems came on line in the 1950s and 1960s engineers across the country were faced with similar problems. But they had no way to communicate regularly with each other.

This created a fertile field for a national organization of cable system engineers that could exchange ideas on improving plant performance and operations. When the trade publication *Cablecasting* magazine called for such an organization in its issue of November 1968, engineers from across the country responded. At the NCTA convention in June 1969, 79 of them got together to form the Society of Cable Television Engineers. Ron Cotten, chief engineer of Concord TV Cable of Concord, Calif., was elected the first president.

Response from management was not favorable. The heads of cable operating companies feared the organization was a prelude to unionization of the cable engineers.

Rex Porter, a cable system operator and salesman for Times Wire, recalled that a group of managers came to the organizational meeting with the express purpose of preventing formation of the SCTE. "A few years later they asked me to help them with the application for SCTE Charter membership since they had been in attendance at the first meeting." By 1978 the society had more than 1,000 active members.

In the early 1980s the group began to face financial troubles. An attempt to hold a national convention and exposition met with opposition from the NCTA, which urged its members not to attend, fearing it would cut into the NCTA convention revenues.

By 1983 the group was $60,000 in debt and was forced to move out of its headquarters in Alexandria, Va. To help get the organization back on track, the SCTE hired as executive vice president Bill Riker who had been director of engineering at the NCTA, director of engineering at Showtime Networks, headend and microwave engineer for MacLean Hunter Ltd., and chief technician for cable oper-

INSTANT SIGNALS: HOW CABLE TV CHANGED THE WORLD OF TELECOM

Service Electric's biggest system was in Allentown, Pa., and it was facing competition from another cable company. Walson wanted something that would give his system an edge over the competition, and he figured HBO might be the answer. But the service planned to carry some sporting events that were subject to blackout rules in the Philadelphia area, including Allentown. So Walson offered his system in Wilkes Barre.

Thompson remembered that, before he would sign the deal, Walson wanted his picture taken with the chairman of Time Inc. So they brought him up to New York where he was photographed with Andrew Heiskell. HBO had its first affiliate.

It was a soggy one.

Wilkes Barre had been devastated by a huge flood just a few months before. People were still living in trailers while they attempted to restore their waterlogged homes. Thompson recalled having to shovel the mud out of the office HBO rented in the downtown area. The water mark on the walls was six feet above the floor. But at least it was a real live system.

The agreement between Service Electric and HBO stated that the cable operator would provide the channel and

ating company AmVideo Corp.

Working out of a garage at SCTE president Tom Polis' construction company in West Chester, Pa., Riker and the SCTE board began to pull themselves out of the financial morass. Riker repaired relations with the NCTA, and the 1984 SCTE Expo was a financial success. Suppliers found they could introduce new products and court the top management at the NCTA convention but use the SCTE Expo to get down in the trenches and show the folks who actually used the equipment how it worked.

In 1985 the SCTE began a program to certify cable television engineers and technicians, later expanded to include installers. It established a group to work on industry standards so that equipment made by different manufacturers would be compatible at some basic level. (This same goal would be a prime mover in the establishment of CableLabs later in the decade).

By 1997 the organization had grown to serve more than 15,000 members, changed its name to the Society of Cable Telecommunications Engineers, helped launch a British version of the Society, seen its Expo grow to attract more than 8,000 attendees, and launched a conference on new technologies attended by more than 1,300 in 1997. In 1996 it moved into a new, permanent headquarters office in Exton, Pa., just six miles down the road from Tom Polis' garage.

Technically Inclined:
Bill Riker held various engineering positions at Showtime, MacLean Hunter and AmVideo Corp. before becoming executive vice president of SCTE in 1983.

billing services, HBO would deliver the programming and the two would share the marketing costs. They agreed to split the revenue, with $3.50 going to HBO and $3 to Service Electric. The $6.50 retail rate was set, Levin recalled, high enough to pay for programming and distribution, but low enough so that it wouldn't exceed the monthly bill cable customers were already paying.

The deal defined future agreements between cable operators and pay television networks, and it made HBO and Service Electric partners. Had HBO simply leased a channel, for example, the two would have had a much more traditional buyer-seller arrangement, and the cable company would have had no real incentive to sell the service. Conversely had HBO simply sold its programming to the system for a flat fee, Service Electric would have kept all the revenue and the network would have had no interest in how many subscriptions were sold.

For all the friction that would later develop between cable operators and programmers, the HBO-Service Electric partnership model remained largely unchanged for the next three decades and guaranteed that for better or for worse, the fates of operators and programmers would be inextricably linked well into the 21st century.

The network-affiliate model, Levin recalled years later, was a prime ingredient in HBO's success.

All the other services, he said, "talked about leasing a channel and delivering movies only by cassette. We were a part of the thinking of Sterling Manhattan Cable. We knew that the cable operators understood their customers. We also came out of Time-Life Broadcasting. So we adopted the network-affiliate model that is used in broadcasting.

"We also decided to be a network, to transmit rather than to bicycle tapes. We wanted to do live sports, and this set us apart from the other pay services. All this is brilliant in retrospect. At the time it was going against the grain."

The final key to HBO's success was the decision to sell subscriptions to the service rather than to sell on a pay-per-view basis. This also went against the grain because at the time the conventional wisdom, grounded in the experience in the 1950s in Bartlesville, Okla., was that viewers would pay only for the individual programs they watched, not for a service that delivered a package of programming, much of which they might not watch.

"At Time-Life everything was sold by subscription," Levin recalled. "We understood the power of a subscription," which remained in effect unless the subscriber made an affirmative decision to end it. Also, a subscription service was "so much easier to deliver by a midband converter," while other services required much more complex technology.

With a monthly subscription, "we

First Subscribers:
Gerald Levin (far right) welcomes HBO's first subscribers. Service Electrics's Hoyt Walter is at the far left.

didn't have to rely on credit cards or a black box. And we had the psychology of it, the inertia of a disconnect."

By launch date, Nov. 8, 1972, 375 customers in Wilkes Barre had signed up to take the service, sight unseen.

Dolan later recalled the launch came on a cold, rainy day. The microwave dish blew down, and an employee had to be sent to hold the tower up by hand in the freezing rain. Dolan wasn't sure they would be able to get the transmission going.

Thompson and Munro were on their way to Wilkes Barre for the formal launch ceremonies, but got stuck in the sleet and freezing rain for three hours on New York's West Side Highway. Finally they ditched the car, hopped the barrier and called Wilkes Barre to tell them to go ahead without them.

Levin, who went on-air to introduce the service to his small group of charter subscribers, later remembered all the glitches that day. "We had some problems with our microwave transmitting. It wasn't clear that we were going to be able to make it. I went on the air a couple of minutes before we went to a hockey game, Canucks against the Rangers, around 7:30. I'd written a little script for myself, never having appeared on camera before. After that I went upstairs to my very small office at what was then Sterling Manhattan Cable where I had set up a little lamp and a chair in front of my TV to watch the first night of Home Box Office. There was no other place to get it in this area where you could simulate home conditions.

"The sense of being quite alone, with very little acknowledgment that this was

Time Triumverate:
Dick Munro, Gerald Levin and Nick Nicholas were the big guns that made HBO a success, and helped jumpstart the cable industry.

happening, was overpowering. It wasn't clear how long this network would last. We had only bought programming for a couple of months, and our corporate sponsorship was only there for a relatively short period of time, unless we could meet certain benchmarks. But just to see a movie all the way through without any commercial interruptions, in its original artistic integrity was, I thought, a profound experience that I hoped people at home would also feel."

They would. But it would be several years before that profound experience was transformed into a positive bottom line, and HBO nearly died on the vine in the meantime. It wouldn't really be successful until it was able abandon the costly and inefficient microwave and make use of a new transmission system, one that could reach the entire nation with its signal: a satellite.

The Satellite Breakthrough

Home Box office was not an instant success. In fact, it almost didn't make it through its first year. On more than one occasion the small HBO staff discussed what would happen if Time Inc. decided to shut the project down.

The service suffered from multiple ills. Most serious was the product itself. Studios were reluctant to provide films, fearing the wrath of theater owners. Lacking a wealth of titles to fill up its schedule, HBO had to repeat the films it did have over and over.

"We were constantly running out of movies," recalled Jerry Levin, who was running HBO just a few months after its launch in Wilkes Barre, Pa., in November, 1972.

Even when a studio was willing to do business with HBO, it was limited by law as to what it could sell. The Federal Communications Commission had ruled that no film could be sold to a pay service unless the film was less than two years old or more than 10 years old. This was designed to protect the syndication window for the broadcasters who generally would rent films for showing a couple of years after they had appeared in theaters.

Scheduling also was a problem. Films started whenever the last one ended. Viewers accustomed to having shows start on the hour or half-hour found the scheduling confusing.

Churn was enormous in part because of overselling by contracted salesmen who were paid a commission on each sale made and therefore encouraged to exaggerate the benefits of HBO. They suffered no penalty if a customer canceled a month or two after signing up. Some of them promised things HBO could not deliver, particularly first-run motion pictures. Some subscribers were shocked by the nature of some of the racier films shown on HBO, films that would not be shown on the broadcast networks without severe editing.

The HBO team had adopted the subscription model, in part because they thought it would make subscribers more likely to continue to pay, as was true in the magazine business. This proved incorrect, as the service found out the summer after it launched, when subscribers disconnected in droves.

"We didn't realize you had to continue to market," Levin recalled, particularly in the summer months when customers had alternative ways to spend their entertainment dollars.

The difficulty in selling HBO was compounded in older cable systems where for years customers had been getting a dozen or so channels for a price

Dishing It Out:
10-meter dishes like this ushered in a new era of nationally delivered program networks that powered cable's growth in the 1970s.

of about $8 a month and were now being asked to pay another $8 a month for a single additional service.

Cable operators also were not accustomed to marketing. Most had been trained as engineers or financial people and knew little about selling. That didn't much matter in the 1950s and 1960s when cable was a service highly desired by consumers who had no other way to receive television signals. But the lack of understanding of how to entice the customer to buy was a critical shortcoming when it came to selling pay television services.

Hardware also presented some major problems. Until the advent of pay television, cable systems simply delivered the same package of signals to every home that signed up. With HBO it became necessary, for the first time, to deliver different sets of signals to different subscribers.

The early pay television systems, including HBO, used converters to accomplish this. The signal would be sent out off-frequency and then could be converted back, using a device that sat on top of the set, to a frequency the TV set could recognize. Customers with these "set-top converters" could get HBO. Those without converters could only get the "basic" package of broadcast and local origination channels.

Converters also could accomplish other tasks, including allowing far more than 12 channels to be delivered to the home. The 12-channel limit had been set by the tuner in the TV set, which could handle only 12 channels. Use of the converter also reduced "ghosting" caused by interference from signals being broadcast on the same frequency as those being sent over the cable system.

But converters were expensive. To meet consumer demand a cable system might need to have several thousand on hand, at a cost in excess of $30 each. Operators then had the problem of reclaiming the converters from those homes where customers dropped the premium service.

This issue was solved with the advent of the trap, a simple device costing only a couple of dollars and placed on the drop cable which ran from the utility pole into the home of each subscriber. The trap would scramble an individual signal. The strategy was to go into a town before the launch of HBO, install a trap for every subscriber and then remove the traps from those homes which signed up for HBO. But it didn't take very long for enterprising consumers to recognize that if they wanted HBO for free, all they had to do was remove the trap themselves.

Traps were useful only where the system could add HBO without breaking the 12-channel barrier. If a cable system was already offering 12 chan-

The Deal in PA.:
HBO found Pennsylvania fertile ground for early affiliation contracts. Jerry Levin is flanked by Keystone state operators John Walson (left), of Service Electric Co., and Bark Lee Yee (right), of Twin County Cable TV.

nels and could not drop one, adding HBO meant using converters.

These issues were new ones for the cable operators, who patched together makeshift premium television marketing plans as best they could. At the TelePrompTer system in Elmira, N.Y., for example, the system manager dumped HBO in the lap of Tony Lynn, a recent graduate of the Newhouse School in Syracuse who had been hired to do local origination programming. Lynn, later became president of Playboy TV, recalled that when the system decided to launch HBO, "we had no marketing department at all.

"Les Read came up from New York to tell us how to sell HBO, and I became the pay TV marketing manager."

"We would hire these Bedouin trap installers," Lynn recalled. Several dozen would check into a hotel and then trap the whole town in a week. The head of one such group became known, Lynn recalled, as "Trapper John."

HBO also had competitors, including a service called Channel 100 started by Geoffrey Nathanson in Los Angles. This service worked with an in-home box. Subscribers would purchase a card that allowed them to activate the box so they could watch the movies. Time Inc. president Jim Shepley was acutely aware of what Nathanson was up to. "He had a friend who would give him information about what Nathanson was doing," Levin recalled. "Here is the president of Time Inc., and in comes this kid (Levin). He was very critical and could be very intimidating."

But the biggest problem for HBO and all the other pay services was distribution. Microwave transmission was enormously expensive and worked from point to point rather than on a broad-

An HBO Original:
On March 23, 1973, HBO made cable history by producing its first original program, the Pennsylvania Polka Festival.

cast mode. A cable system could receive the signal only if it was along the microwave route.

Claus Kroeger, an assistant general manager for a group of New England systems owned by Cox Cable, recalled how difficult it was to manage the microwave transmissions.

"The towers were all up on a mountain top," he remembered. "If the signal went out in the winter somebody had to snowshoe up to the top carrying all the equipment while it was 10 degrees below zero with the wind howling. And if any one tower went out, all the systems down the route would be affected.

"It was an immense amount of work to keep that signal going."

His systems were classic operations that had been running for many years, mostly without price increases. And the consumers were all New Englanders who took pride in their reputations as skinflints. Moreover, Kroeger recalled, they just felt that movies should be seen in the theaters, not at home. For all these reasons there was considerable consumer resistance to the out-of-town, fast-talking marketing crews that swept down to sell door-to-door when HBO launched.

Kroeger tried lots of marketing ideas. He put an HBO promotion on the message wheel. This was a device that would rotate different print advertisements every few minutes in front of a stationary camera. The systems decorated their offices with giant movie posters touting upcoming HBO films. He put popcorn in the lobby and offered lollipops to the kids. He tried direct mail and newspaper ads.

But nothing really worked very well. "We got to 10%-12% penetration and just stalled there," he remembered.

The troubles added up. A year after launching, HBO had about 8,000 subscribers nationwide, down from the 12,000 it had had at the peak. Then Time Inc., majority owner of HBO and one of the nation's largest cable operators, got cold feet about the business.

HBO had started as an outgrowth of the Sterling Communications partnership Time had with Charles Dolan, a partnership which also operated the enormously expensive New York City cable system. In 1971 the system lost $2.5 million and in 1972 $4 million, according to the *New York Times*. Time Inc. had been loaning money to the enterprise in return for convertible debentures that could be transformed into stock. By the beginning of 1973 Time's equity and convertible debentures entitled it to own 80% of Sterling.

According to James Heyworth, then HBO's chief financial officer, Time's accounting firm advised the company that it could no longer list Sterling as an off-balance-sheet asset. It needed to consolidate Manhattan Cable and HBO into its regular operations.

So Time sent Dolan packing, with a nice check for his remaining share of the business, but with a bitter sense that he had been booted out of a business he had spent the better part of a decade building. "It was a difficult period," was all that Dolan would say 25 years later.

Then Time decided to get out of the cable business altogether. Its stock had come under attack on Wall Street, and the company was shedding non-core assets such as its paper mill and broadcast stations. It even shut down one of the jewels of its publishing stable, *Life* magazine.

Time Inc. agreed to sell Sterling — the Manhattan cable system, systems in Long Island and a controlling interest in HBO — to Warner Communications for $20 million. Time traded its remaining systems to ATC for roughly 9% of ATC's stock.

But the deal for Sterling fell through when Warner took a look at the New York City franchise requirements. Time offered Manhattan to ATC, but Rifkin also balked at the enormous cost to build and manage the system.

Dolan got word of the Time-Warner deal while on a trip to Las Vegas. He flew back to New York and went to see the folks at Time Inc.

Grabber: *HBO's popular programming guide also was used as a promotional piece for a 1975 convention.*

"I asked them: 'How about taking a bid from me?' " Dolan was told the deal was done and was given a copy of the contract, signed by Warner but not yet by Time Inc. "So, with the contract in my hand, I went in to see (Time president) Jim Shepley."

Shepley, irked that Warner was not going to take Manhattan, said he was willing to split off the Long Island system and give Dolan a chance to bid. "Shepley told me 'The price is $900,000. Show up here in the morning with a check for $100,000 and a note for the balance and you can have it.' "

Dolan did just that, forming Cablevision Systems Corp., which would operate the systems on Long Island and become one of the nation's largest and most innovative cable operations. (And Dolan never lost his love for movies or his faith that showing them on cable would produce profits. He founded American Movie Classics and Bravo).

So Time Inc. was forced to keep its white elephant cable system in Manhattan and with it the fledgling Home Box Office. The Time brass began to put the heat on Jerry Levin, telling him that unless considerable progress could be made toward profitability the future of the pay service was in serious doubt.

Shelpley gave Levin a benchmark: Get 20,000 subscribers by June of 1974 or face a shutdown of the service. "We just made it," Levin recalled, and some who were there at the time remembered that the goal was achieved in part because the service ignored a couple of hundred disconnects. In any case Levin remembered that the HBO staff held a victory celebration at a small Italian restaurant in New York.

But their troubles weren't over yet.

The Master Plan:
Nick Nicholas and Jim Heyworth plot HBO's affiliate growth throughout the U.S. in the mid-1970s.

HBO continued to operate in the red. And it was clear that this small, regional pay service delivering programming to a few thousand customers in the Northeastern U.S. didn't fit the image of Time Inc., publishers of the mighty *Time*, *Sports Illustrated*, Time Life Books and other huge nationwide publications.

"To sustain Time Inc.'s interest, we needed a big idea," Levin recalled. He set to work to find one.

Dolan, meanwhile, took over in Long Island and began to look for financing to upgrade and expand the system and bid for other franchises. The best place, in fact just about the only place, to find money for cable systems in the early 1970s was from the Jerrold Electronics Division of General Instrument Corp., still the leading supplier of hardware to the industry and increasingly reassuming the role it had played in the late 1950s as a major source of funding for cable systems as well.

In March of 1973 the company had set up the GI Credit Corp. to provide financing for cable system construction. Backed by a $25 million line of credit from Chase Manhattan Bank, the structure was similar to the automobile com-

The Young Doctor: A youthful John Malone, fresh out of McKinsey & Co., helped turn around Jerrold Electronics for General Instrument in the early 1970s.

panies, which sell a car to a customer and then loan him the money to purchase it through an in-house finance operation.

When Dolan completed arrangements for the first infusion of cash from Jerrold, the manufacturer's new 32-year old president came out to ink the deal. Dolan remembered him as looking even younger than his age, with a very matter-of-fact, straight-to-the point style.

His name was John Malone.

Malone grew up in the upper-middle class town of Milford, Conn. He graduated Phi Beta Kappa with a degree in electrical engineering from Yale in 1963. At Johns Hopkins University in Baltimore he earned a Masters of Science degree in Industrial Management and a Ph.D. in Operations Research. Otherwise a decidedly unpretentious man, he does prefer to be known as Dr. Malone.

While still pursuing his doctorate, Malone went to work for Bell Labs, the think tank for AT&T. He was assigned to do a study of how AT&T could create more wealth for its shareholders. He prescribed a cure that included more debt and less equity, a pretty radical prescription for the queen of the Blue Chips.

After he presented the plan to the AT&T management committee, he recalled later, "I was waltzed out of the room by the AT&T chairman. He told me, 'Kid, that's very nice, but don't expect too much. If you can see one change in Bell during your entire career, you have accomplished a lot.'"

Malone chafed under the strict Bell system, and soon took a job at McKinsey & Co., the high-prestige business consulting firm. His first day on the job he reported to an office he would share with another McKinsey hot-shot, Lou Gerstner (who later became CEO of IBM), to find a note on his phone that read: "If you can't be in Montreal for lunch with the chairman of Bell Canada this afternoon, please call."

So he sped out to LaGuardia, hopped a plane and spent the rest of his first day with McKinsey trying to help Bell Canada reorganize. He worked on a variety of accounts including development of a new organizational structure for General Electric Co., where he met two rising GE executives, Jack Welch and Bob Wright.

But all these experiences, Malone later recalled, had been in a rarefied atmosphere, an intellectual arena of philosophical and broad strategic issues. "It was a gentleman's club to the extreme," he remembered. "It was all so disconnected from the real world." He started to get itchy to get his fingers dirty running a real business.

When McKinsey was asked to look at the operations of General Instrument Corp., Malone first encountered cable television in a major way. He also encountered GI's chairman: a gruff, hardened, old curmudgeon named Moses "Monty" Shapiro.

Malone fell in love: with Shapiro, with GI and with cable television.

GI had just purchased Jerrold Electronics Co., founded by Milton Shapp, and it was going very badly. Costs were soaring, cable companies were on the verge of bankruptcy, construction had stalled and, while the government had relaxed its distant signal rules, it showed no real signs of lifting a host of other onerous regulations.

Malone took Shapiro's offer to become head of Jerrold and a vice president of GI. He set to work to slash costs, outsourcing many of the company's operations to cheaper vendors and moving its expensive manufacturing operations from Philadelphia to Mexico and Taiwan.

He instituted a policy of "key account marketing," which called for him to visit personally on a regular basis all of his biggest customers, all of whom soon owed Jerrold huge sums of money thanks to the GI financing plan that Malone instituted and without which many cable operators would have had to halt construction and perhaps operations as well.

Every few months he would do a swing around the country, stopping off in Tulsa to see Gene Schneider and in Denver to see Bob Magness, among others. He would stay at their houses, enjoy a cocktail or two with them, get to know their families and become more of a partner and a friend than a simple vendor selling some products.

These businessmen were very different from those he had dealt with in such huge corporations as Bell Canada, GE and AT&T. These cable guys were worried about how to make next month's payroll and how to pay the banks. It was a long way from Yale, Bell Labs and McKinsey. These guys were living in the real world.

So was Shapiro who, Malone recalled, would call him regularly to chew him out for one alleged sin or another. Malone would put down the phone during Shapiro's tirades and, 30 years later, would swear that the receiver would actually move across the desk so strong were the vibrations from Shapiro's voice.

One time Shapiro sent down a group of auditors to study Jerrold's books, and Malone locked them out of the office. "So he called me up again to chew me out." But Malone did listen to Shapiro's advice, bits and pieces that the GI chairman called his "Talmudic wisdom."

"The question you have to ask is 'if not,' " Shapiro would advise him, always pressing for action rather than seeking reasons not to act. Thirty years later Malone would still refer to Shapiro as "my mentor."

Malone over time adopted some of Shapiro's techniques and mannerisms, including a habit of attempting to intimidate people when he first met them.

When this reporter first met Malone in 1981, it was just after I had printed a story about a deal his company had done, including some details that had not been part of any press release. "It's a pleasure

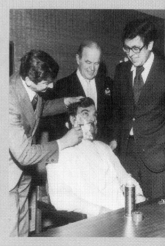

Marketing Stunt: TelePromptTer's Jeff Marcus vowed not to shave until the MSO reached the one million subscriber mark in 1974. When the plateau was achieved, Bill Bresnan did the honors as owner Jack Kent Cooke (rear) and Marc Nathanson (right) looked on.

Matchup:
Bob Magness lured John Malone to TCI in 1973. Together the two executives weathered cable's rough financial storm and built TCI into an MSO and programming powerhouse.

to meet you," I said. He glared at me with cold, gray eyes and responded: "Where do you get that shit?"

"Excuse me?" I said.

"You know, that shit you print," he growled. "You know what we do with publications like yours? We buy them and we close them down."

I noted that we had attempted to get his comments on the story but had been unable to reach him. I suggested that had he commented we certainly would have printed what he said, and perhaps altered or even killed the story. I recommended that he authorize someone else to speak for his company when he was unavailable. The story, by the way, had been confirmed off the record by one of Malone's own lieutenants and was accurate.

After that first meeting Malone was unfailingly courteous and kind to me, even when he was rushed or under pressure. I later figured his opening bluster was just a test he applied to new acquaintances from time to time to see if they would buckle and could be bullied.

Malone also had an unusual approach to business. Most Americans, paritcularly those who spent too much time watching sporting events, tend to see the world as a series of contests in each of which there must be a winner and a loser.

Malone, instead, views the world as a series of problems which can be solved much as a mathematician solves an equation. In mathematics the object is not to defeat the other guy. The object is to arrive at a solution. The best solutions, those which are deemed to be "elegant" are often the simplest, as in the most famous equation, $E=MC2$. One of Malone's favorite pieces of Talmudic wisdom is "KISS," which stands for "Keep it Simple, Stupid."

He also believes that there are many ways to solve a problem and that if one is blocked, another can be found. "I'm not like most people," he explained. "I don't think that there is one right and one wrong branch in the road."

Malone's gruff style earned him a reputation as a ruthless competitor. He appeared at times to revel in his self-generated image as a tough guy who would run over his grandmother to gain a few extra points on the cash flow margin. The shareholders of the companies he ran sure loved that notion.

But it's difficult to find many in the industry who, even off the record, will say that Malone set out deliberately to ruin them or that he acted unethically or unfairly. About the worst that is said of Malone came from the CEO of another cable company who remarked: "If John is going to rape you he tells you in advance and explains why."

And dozens of his colleagues and peers relate stories about how Malone was willing to lend a helping hand to get a new project started or to bail somebody out of a tight spot even when it didn't appear to do him any particular good.

Eighteen months after Malone took over the company, Jerrold had turned around. Its profit margins had moved to the 50% range from the mid-20s that Malone had inherited and it was contributing much of the parent company's revenue and all of its cash flow.

But Shapiro was nearing retirement age, and Malone figured the company would never consider a 33-year-old as a successor. So he began to look around for other opportunities. He talked with

A Marketing Organization

Unlike car dealers or grocery stores or movie theaters, cable systems don't compete directly with each other. As a result, system operators have always been willing to share information without fear of giving away trade secrets to a competitor.

As HBO launched on the satellite, followed quickly by WTCG and MSG Network, cable operators were propelled into an era in which marketing would be critical to their success.

Yet few systems in the early 1970s had marketing personnel, and those that did often employed junior people often just out of college.

One of the first full-time marketing executives in the business was Greg Liptak, who started out at the Cox systems in the suburbs of Cleveland, moved on to work for Gene Schneider's flagship system in Tulsa, Okla., and by 1975 was vice president of marketing for Communications Properties Inc., the nation's eighth largest MSO.

When Home Box Office announced its plans to distribute via satellite at the 1975 NCTA convention in New Orleans, the torrential rains that devastated the convention arena did nothing to dampen the excitement of Liptak and a small group of marketing colleagues.

As the conventioneers headed for home, Liptak recalled, he found himself in the damp and deserted hallway outside the convention hall where he ran into George Sisson, president of Colony Communications and a member of the NCTA board. The two were joined by Tom Willet of Continental Cablevision and David Lewine from TM Communcations.

The talk turned to formation of an organization that would enable marketing executives to meet and exchange ideas. Cable marketing executives had gotten together before on an ad hoc basis. One such meeting took place in 1972 when the NCTA sponsored a marketing meeting. A couple of years later TelePrompTer marketing vice president Marc Nathanson held a similar gathering. But

C T A M

nothing permanent had ever been set up.

Liptak sent out invitations to a meeting to be held at the O'Hare Hilton in Chicago, and the result was the establishment of the Cable Television Marketing Society (later changed, at Nathanson's suggestion, to the Cable Television Administration & Marketing Society). Liptak served as the first president, with Telesis marketing vice president Gail Sermersheim as vice president.

Each year since, the CTAM meetings, conferences and related activities have proved to be the focal point for discussions of marketing and operations issues facing cable systems across the country.

Jerrold Veteran:
Ex-Jerrold employee J.C. Sparkman ran TCI's cable operations for more than two decades and was one of the many familiar faces John Malone encountered when he left Jerrold to join TCI.

TelePrompTer, Warner Communications and his friend in Denver, the gap-toothed, homespun, nearly bankrupt, ex-Oklahoma cotton-seed buyer, Bob Magness.

TelePrompTer he dismissed because of the battle for control raging between Irving Kahn and Jack Kent Cooke. Warner chairman Steve Ross offered to double Malone's salary.

"But I didn't relish the thought of the New York City commute, and I wanted a better place to raise a family," Malone remembered. Even when the famously generous Ross offered to supply a company limousine to take him to and from work each day, Malone declined.

He opted instead to take a substantial pay cut to join a company that had $132 million in debt with only $18 million a year in revenue and that faced almost certain bankruptcy.

Asked later why he chose TCI, he said it was partly because he liked the people so well, in particular one former Jerrold employee, J.C. Sparkman, who had become TCI's head of operations. When he got there, he knew, he would not have to make any changes in people. He also liked the idea of working for a company where cable was the only business rather than a company such as Warner where cable was one of many businesses.

Most of all, he said, he like the challenge.

But just before Malone left Jerrold, he needed to hire somebody to head up a new division — the terminal products and services division — to try to catch up with competitors in the hot new market for interactive devices. The fellow he hired came to the cable industry from a longer distance — both geographically and culturally — than any other top executive. Yet he was to be the only person ever to achieve the cable industry hat trick: serving as the head of the leading hardware supplier, as an executive of the largest MSO and as CEO of a major programming service. John Sie was born in Nanking, China, in 1936, the son of a diplomat and scholar. His family moved from Nanking to Shanghai just weeks before the Japanese Imperial Army entered Nanking and slaughtered tens of thousands of innocent Chinese civilians. Sie grew up in a China that was occupied by Japanese troops. His father was in Italy, serving as ambassador to the Vatican for the Chinese nationalist government of Chiang Kai Shek.

After World War II, Sie's father returned to China, helping to get his family on the next-to-last boat that left Shanghai for Taiwan before the Chinese Communist takeover of the mainland.

In April, 1950, Sie's father had been appointed ambassador to Belgium and the family was in the United States, en route from China to Belgium, when the job fell through.

"He asked us what we wanted to do: stay in the U.S, return to China or go on to Europe," Sie recalled. The family voted to stay.

Sie enrolled as a sophomore in public high school in Staten Island, N.Y. He didn't speak a word of English. He would copy down the lessons from the blackboard every day and then translate them at night. His first year he averaged a C. Second year he made a B and from then on it was all As.

"Two things made me excel," he recalled later. "In China you didn't want to lose face for your family. And my uncle had always told me not to worry if I came to America because all Americans are dumb. So I figured if they were all so dumb I better not fail or I would really lose face for my family."

He wanted to be an artist, but his father steered him to engineering where there was a better chance to make a living. He won a full scholarship to Manhattan College in the Bronx and received his masters degree from the Polytechnic Institute in Brooklyn. He went to work at RCA, specializing in advanced amplifiers.

Pretty soon he realized that while there were plenty of Chinese-American engineers and technicians, there were very few Chinese-American executives in management. And it was there, he saw, that the real power lay. So he gave up on becoming a Ph.D. and instead joined a group of his fellow RCA employees to start a company developing advanced microwave technology. This was sold to Raytheon in 1964, and Sie worked there until 1970 when Raytheon began to retrench and Sie got the entrepreneurial bug again.

He set up a company to create synthetic diamonds, becoming a certified lapidary in the process. He worked at this until similar operations in Israel and Brazil began to produce cheaper versions and Sie ran into the opportunity at Jerrold.

"I remember he (Malone) didn't understand the word 'lapidary' on my resume," Sie recalls. But the two hit it off when they began to talk about advanced electronics, particularly the Asaki diode that had been Sie's specialty at Raytheon.

"He (Malone) wanted to know how you created a negative resistance from a two-terminal device," Sie recalled.

He also recalled how quickly Malone made decisions, including the decision to hire Sie to head up a new division for Jerrold.

Among Sie's first customers was Service Electric Co., which was using Jerrold converters to offer its customers the newly launched Home Box Office pay service. Sie, a tireless worker, soon got to know all of the other entrepreneurs in the cable business, a group that even in the face of hostile government regulations, opposition from the broadcasters, soaring interest rates and

Hat Trick:
John Sie, shown here in his Showtime days, is one of the few cable executives to have held senior positions at an equipment company (Jerrold), an MSO (TCI) and programming networks (Showtime and Encore).

other obstacles, continued to find a way to survive and grow.

Among those who did so were two products of small town America — one from the coal mining regions of Pennsylvania and the other from the Catskill mountains of upstate New York. Each grew up in a lower middle class family where, while nobody ever starved, nobody ever had an extra nickel either. Each got his start in the military, learning valuable skills. Each started his career without a dime. Yet each would build a company that would be worth, by the end of the century, more than a billion dollars. One was Glenn Jones. The other was Alan Gerry.

Of all the mavericks in the cable industry, Jones may be the most unusual. The son of a coal miner and truck driver, Jones turned down the chance of a scholarship to Yale and got a job in the local steel mill after high school. One day his father woke him up before the swing shift, said, "Get dressed, you're going to college." Jones was driven west to Findlay, Ohio, where his father had enrolled him in Findlay College. Affiliated with the Church of God, Findlay didn't look kindly on the young Jones when it was reported in the newspaper that he was playing the piano for a local vaudeville show and assisting the magician Theo the Great.

His father hauled him back home and enrolled him in the closer Allegheny College from which Jones graduated with a degree in economics. With the Korean War still going, he joined the Navy and volunteered for the underwater demolition unit, the folks who defuse unexploded bombs. He knew he had weak eyes and memorized the eye chart the day before the exam. But they switched charts on him, and he flunked. When it turned out there were no other volunteers, they lowered the standards and let him in anyway. Asked later what drove him to volunteer for such duty, he cited the $120 a month additional pay for hazardous duty.

So the nearsighted lieutenant junior grade was shipped to the Far East where he donned an underwater suit to dismantle sea mines, unexploded bombs and other ordinance left on the beaches and ocean floor after Korea and World War II.

After his stint in the Navy, Jones found a job with Martin Corp. in Denver, testing explosive devices needed to launch the intercontinental ballistic missiles Martin was building. He also went to law school, graduating from the University of Colorado.

One evening while hanging laundry, Jones met a neighbor in his apartment

Glenn Jones: Then and Now
Glenn Jones did a stint as a Navy diver in the demolition unit before making his way to Denver in the 1960s. He parlayed a small investment in the Georgetown, Colo., cable system into a cable programming and MSO empire. Thirty years later he displayed his Navy gear in his state-of-the-art office at Jones Intercable headquarters.

building, Carl Williams, who was a partner of Bill Daniels. Williams gave the young attorney some legal work and liked what he saw well enough that pretty soon Jones was handling many of the Daniels transactions.

In 1964 Jones read the book "Conscience of a Conservative" by Barry Goldwater and decided to change the world. Jones sought and won the Republican nomination for Congress from Colorado's First District, encompassing the solidly Democratic city of Denver and represented by a popular, long-time incumbent. Jones was buried in the Johnson landslide and left $40,000 in debt.

While he never ran for office again, Jones' ambition to change the world never really left him. He just shelved it while he focused on getting out of debt and bringing in enough income to keep his family fed and clothed.

He found legal clients wherever he could, operating out of cafes and restaurants because he couldn't afford an office. One client was the Southern Baptist Convention, which from time to time had legal squabbles with its member churches. Jones was sent to resolve them, and the church, so as not to be embarrassed by the old junk heap their attorney was driving, gave him a brand new Volkswagen, brought over from Germany by a missionary.

All the while he kept an eye out for cable systems that might be for sale, recalling the millions that had been made by Williams and Daniels. In Georgetown, Colo., a mining town of a few thousand souls, he hit pay dirt. The system was owed by the local plumber, and it was leaking like a sieve. Built with open, 350 ohm wire strung on trees and bushes, with a live, 440-volt line running up the mountain to

Pioneer:
Jones found innovative ways to finance investments in cable, including limited partnerships.

the antenna, the system was more trouble than it was worth. It had fewer than 100 subscribers, most of whom had been refusing to pay their bills for months.

Jones offered to buy the system for $12,000, with $1,000 down and the rest secured by his signature. Finally, after wooing the plumber for weeks (including spotting his truck in front of some home where he was working on the pipes and then bringing him fresh coffee) Jones won out. When he couldn't raise the down payment, Jones borrowed $400 against his new Volkswagen, and promised the plumber he would have the rest shortly.

The deal was made. Jones bought on credit a small state-of-the-art television monitor, the kind used in TV studios. He took this with him as he went door-to-door in Georgetown to collect the past due bills from his new customers.

When the customers complained about poor pictures Jones hooked up his top-of-the-line monitor to the cable and — lo and behold — the picture was far better than on the larger, less efficient TV sets most of the cable subscribers had. "Must be your TV set," Jones would say as he collected his funds.

Within a few weeks he had the rest of the down payment.

Jones continued to bootstrap his way along, winning the franchise to build the system in neighboring Idaho Springs, Colo. The only hitch was that the city council didn't much like the name of Jones' company: Cowpoke Cable. The council members felt that had a bit of a fly-by-night ring to it. On the way to the final council meeting Jones passed one of the local mines, The Silver King, and decided to call his company by the more distinguished name: Silver King Cable. (That lasted a few years until a company financial officer, a native of Czechoslovakia, found it difficult to pronounce, leading many investors to write checks to Silver Kink Cable, and Jones renamed the business after himself.)

As Jones expanded he continued his unorthodox ways. He would look through Warren Publishing's *Television & Cable Factbook* and find systems that had multiple owners. Then he would get out a contour map and fly over the areas the system served to check out the lay of the land. "I must have looked a bit strange with my contour maps, the *Factbook* and binoculars on these flights," Jones later recalled.

If the area looked promising Jones would drive up to see the owners. One system in California, it turned out, was owned by a group of businessmen headed by the local Republican County Chairman. To buy it, Jones needed $20,000 down. On a single day he borrowed $5,000 each from four different banks in Denver, all on unsecured, personal 30-day notes, and bought the system. As soon as the ink dried on the sale document, Jones drew out the $23,000 in cash the system had on hand and used it to repay the banks. It was bootstrap financing at its most basic.

But Jones didn't really leap into the front ranks of the cable industry until he hit on a new way to finance acquisitions, a device that had been used with great success in the oil and natural gas business: the public limited partnership.

It worked this way. Investors would put money into a fund to finance the acquisition or construction of a cable system. Jones would manage the system for a fee. For the first few years the system would operate at a tax loss, and the losses would be passed on to the investors. Then, when the system began to turn a profit or was sold, the proceeds were split among the investors, with Jones taking 25% and the investors 75%.

In the 1960s and 1970s the tax bite for high-income individuals was often in excess of 50%, and doctors, dentists, lawyers and others with high salaries were desperate to find ways to shelter some of that income from the IRS. Cable was a perfect vehicle, particularly for those nearing retirement who needed tax losses in their peak earning years and income after they retired.

Jones, together with his brother Neil, an oil and gas executive, finished raising the money, $150,000, for the first fund in 1972. The money was used to purchase a system in Attalla, Ala. Four years later, in 1976, the system was sold for $500,000, producing a gain for the partnership of $350,000. Jones kept $50,000, leaving the partners with an increase on their investment of nearly 200% in four years. In addition, the depreciation allowance and other tax benefits from the investment had meant that the government, in effect, picked up more than half the original investment. It was a pretty sweet deal.

Soon Jones had turned over the operating side of the business to Robert Lewis, who had worked in the industry for 18 years, including a stint as president of CableCom General, the successor company to VuMore. The Jones brothers concentrated full time on the financial side of the business, constructing a sales force that combed the country for potential investors.

At its height, the Jones money machine was holding a seminar every two weeks or so for high-powered limited partnership brokers from Wall Street houses such as Prudential Bache, Shearson Lehman and Dean Witter. In 10 years Jones raised $1.2 billion for his limited partnerships.

The system worked until the mid 1980s when the tax laws were changed to limit the writeoffs available to limited partner investors. By then, though, Jones had made enough to buy out many of the partnerships himself (including the Alabama system that had been purchased and then sold by the first fund). Thirty years after borrowing $400 against his VW to buy his first system, Jones had built the eighth largest cable company in the country, serving nearly 1.5 million customers and worth in the area of $3 billion, give or take a hundred million or so. And Jones returned to his original plan of trying to change the world, using the communications medium that had made him rich.

While Jones was negotiating the loan on his Volkswagen, 1,500 miles to the east another entrepreneur was building his cable company, only doing it with means much more conventional than the exotic limited partnerships Jones would construct.

The story of Alan Gerry is as salt-of-the-earth, American, Horatio Alger as

The Beginning:
Alan Gerry, front row, left, graduated from a TV service institute in 1950. In 1956, on a tip from a Jerrold salesman, he got into cable.

they come. The son of a frozen food distributor who struggled to keep his family fed in the late Depression, Gerry grew up in Liberty, N.Y., about 90 miles from New York City in the heart of the Catskill vacation region. The 20,000 population of the county would swell to 200,000 or more in the summer when sweltering New Yorkers would swarm to the "bungalow colonies" to find relief from the city heat.

After high school Gerry joined the Marine Corps and was trained in communications. After getting out in 1948 the 19-year-old worked at various jobs — on a coal truck, chicken farming, construction — before taking a course in television at the Delehanty Institute in New York, paid for by the GI Bill.

He came back to Liberty, put an ad in the paper that read "Got a TV problem? Call Alan Gerry" and, as he later remembered, the phone never stopped ringing.

Gerry worked like a dog: nights, Saturdays, Sundays running the repair business out of the basement of his home. It didn't matter when a TV set went out,

Gerry would be there to fix it in an hour. He started to sell his own TV sets and would routinely double the manufacturer's warranty on every set he sold. "We wanted to make certain the customer was very, very satisfied," he recalled with an intensity undiminished 30 years later.

Among the jobs he took on was to build huge towers on which he mounted antennas for the TV dealers in the county so they could get a clear signal to show prospective TV set buyers.

"One cold day in February I was up on one of these towers, 80 feet up, changing out a tube and this fellow stops by and yells up at me," Gerry recalls. It was a salesman for Jerrold Electronics, Walter Goodman, who happened to be passing through Liberty and noticed the big tower and the fellow on top.

"How come there's no cable in this town?" Goodman yelled at Gerry.

Thriving Business:
Alan Gerry built his TV service business in Liberty, N.Y., into one of the most profitable cable companies. Below, an early cable TV bill from Liberty Video.

"Cable, what's that?" Gerry shouted back. After coming down off the tower and hearing the Jerrold pitch, Gerry was hooked. But he didn't fall for the "yellow dog" deal in which Jerrold would get equity in the system and a per sub fee in return for helping Gerry to get started.

"They talked about it," he recalled. "I wasn't a Ph.D., but it just didn't make any sense to me."

Instead, he scraped together $1,500, rounded up seven or eight local businessmen to form a pool of $20,000 and borrowed another $20,000 from the local bank and began to string wire. It was 1956.

"The state-of-the-art in those days was five channels," Gerry recalled. Since there were seven channels available out of New York City, Gerry would pick and choose which to put on.

"I didn't always please everybody," he recalled. "I remember one day I was out mowing the lawn and this good old gal comes driving by, Mrs. Blatchley. She raised bloody hell because I had taken off the wrestling matches to show a Yankees game. Boy, was she ticked off that she couldn't watch her wrestling."

And boy, was Gerry close to his customers. He stayed that way for the rest of his career, never moving from the home where he had repaired those TV sets in his basement (he did put on some additions) and still passing on his way to work every day the two-room schoolhouse where he had learned to read and write and planted a line of scrawny pine trees one Arbor Day that later would tower a hundred feet above the road.

After wiring Liberty, Gerry began to look at franchises in nearby towns but had a tough time persuading the banks to loan him more money. His partners didn't help much either, taking a dividend every time there was a little extra cash in the till. Gerry wanted to use every dime to expand. Eventually he bought them out.

One day in the mid-1960s a fellow came by to see him with an offer to buy him out. It was Fred Lieberman, then of GenCoE and later a founder of Communications Properties Inc., the Texas cable company that eventually was sold to Time Mirror Co.

Lieberman asked Gerry if he would consider selling and offered to pay him $250 to $300 for each of Gerry's 1,000 subscribers. "Would you mind putting that in writing?" Gerry asked. Lieberman did and Gerry promptly took the offer to the local bank which agreed to lend him $100 per subscriber, raising his line of credit to a cool $100,000.

It would be the closest Gerry would come to selling for the next 30 years.

Gerry's company, Cablevision Industries, never went public, never sold equity, never took on a big company as a partner, never did a deal with venture capitalists, never used limited partnerships or other exotic financing schemes. He built his company, which eventually became one of the nation's top 10 MSOs with more than 1.5 million subscribers, in the most prosaic way imaginable. He borrowed money, first from banks and then from insurance companies and other lenders.

At times he was heavily in debt, always among the most heavily leveraged cable companies in the country. But he paid back every nickel and never violated a covenant or missed a payment. The more he borrowed and the more he paid back on time, the better his credit became and the more he expanded, always paying close attention to the needs of his customers just as he had back when he was repairing all the TV sets in Liberty.

"Our hallmark was always excellent service and excellent pictures," he recalled with evident passion in his voice. "You read about cable systems going out. Not our systems!" Gerry hired crews of people to "run the lines," looking for amplifiers and other equipment that might cause problems and fixing them before they went out.

The tactics produced extremely loyal customers, shielding Gerry's systems from the churn that plagued so many other cable companies. He also operated in communities where the population was more stable than in larger cities and where word of mouth worked better than more expensive forms of advertising to let people know about a good service. For years he had some of the highest penetration numbers of any cable company in the country.

He had lots of chances to sell out after the Lieberman offer. But he never did. Not until 1996, when he finally merged his company into Time Warner Cable.

Why? "Because I loved it so much," Gerry said later. "I was passionate about it. It was the center of my whole being, building that company. The first time I saw my name on Kagan's list of the top 100 cable operators it was like I had been nominated for an Academy Award. The same was true the first time I was interviewed for a cable magazine or the first time I was on a panel at a Kagan conference. It got to the point where there wasn't enough money in the world to take that away from me."

But Gerry, Jones and the other cable operators struggling to expand in the 1960s and early 1970s would never have become as big as they did, or even survived in some cases, had it not been for a new technological marvel that emerged to save their skins: satellites.

In 1958 Americans had awakened to the most dramatic and disturbing news to hit the country since the death of Franklin Roosevelt: the Soviet Union had launched the world's first satellite, Sputnik.

The nation raced to catch up. It beefed up mathematics and science courses in schools, diverted hundreds of millions of dollars into the military and launched a space program that the newly elected John F. Kennedy promised would put a man on the moon by the end of the 1960s.

But a handful of cable operators and suppliers understood that these satellites also could revolutionize the communications business. A science fiction writer, Arthur C. Clarke, in the late 1940s had proposed that a satellite placed 22,300 miles above the ground would orbit at a speed matching the rotation of the earth. The satellite would therefore appear to be stationary to anyone standing below it. Such a "geostationary" satellite could be used to beam radio, television and telephone signals sent from a single point to a vast area, called a footprint.

At TelePrompTer in the mid-1960s, chairman Irving Kahn and his chief lieutenant, Hub Schlafley, began to talk about the potential of satellites and take the first steps to make them a reality. TelePrompTer even filed an application with the FCC for permission to launch its own satellite.

But there were some pretty big hurdles in the way. First, the United States in 1972 still had no communications satellites. The FCC and other agencies of government couldn't decide who should operate the satellites.

In 1962 Congress had established a quasi-public company, Comsat, to develop domestic communications satellites and to represent the United States in Intelsat, the organization that was developing international satellites.

Comsat developed many of the early experimental plans for domestic satellite communications and argued it should be granted the sole franchise for development of domestic birds. The FCC seemed inclined to go along with the notion that a single entity should be given the right to handle all satellite business, the same way AT&T handled all telephone business.

But the country had a president, Richard Nixon, who was deeply distrustful of the federal bureaucracy, which he felt, with some justification, was riddled with appointees of Presidents Kennedy and Johnson. Nixon had established a new office in The White House, called the Office of Telecommunications Policy, headed by a man named Clay Whitehead (and employing, among others, a young staffer named Brian Lamb and an attorney named Antonin Scalia).

Whitehead advocated what he termed an open skies policy under which any entity with the financial and technical means could apply to launch a

Bird Man:
Nixon administration official Clay Whitehead spearheaded efforts to open up satellite distribution for media companies. The government policy encouraged cable operators and programmers to use satellite technology to deliver programming.

domestic communications satellite.

While the U.S. was debating how to proceed, Canada launched its own satellite, the Anik I. And Hub Schlafley, by then president of TelePrompTer, decided to demonstrate how the signal of the Anik satellite could be used to reach all the TelePrompTer systems in the U.S. He put out a request for proposals for a company to build a portable earth stations.

At that time a major shareholder in TelePrompTer was Hughes Aircraft, itself a giant communications company. But when Hughes learned that the budget was only $100,000, it passed.

Into this scenario walked Sid Topol, a short, energetic engineer from Boston, who had spent 22 years developing new communications systems for the Raytheon Corp.

Born in Boston to Polish immigrant parents, Topol worked as a young boy in his father's fruit and produce distribution business. In the ninth grade he won admission to the Boston Latin School, the oldest and many believe the most prestigious and academically rigorous public school in the country.

In the Army in World War II Topol was placed in an accelerated program to learn communications systems, including stints at Harvard and MIT. He emerged from the war trained in some of the very latest techniques: radar, microwave and other technologies.

After the war he returned to Massachusetts, graduating with a bachelor of science in physics from the University of Massachusetts. At Raytheon after the war he worked in the antenna division and then moved over to microwave design. Among his biggest customers was Tele-Communications Inc., where chairman Bob Magness was using Raytheon

Breaking Out Of The Pack:
Sid Topol took a gamble on building 10-meter earth stations for cable operators who wanted to receive satellite signals.

equipment to build a microwave network to serve his far flung collection of cable systems in the Western United States.

In 1969 Raytheon was approached by a small company based in Atlanta, Scientific-Atlanta, about the possibility of a merger. S-A had been founded in the early 1950s by a group of professors at Georgia Technological Institute to manufacture antennas. Topol looked into the company and recommended against a merger.

The S-A board continued to express interest, however, this time in Topol himself. In 1971 he agreed to become president of the company and moved to Atlanta to take over. S-A made a good deal of equipment for the cable industry, having just purchased the amplifier business of Spencer Kennedy Labs. But it was by no means in the same league as Jerrold or even Ameco.

S-A's chance to break out of the pack of cable equipment manufacturers that trailed Jerrold came when Topol heard

DISTANT SIGNALS: HOW CABLE TV CHANGED THE WORLD OF TELECOM

Demo: The 1973 NCTA convention featured a live satellite hookup from Washington, D.C., to Anaheim, Calif. House Speaker Carl Albert spoke to convention attendees live by satellite. A little over two years later, HBO would stage cable's first satellite-delivered programming, the Thrilla from Manila.

that TelePrompTer was looking for a company to build satellite receiving equipment. "I had learned the budget was $100,000, and although it was going to cost us more to build that unit than $100,000, we bid $100,000.

"Hub was not convinced that Scientific-Atlanta was a company that was in the communications business," Topol recalled, "and preferred a bigger company such as ITT or Raytheon." But Topol persisted. "We convinced him that we were capable, that we were eager, that we were enthusiastic and we bid $100,000 and we won the job somewhere in the fall of '72."

Teleprompter wanted the system ready for a demonstration at the NCTA convention in Anaheim, Calif., in the spring of 1973. So S-A set to work and had an earth station ready in time. Home Box Office, the fledgling pay television service, agreed to provide the programming for an initial demonstration in its suite at the show. The next day TelePrompTer president Bill Bresnan introduced satellite-delivered television programming to the NCTA national convention.

Live, via the Canadian satellite Anik I and a Scientific-Atlanta earth station, the convention heard an address by Speaker of the House of Representatives Carl Albert.

Bresnan was nervous as he introduced Albert. "I just wondered if it was going to work," he recalled. "But we flipped the switch and there he was, just fine."

It was the first demonstration ever of the possibility of live satellite transmission of television programming.

But nobody bit. Most cable operators were impressed with the technology but didn't see what it would mean for their business, which was still primarily delivering local broadcast signals. "What does this have to do with me?" was the general response, recalled S-A's Alex Best who helped set up the downlink.

TelePrompTer and S-A continued to haul their earth station around the country, demonstrating the feasibility of satellite transmission at a variety of different locations. But the effort generated "no real excitement," Topol later recalled. The earth stations were still enormously expensive, in the six-figure range. There were no U.S. satellites to deliver the signal and no cable programmers with a signal on the satellite. It made for a tough sell.

When RCA won permission to launch its domestic satellite in 1974 Topol wrote to HBO chairman Jerry Levin, a group of cable operators including Rifkin and Rosencrans, and RCA asking for a meeting to set the stage for the debut of satellite-delivered cable programming. The meeting was set for the Time Inc. headquarters in New York.

Topol opened: "Listen, gentlemen, I think we have everything in place now for satellite-delivered television programming. HBO, you have the programming. Monty Rifkin and Bob Rosencrans, you have the subscribers. Sid Topol, you have the capability of building the earth stations. RCA, you have the satellite. What's preventing us from going moving ahead?"

Rifkin's initial idea was to build 10 earth stations and then feed the signals from them to other cable systems via microwave. In the end it proved more economical to put an earth station at every cable headend.

Levin pledged to ask the Time Inc. board for permission to lease a satellite transponder at a cost of about $8 million for a five-year period and to build an uplink. Levin must have had some magical ability in front of the Time Inc. board. Over and over he had persuaded them to keep supporting HBO, and here he was, back again, asking for a multimillion dollar pledge to try out a completely untested technology to support a business that had yet to show a dime of profit. But he did it.

The next thing he needed was a commitment from some cable operator willing to install an earth station and receive the HBO signal.

But in 1975 most cable companies were deeply in debt and few had any funds to spend on an expensive, untested system to deliver pay television programming. They were having a tough enough time making their bank payments without shelling out $100,000 for an earth station to bring in a signal from HBO which was selling to only about 15% of their customers in any case.

An exception was Bob Rosencrans.

A soft-spoken, amiable native of Woodmere on Long Island, Rosencrans had graduated from Columbia College and Columbia Business School. After a stint in Korea he got involved in the closed-circuit television business and eventually went to work for TelePrompTer.

In the mid-1960s, Rosencrans left TelePrompTer to found his own cable company, which he named Columbia, after the river which flowed nearby one of his first systems.

He took the company public in 1968 and merged it with a Texas-based cable company, International Cable, headed by Ken Gunter.

Unlike many of his colleagues, Rosencrans proceeded cautiously when it came to finances. "We nursed what we had," he said, never taking on great amounts of debt. By 1969 he merged again, this time with a group of cable systems owned by United Artists Theaters to form UA Columbia Cablevision. The mergers were all stock for stock, avoiding taxes and debt. He locked in long term fixed debt at about 10%, forming an alliance with Home Life Insurance Co. to provide the funds needed to expand.

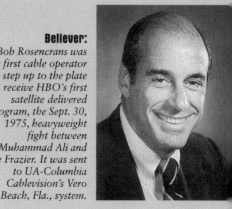

Believer: *Bob Rosencrans was the first cable operator to step up to the plate to receive HBO's first satellite delivered program, the Sept. 30, 1975, heavyweight fight between Muhammad Ali and Joe Frazier. It was sent to UA-Columbia Cablevision's Vero Beach, Fla., system.*

"When the (interest) rates went up" in the early 1970s, Rosencrans recalled "we were untouched."

Rosencrans pursued franchises in the suburbs of New York, hiring a husband and wife duo — lawyer Bill and TV producer/public relations executive Kay Koplovitz — to head the franchising team.

He also experimented with pay television. He tested the Channel 100 system developed by Geoff Nathanson and imported games from Madison Square Garden to run on his systems in New Jersey and New York State.

So when Jerry Levin called in 1975 seeking a cable operator willing to install earth stations to receive HBO, Rosencrans didn't have to think about it very long. He had been at the meeting with Topol and Levin in New York and knew what the deal was.

"He (Levin) called on a Friday, I remember, and I called him back Monday to say we would take seven or eight earth stations at $100,000 each. It was very easy to do. A management team ran the company, and we weren't laden with debt."

Levin was astounded. "It was so easy," he recalled. Rosencrans didn't have to get anybody else's approval or do a study. He simply followed his gut.

"Bobby is the symbol of the cable industry," Levin said 30 years later. "He is an instinctive player. He didn't run it through a committee or do a business plan. It just seemed right, so he went ahead."

Rifkin committed ATC to take at least two downlinks, one for the system in Jackson, Miss., and another for Orlando, Fla.

At the NCTA convention in New Orleans in April 1975, the three unveiled their plans. It was, reported *Broadcasting*, "the best piece of news cable TV has heard in two years.

"An industry chastened by a two-year diet of humble pie began to act like a winner again," the magazine noted.

Still, Levin had to keep the folks at Time Inc. from turning back. The NCTA announcement generated enormous press, including articles in the *New York Times* and *Wall Street Journal*. Time Inc. president Jim Shepley began to get calls from some friends saying that the company was going too far out on a limb. Levin remembered the heat Shepley was taking. " 'This is crazy,' they would tell him," Levin recalled. " 'You've got this kid running this thing.'"

It helped enormously when TelePrompTer president Russell Karp agreed to purchase 65 earth stations to serve his vast cable empire. It was a start to get the handful of systems Bob Rosencrans had. It began to look like a real business when TelePrompTer came on board.

Holding the Time brass in line, Levin then tiptoed down to Washington to get

Heavy Hitters:
Hollywood studios opposed loosening restrictions on pay TV and would bring in some added firepower to their lobbying efforts. MPAA president Jack Valenti (seated at left) had Charlton Heston testify at the 1974 FCC pay cable hearings. Warren Beatty is seated at right. Cable prevailed.

the necessary FCC approval for his plan. The HBO chief assumed it would be blocked by somebody — the networks, the theater owners, a rival pay service — any one of a dozen or more who might want to delay or kill the HBO plan. Any objection would have triggered a process that could have lasted years, time which HBO did not have.

"I assumed the networks, particularly CBS, would object," Levin recalled. "All they needed to do was object and it would have been held up by a year. Time Inc. would have lost its nerve."

But nobody did object, and Levin received his approval.

Then came another indication that the stars were at last lining up right for HBO. Levin ran into fight promoter Bob Arum of Don King Productions, who offered to supply the upcoming heavyweight championship fight — Muhammad Ali versus Joe Frazier, the Thrilla from Manila — for a price that Levin later described as bupkis. That set the date by which HBO would have to launch its satellite service: Sept. 30, 1975.

Rosencrans suggested his Vero Beach-Fort Pierce, Fla., system as the site of the first satellite-delivered programs. Vero Beach was an up-to-date system with about 10,000 subscribers and was out of the way of ice and snow storms that had nearly doomed the launch of HBO on microwave in 1972. HBO's chief finance officer, Jim Heyworth, had just returned from Florida with optimistic news about a possible microwave link up the backbone of the state. A dish in Fort Pierce could supply more than one cable system, they reasoned.

"The equipment went in the 28th or 29th of September," Rosencrans recalled. The company had planned a huge

Turning Point:
HBO's September 1975 transmission of the Muhammad Ali - Joe Frazier heavyweight fight from the Philippines set the stage for satellite delivery of the network.

party and unveiling for Sept. 30. Rosencrans hired Kay Koplovitz to stage the event, and she brought down all the major press, members of Congress and top executives of the cable industry. Another downlink was set for the ATC system in Jackson, Miss., and the Scientific-Atlanta headquarters outside Atlanta where the Southern Cable Association was holding its meeting.

Among those on hand for the Florida launch was Irving Kahn, Rosencrans' old boss at TelePrompTer and one of the first to talk about the potential of satellite-delivered television. Kahn, only a year out of prison where he had served 20 months for bribery, had not been formally invited to the event. But Rosencrans made him welcome nonetheless, inviting him to the dinner after the launch. "He had always dreamed about it," Rosencrans said, and here was his dream coming true.

The signal was uplinked from Manila to California via an Intelsat satellite. It was sent across country via AT&T landlines and then sent by microwave to HBO headquarters in New York and then to Valley Forge, Pa., where it was uplinked once again to Western Union's Westar satellite and back down to the dishes in Florida, Georgia and Mississippi.

(HBO had planned to use an RCA satellite, Satcom I. But it would not launch until December, so for the first couple of months HBO sent its satellite signal via the only available domestic bird: Westar.)

In all, the signal traveled 93,000 miles, noted *The New York Times*, which added: "How long did it take to make the connection and travel that extraordinary course? Almost, literally no time at all. Give or take a millisecond or two, the reception in Florida of the Manila fight was instantaneous."

The pictures were so good, Rosencrans recalled, "It looked like they were fighting in the next room."

The next day ATC launched its satellite service in Jackson, Miss. The world had changed.

"In short order we put them in all our larger systems," Rosencrans said. Other cable companies followed suit, particularly after the FCC ruled that smaller dishes were permitted and the price began to decline.

The economics were so compelling that the old idea of installing a few dishes and hooking up multiple systems to each downlink via microwave was quickly abandoned.

"It was just cheaper and easier to duplicate the dishes," Rosencrans said. Charging $8 a month retail for HBO returned about $40 per year per subscriber in incremental revenue to a cable system after the split with HBO. By adding 1,000 subscribers, the system could pay for the dish, even a $100,000 dish, in just over two years. And there was no cost after that, whereas the price for microwave transmission was an ongoing expense.

HBO also moved to solve some of its other problems as well. Levin was promoted to chairman, and Time Inc. brought in Nick Nicholas, a hard-headed financial expert who had been running Manhattan Cable, to help deal with the explosive growth ahead.

Tony Cox came over from *Sports Illustrated* to head up field marketing. Austin Furst, who had been looking into pay-per-view for Time Inc., was named head of programming. And a young attorney from the William Morris talent agency, Michael Fuchs, was hired to develop original programming.

Fuchs took the job after his boss at Morris turned it down. He later recalled: "I was at William Morris, and Neal Pilson was my boss. He came to me one day and tells me he'd been offered two jobs: one at CBS Sports and one at HBO. He said he was taking the job at CBS Sports because it was more aligned with what he wanted to do while

Programmer:
Michael Fuchs left the William Morris Agency to join HBO's programming department. Over the next decade and a half, Fuchs help build HBO into a programming powerhouse with theatrical motion picture purchases, original movies, series and documentaries and comedy and concert specials.

the HBO job involved more programming and seemed like something I might like. That was like telling a kid he could have all the chocolate sauce he wanted on his sundae. I literally went down the hall and called Austin Furst and I told him 'I hear you're looking for talented people. Why don't I save you some time and money.' I think I got the job based on that comment because I could hear Austin say 'I love it.'"

Furst reorganized the programming schedule, setting movie start times on the half hour or hour and developing "interstitial" programming to fill the gaps. Among those he hired was the former local origination manger from the TelePrompTer System in Elmira, N.Y., Tony Lynn and a young film buyer for a broadcast station in Washington, D.C., Jim English.

Fuchs began to court Hollywood producers and studios. In particular he pitched the stars and big-name directors, offering them the type of creative freedom that would never be allowed on broadcast TV. HBO could present uncut productions, free from advertising breaks. It would allow off-color language and nudity. It could provide major chunks of prime time, and devote that to live concerts or events. It didn't shy away from politically controversial topics. It was an attractive proposition for many in Hollywood, tired of the tyranny of the broadcasters, even if the money wasn't as good as what the networks would pay.

And Cox put together a team of hungry young "affiliate representatives" to market HBO to the cable operators and help them in turn to market to the consumer.

Among them was the manager of a group of small TelePrompTer systems, Bill Grumbles. Grumbles' father had

Marketer:
Bill Grumbles laid siege to the HBO offices in New York to persuade Tony Cox to give him a job. Later he became president of Worldwide Distribution for Turner Broadcasting System.

been a broadcaster in Memphis, Tenn. When Grumbles went to Southern Methodist University he took an investment course and found that the stock of a company called TelePrompTer was skyrocketing. The two experiences led him to seek a job in the business, first with Ben Conroy at CPI and later with TelePrompTer, running a group of small systems in Texas.

When HBO came along "I thought it was the coolest thing I had ever seen in my life," Grumbles said 25 years later with an enthusiasm undiminished by time.

He got an appointment with Cox and managed to "blow the interview" as he later recalled. Learning that Cox had decided not to hire him, Grumbles laid siege to the HBO headquarters and didn't quit until he had another chance. Finally Cox told him "I have to admire your persistence and your gumption. I'll see you around."

HBO sent him to open the regional office in Kansas City in January, 1978. He combed through the *TV Factbook* looking for systems with 10,000 subs or more. When the price of a dish dropped to around $30,000 HBO began to be affordable to systems with as few as 2,500 subscribers.

HBO provided the field reps with a kit to cover all aspects of their job. A one-sheet pro forma showed the operator that by charging $9.95 retail for the service and sending HBO $4.25 the service could generate in excess of $60,000 per year for the system. After paying off the cost of the dish, nearly 90% of that could be taken to the bottom line every year.

The HBO field reps were, as Grumbles later described it, "a different breed of cat" from the operators or from the equipment sales people. On the equipment side, once the sale was made, except for service, the salesman's job was done. For the HBO rep the sale meant the job was just beginning.

HBO handled all the launch plans, providing movie posters, customized press releases, direct mail pieces, newspaper ads, training for customer service representatives, and the first of what would become a flood of HBO premiums: key chains, T-shirts and lighters. "We called it the 'womb-to-tomb' plan," Grumbles recalled.

"The whole town would get real excited," Grumbles recalled. "We'd get the mayor out for the launch and make a big deal of it."

Levin attended many of the early launches. He recalled in particular the excitement that the installation of the dish itself would generate in towns such as Laredo, Tex., where HBO launched in 1976. "When they installed the earth station, it was like the astronauts had landed," he remembered. "It was a huge attraction. People would come just to look at it. It was such a galvanizing thing from a marketing point of view."

The systems were working out the marketing and operational kinks as well. Kroeger, at the New England Cox Cable systems which had had such difficulty selling HBO as a microwave service, recalled that the first step was to end the practice of roving bands of salesman and recruit a permanent sales force from the local towns. They instituted tap audits to find out who might be stealing the service. And more and more systems moved to computerized billing systems, replacing the old coupon books and card files. Some, like Cox, did their own in-house operations while others signed on with a new Sacramento-based company, CableData, that within a decade would handle half the billing for the entire industry and become, in the process, the single largest customer for the US Postal Service.

But HBO's biggest marketing tool was the guide. Most local newspapers refused to carry listings of HBO programming.

Selling Pay TV:
Viewer's Choice senior-VP Bob Bedell and Showtime president Matt Blank were two of the many executives who worked at HBO in the early days of the network and its sister service, Cinemax.

The service developed its own guide which it mailed to subscribers every month, inserting it in with the bill.

"The (arrival of) the guide was a terribly exciting event every month," Grumbles recalled. "People would collect them like baseball cards." By enclosing the guide in the bill, HBO radically reduced churn, because those who were thinking about disconnecting when the bill arrived would find something in the coming month's attractions to keep them paying.

The results for the cable industry were phenomenal. Not only did HBO provide a new source of revenue from existing subscribers, but it pushed those who had not signed up for cable to become customers as well. In the suburban markets where off air-signals were plentiful "penetration had been 10% to 15%," Rosencrans recalled, and "in no time we moved up to 30% to 40%." It didn't take him long to realize what this meant. "All of a sudden the big markets began to look feasible."

The results for HBO were astounding as well. By the end of 1977 HBO had more than 1.6 million subscribers, from a base of 20,000 (minus a few disconnects) in June of 1974. Paul Kagan Associates estimated that in 1977 HBO generated gross revenue of $124 million of which it kept more than half (the rest went to the cable operator). In the fourth quarter of 1977 HBO moved finally into the black.

The satellite launch of HBO did more than just rescue a struggling regional pay TV service from the brink of insolvency or give the cable industry the engine it needed to drive growth in the suburbs and cities. When HBO went up on the bird it fundamentally changed the nature of the U.S. television business, and ultimately the entire world of communications.

"We brought the glitz of Hollywood to the plains of Kansas," Grumbles recalled. "We were the entertainment industry, and we took what was essentially a retransmission business and transformed it into an entertainment business."

In the process HBO groomed a generation of executives who would go on to help launch dozens of other services, having learned far more about how to win over the American consumer than they could have in any college or graduate school.

Typical of the HBO shock troops was Matt Blank, a native of Queens and graduate of the University of Pennsylvania who joined HBO in 1976 after a stint as a marketing trainee at Phillip Morris Co.

"I was raised in New York City and went to school in Philadelphia. I had never seen anything of America," Blank recalled. "A week after I started I was in Lewistown, Pa.," managing an HBO launch.

Blank found the contrast with Phillip Morris sharp and exhilarating. At the big company there were separate departments to handle everything — advertising, ad placement, marketing, promotion. At HBO he did everything: made up the print ads, ran the slicks down to the paper, cut the radio spots, bought the time.

In Hickory, N. C., Blank recalled, he met up with TelePrompTer vice president of marketing Marc Nathanson and the system manager to get ready for a launch. Then they found there was a problem with the satellite dish.

The three drove out to the headend

Local Ads

He raced cars, dropped out of school, figured to die young. Somewhere along the way, Bill Killion invented the local cable advertising business.

"My thing," says the soft-spoken Killion, stressing the singularity of the word, "is putting advertising on cable."

Which hardly seems novel now. But to understand the contribution Bill Killion made, you have to understand that he was talking about putting advertising on cable well before the late 1970s satellite communications revolution, before Ted Turner launched Cable News Network and 39 systems inserted their first local spots on CNN's network feed, and before anyone anywhere figured small-time merchants would put up money to see themselves on television.

Killion was talking about putting advertising on cable at a time when the only reaction he got was silence and puzzled stares. At trade shows, speeches, conferences, Killion told people the day would come. One day, he brought a program switching device to the Western Cable Show in California, boasting that it would allow cable companies to place local commercials within the channels that carried the only cable-original programming of the day, channels like the precursor to USA Network, Madison Square Garden Network. It was a new idea and a new system. "But there was no interest in it at all," Killion said. So he went back to the place where many American inventors go, back to where the drawing board was, back to his garage.

The garage where he started was in El Cajon, Calif. Killion was there out of choice, having quit a technical writing post at a southern California military contractor and broadcast electronics firm, Dynair, in 1974. The idea was to operate his own small electronics manufacturing business backed by a local bank which had promised him financing.

But without a solid business plan and without his Dynair connection, Killion's backing fizzled. Living off his wife's salary as a law firm assistant, Killion scraped up enough components and solder to build 100 VCR controllers, filling an order he won from a hotel group that wanted a way to control in-room movie telecasts. Killion sold them for $650 a piece to Holiday Inns and apartment building owners, and later to cable TV companies for automating crude pay-TV channels. The orders kept him afloat. He named the product Channelmatic.

Killion was selling product, but living lean. A big celebration was "a hot dog and a six-pack," he said. But two things happened in 1977 that would have an ultimate impact on Killion and the spot cable advertising business. Home Box Office had begun transmitting its commercial-free movies to cable systems by satellite. And the microprocessor, the brains behind the personal computer, became affordable.

Killion introduced a microprocessor to his VCR controller, tied in a tone encoder and was shortly outfitting cable systems with devices that automatically switched among program sources so subscribers wouldn't see a dark screen after HBO's feed ended for the day.

The applications for advertising were obvious to Killion. Switching from a satellite feed to a commercial stored on a VCR was conceptually no different from switching between two satellite sources. There were few people besides Killion who made the connection. In Hawaii, Oceanic Cablevision engineer Jim Chiddix began inserting local cable commercials using Killion's invention. USA Network president Kay Koplovitz was intrigued, says Killion, but resisted the idea of cue tones that would be audible to viewers. UA-Columbia engineers took more interest, inviting Killion to San Angelo, Tex., where he received a breakthrough order for 10 systems.

Killion was gleeful over the first big sale. He bought a house in Alpine, Calif., out of bankruptcy and hired six employees to assemble components. But he soon turned pessimistic. "After the UA order, the cable business sort of dried up for us," he says.

Worse, hotel customers were proving to be a collection problem, leaving Killion drained of cash. Their bank account empty, Killion and his wife Sally decided to make one last stab before calling it quits. They charged airline tickets on their credit cards and headed to a cable convention in San Antonio, hoping for a turnaround. The mood was grim. "If we went home without orders, we would have been broke," Killion says.

In Texas, with the stakes high, Killion watched in horror from his exhibit floor booth as show-goer after show-goer passed him by without a second look. "Nobody stopped. We didn't know why," he says.

Discouraged, Killion surveyed the otherwise busy show floor, stopping at the distant sight of a man wearing a cowboy hat. As the individual neared, Killion recognized him as an old acquaintance from his Dynair days. But that wasn't what intrigued Killion. It was the hat. Killion was taken by the image. With nothing to lose, he left the show, found a nearby shop — the Paris Hattery — and watched as the proprietor handcrafted a facsimile of the wide-brimmed Stetson Killion recognized from Clint Eastwood westerns.

Whether the crowds that gathered around Killion upon his return that afternoon were drawn by the sight of a tall man in a cowboy hat, or were simply moved to check out the Channelmatic displays they had previously overlooked, doesn't matter to Killion. He sold dozens of commercial insertion system orders that day and went home delirious — and still in business. Killion, who was diagnosed as a child with rheumatic fever and expected to die before adulthood, was not one to tempt fate. Since that day, he says, "I was never without the hat."

Or without a paycheck. Killion hadn't paid himself a salary from Channelmatic from 1974 to 1981. But shortly after Texas, he began to realize the fruits of his invention. On June 1, 1980, 39 cable systems, all using Channelmatic gear, aired their first local commercials the day Ted Turner flipped the switch that turned CNN into the world's first around-the-clock news network.

Killion's biggest break came in 1981. Thom McKinney was orchestrating Group W Cable's expansion of local advertising. After meeting with Killion, McKinney took a headlong plunge, agreeing to buy 55 random-access commercial insertion systems recently developed by Channelmatic. Group W got its advertising business rolling. Killion got rich.

Inventor:
Bill Killion, above, developed the technology for inserting ads on local cable signals in his garage.

Something Big:
Ted Turner bought a struggling Atlanta UHF station in 1970. He changed the call letters to WTCG, which stood for "Watch This Channel Grow," and proceeded to build a cable programming empire.

The HBO reps were young and enthusiastic. They were mostly single and worked seven days a week. Conventions, Blank recalled, were "one giant party. It wasn't just a business experience, it was a cultural experience." Working at HBO in the late 70s, he said, was the "relative equivalent of going to Woodstock."

"I view my time at HBO not as part of my career, but as part of my education," Blank said later.

But while HBO was the first on the satellite it wasn't the last. As Jerry Levin and Bob Rosencrans met in Vero Beach to launch the first satellite transmission of HBO, their progress was viewed with great interest just to the north. In Atlanta the owner of a small, nearly bankrupt UHF broadcast television station had decided that the satellite would be his path to the future as well, a future in which Channel 17, WTCG (for Watch This Channel Grow) could catapult itself from fourth place in a four-station market to a major regional if not national programming force with a distribution as wide as that of the mighty broadcast networks themselves.

in the middle of a pasture. A shabby fence surrounded the dish, and three cows had trampled the fence down and were resting their chins on the dish, interfering with the satellite signal.

"There we were, up to our ankles in mud, dressed in our blue suits and Gucci loafers, trying to move these three cows," Blank recalled.

Blank found the cable operators eager to help educate him about the business. After a dinner at the Pennsylvania show one year, Adelphia Cable founder John Rigas stayed with Blank at the table long after everyone else had left, tutoring the young HBO exec about how cable worked.

"He didn't even notice the time," Blank recalled. "He was so proud of what he did for a living he just wanted to sit there and tell me about it."

The broadcaster was also a sailor and a friend of Time Inc. president and fellow yachtsman Jim Shepley. Through Shepley he had met Jerry Levin. In September, 1975, he read in *Broadcasting* about HBO's satellite plans and picked up the phone to ask Levin about it.

"Jerry," he said. "This is Ted Turner."

The Programmers

Ted Turner really isn't a businessman. He is more a force of nature, closer to Hale-Bopp or El Nino than to any business school graduate.

Raised in Savannah, Ga., Turner was the son of an alcoholic father who believed the best way to build character in the boy was to beat him regularly and praise him never. Young Ted was sent away to boarding school in the fifth grade and never really returned home except for summer vacations until he was tossed out of Brown University for having a woman in his room.

A magnificent sailor, he had led the Brown sailing team to a national championship by sailing across a stretch of ice to overtake a boat from the University of Michigan. So uncanny were his instincts that one of his crewmates was convinced Turner could tell ahead of time when the wind would shift and in which direction.

That same fearless bravado epitomized his business career.

Turner's father committed suicide when Ted was 24, leaving the young man with a small outdoor advertising business the senior Turner had built up during years of battling rivals for choice locations on two-lane southern highways.

All his life, Ted Turner loved being the underdog. Over and again he would portray himself as the Vietcong fighting the U.S., the Confederates surrounded by Union soldiers, or Hannibal outwitting the much larger Roman legions. Success bored him. Each time his business reached the brink of profitability and stability, Turner would risk everything, including his own personal fortune, to take on some seemingly impossible task and regain his status as the underdog.

One early example came when he decided to become a broadcaster. He had finally managed to get his outdoor advertising business on an even keel when he jumped into a new business he knew almost nothing about.

In 1970 he purchased for $2.5 million WJRJ-TV Channel 17, a UHF station in Atlanta that was losing nearly $1 million a year as the fifth-ranked broadcaster in a five-station market. Within a year he had bought another bankrupt UHF in Charlotte, N.C. He renamed the Atlanta station WTCG for Turner Communications Group or, as Turner loved to say, Watch This Channel Go, and the Charlotte station WRET, for himself, Robert Edward Turner.

He hired a group of young sales people, including two who would remain with him for years — Gerry Hogan and Terry McGuirk. But Turner led the troops. An amazing general he was.

He would do anything to get a sale. He would come into the office of a po-

Sales Tornado: *Once Ted Turner discovered the potential of satellite distribution for his struggling Atlanta TV station, there was no turning back for Turner Broadcasting System. In the early days, Turner was a tireless pitchman for WTCG, later renamed WTBS, which became the foundation network for his vast media empire.*

Marketing Conference:
Ted Turner confers with Art Dwyer (right), a former marketing executive at Turner and Cox Cable, who went on to help design many of the marketing pitches for Turner's growing stable of cable networks.

tential advertiser, jump up on the desk and blare out his sales pitch, punctuating his performance with wild gestures. If he was turned down he would fall prostrate on the floor, grab his throat, choke himself, gag, and cry out, "You're killing me."

Sometimes the antics worked, but sometimes they didn't. WTCG moved up in the Atlanta rankings only when the other UHF station in town folded for lack of audience.

Turner tried just about everything to draw attention to his station. And in the course of it he developed an instinct for promotion, particularly cross-promotion. He used his billboards to promote his station and vice versa. And he used his own personality to promote them both. His persona became an even more valuable asset after he won the America's Cup yacht race in 1977 — and after his picture began to appear on the cover of such publications as *Sports Illustrated, Time, Newsweek, Forbes* and *Fortune*. These were tactics

that would serve him well in the future.

Above all he learned to market, rather than to sell. Art Dwyer, who worked both for Turner and for Cox Cable in the 1980s, defined the difference: "The marketer creates a need and a desire for the product. A salesman is just trying to sell you a bill of goods. The two best marketers in the business were Chuck Dolan and Ted Turner. They could truly excite you about their product. They talked marketing potential. Neither one ever sold anything, but they never left without getting the order."

As he cast about for ways to keep his company afloat, Turner became aware of the cable systems that were carrying his little UHF station in places such as Dalton, Ga., just outside Atlanta. The biggest problem for a UHF station was that it required a second antenna on the TV set that had to be tuned just right to allow for reception of the weak UHF broadcast signals. Compared to the stronger VHF stations, the UHFs looked like poor stepchildren to most viewers.

Cable leveled the playing field. When picked up and retransmitted by cable systems, the signal of a UHF station could be received by a cable subscriber without the need for a special antenna and could look nearly as good as the signals of the VHF stations.

"I realized that cable converted a UHF station to a VHF that people could get," Turner recalled. "And that was good for Channel 17."

He called his friend Henry Harris, president of Atlanta-based Cox Cable, and over lunch Turner learned everything he could about cable. He asked Harris to introduce him to the cable operators. The two went up to a regional cable meeting at Myrtle Beach,

S.C., and Turner pitched the audience on the virtues of adding WTCG to their channel lineups.

"I formulated a plan to gain the rights to the Atlanta Braves (Major League Baseball team) games and (National Basketball Association) Hawks," and to televise the games on WTCG throughout the southeastern region, via microwave and cable, Turner said. There were no other independent stations in the Southeast at that time, and cable systems in Savannah and other southern cities were eager to have more programming, particularly something that would allow their subscribers to watch the South's favorite (in fact only) pro baseball and basketball teams.

Turner struck a deal for the TV rights to the Braves in 1973 and to the Hawks shortly thereafter. Within a couple of years he would own both teams outright. (Neither team was very good in those days. When Turner took over the Braves he was so excited when one of his players hit a home run in the opening game that the young owner leaped out of his seat, ran onto the field and accompanied the player from third to home base. The Braves still lost the game, but the photo of Ted Turner crossing home plate made nearly every paper in the South.)

Turner's plans for a regional network received a major boost in 1972 when the FCC relaxed its prohibition on the importation of distant signals, allowing each cable system to carry two distant signals so long as they did not "leapfrog" other nearby stations. Since WTCG was the only independent in the Southeast, Turner had an open shot at carriage on all systems in the region.

McGuirk, who had been promoted to Turner's special assistant in charge of cable, recalled that "Ted really laid siege to the cable industry." One early meeting took place in Lakeland, Fla., at the Florida state cable convention.

"Ted had been listed as a speaker, but they had put him 20th on the roster, which was not really acceptable at all," McGuirk recalled. "We forced our way up to second or third."

"It was the first time anybody had really seen him. He had them rolling in the aisles, laughing and having fun and when he was finished with his speech he gave them all an unscheduled intermission."

He preached like an old-time evangelist, waving his arms and asking his audience rhetorical questions. He electrified the room.

His pitch was perfect for the cable operators, who also saw themselves as upstarts competing with the giant broadcasters who were using their clout in Washington to crush the small cable operators.

If Turner dreamed of becoming a general leading a ragtag army against a giant, evil enemy, he found willing recruits in the ranks of the cable operators.

The operators, initially taken aback by this gale wind from Atlanta, lapped

Tutoring Ted:
Henry Harris was the Cox Cable executive Ted Turner turned to in the 1970s to learn more about the cable business.

up his act. None of them had ever seen anything like him, or would again. "It brought him instant icon status," McGuirk recalled.

It also brought him more and more viewers. By 1976 his little station was being carried by systems with nearly half a million subscribers throughout the Southeast.

The big jump in viewers barely moved the bottom line. Advertisers in Atlanta didn't much care about reaching viewers in Birmingham, Ala., or Charleston, S.C. And regional or national advertisers wouldn't even consider buying a station with no discernible ratings even in its home town. But a new kind of advertiser was starting to emerge on the national scene, one who didn't care about ratings, only about results: the direct response, or per inquiry, advertiser. The purveyors of Ginzu knives or K-Tel knitters didn't really care at all whether the order for their product came from Alabama or Florida. All that mattered to them was how many orders there were. And WTCG delivered orders, splitting the revenue with the manufacturer.

While this revenue helped, the relationship between Turner and the cable operators was still hampered by technology. The microwave hops that were needed to bring the signal to cable systems hundreds of miles away were expensive and unreliable and left hundreds of cable systems out of range of the signals.

Then, one day in 1975, Turner recalled, he was reading a story in *Broadcasting* magazine about the plans HBO had to put its signal up on the satellite.

He knew instantly that this would be his path to success. "I was like the hound on the trail of an escaped criminal," Turner later recalled with glee. He hired an NCTA official, Don Andersson, to help him work out his satellite plans and help find a way through the Washington land mines.

He set out to find a satellite. His search led him first to Western Union, which was planning to launch a satellite and whose marketing vice president, Ed Taylor, tutored the young Atlantan about satellite technology. But when Turner and Taylor went to call on Andy Goldman, a TelePrompTer executive who had run some of the Alabama systems that carried WTCG, Goldman told Turner he was on the wrong bird.

Only if he were going to be carried by the same RCA satellite that HBO was using could cable operators pick up both signals with a single dish. TelePrompTer was not about to shell out an additional $100,000 per system to pick up the WTCG signal on a second satellite dish, Goldman said.

But after steering him to RCA, Goldman was encouraging, as were other cable operators. "(TCI President) John Malone told me 'Ted, you'll never get it through the government, but if you do I'll put it on everywhere,'" Turner later remembered.

Malone initially appeared correct

Ted's No. 2:
Terry McGuirk (right) has been Ted Turner's right hand man since the late 1970's when Ginzu knifes and newsman Bill Tush, and not the NBA and CNN, were the hallmarks of the company.

about the government, but Turner wouldn't take no for an answer. Andersson arranged for him to testify before a House subcommittee where Turner's down-home style and self proclaimed crusade against the broadcast networks found a sympathetic audience.

Eventually he worked out a deal with the FCC under which he would separate the programming and transmission aspects of his business. He set up a separate company, Southern Satellite Systems, to lease the transponder and distribute the station to cable systems. He offered ownership of the new company to a number of his friends, all of whom turned him down until Ed Taylor, the Western Union vice president, agreed to buy it for the price of $1.

It took Taylor a while to raise the money to buy a transponder, but at last, on Dec. 17, 1976, the FCC gave the final go-ahead and WTCG launched satellite service on Satcom I. (Shortly after going up on the satellite, Turner changed the call letters of the station to WTBS, for Turner Broadcasting System. To do so he had to buy the rights to the call letters from Massachusetts Institute of Technology, which had been using the same initials for its student-run Tech Broadcasting Station.)

The impact of satellite distribution astounded even Turner, who had entered the venture hoping to become a regional network, but with little thought that he would reach nationwide audiences.

"We had no way of knowing who was carrying us," he said later. Although cable systems in theory were supposed to pay SSS a small carriage fee, there was no way to prevent them from picking up the signal for free if they chose, particularly in the initial weeks and months after the satellite launch.

Upstart:
Ted Turner poses with WTCG equipment in the late '70s; WTCG personality Bill Tush with co-host Jan Hooks, who went on to 'Saturday Night Live' fame.

The only way to judge who was watching was from the direct response advertising mail, the orders for Ginzu knives and K-Tel knitters that poured into the WTCG offices every day. "I would personally go and get the bushel baskets of envelopes every day," recalled Turner who would count the letters from Atlanta and those from outside the city.

(One consequence, he recalled, was that he noticed how sloppy the post office was. "A lot of times they didn't cancel the stamps." Turner ordered that the uncancelled stamps be steamed off the envelopes and reused.)

The letters, he found, were coming from all over the country. "We were looking for an audience in the Southeast," McGuirk recalled, "in Louisville, Ky., and Ft. Lauderdale, Fla., and places like that. That is where we expected the signal to come down."

But within days McGuirk and Turner realized that this was going to be bigger than they had thought. "We were getting mail from Minnesota and Washington State and California. It was magical. We couldn't keep up with the (direct response) orders," McGuirk said.

DISTANT SIGNALS: HOW CABLE TV CHANGED THE WORLD OF TELECOM

Pleading The Case:
Cable operator Ed Allen and Ted Turner appeared regularly on Capitol Hill in the 1970s and 1980s to plead cable's case on issues ranging from copyright to loosening restrictions on distant signal importation.

Because he was getting as much mail from outside Atlanta as from within the reach of his broadcast signal, Turner figured his nationwide audience was at least as large as what he was getting in Atlanta.

He tried to sell that idea on Madison Ave. "I made the calls personally," Turner remembered. "I remember one guy at (advertising agency) Dancer Fitzgerald. After I made my pitch, he laughed so hard his swivel chair want out from under him and he fell back and hit his head so hard it almost killed him."

He called on the A.C. Nielsen Co. as well, to complain that the ratings service was undermining his station. "I'm going to sue you for every Goddamn penny you have," Turner recalled telling them, "for treble damages. You are treating me like Lester Maddox used to treat the black people. I'm going to be like Martin Luther King."

But Turner made no headway. Advertisers wouldn't budge, while syndicators of programming, convinced that WTCG had a national audience, were demanding higher fees for their programming. It was starting to look as if Turner had outfoxed himself when he went on the bird.

As McGuirk later put it "It was cowboys and Indians time."

Typically, Turner struck hard. He set up a rate card based on the audiences he was getting nationwide and announced he would charge everybody those rates, even the local Atlanta advertisers. The rate was 10% higher than what he had been charging. Atlanta advertisers balked. And nobody signed up for a national buy. For six months, Turner made not a single sale to a national advertiser. But he wouldn't give up.

"Then one day an advertiser came to us and bought the whole boat," McGuirk recalled. "Within 30 days the whole thing turned around. They just wanted to make him sweat ... for a very long time."

While TBS was the first advertising-supported service to launch on the satellite, it was still essentially just an Atlanta broadcast station that was now being distributed nationwide.

The first advertiser-supported programming service created expressly for satellite distribution to cable systems was the brainchild of Bob Rosencrans, the head of UA Columbia Cablevision who had been the first to agree to install a satellite dish to receive the HBO signals from Satcom I.

Rosencrans quickly saw the benefits additional programming could bring to cable systems. His subscriber rolls had soared in the first couple of years after HBO launched on the satellite.

For several years he had been running events from Madison Square Garden, imported via microwave, on the UA systems in the suburbs of New York.

"So we began to talk to MSG," Rosencrans said. "The events were already being produced, so there was no programming cost." He told MSG he would distribute their events by satellite, charge cable operators a fee for the rights to carry the service and then split the revenue with MSG. The fee for carriage would be based on the number of subscribes a system served.

It would all be incremental revenue for the Garden, Rosencrans reasoned, and wouldn't hurt its gate since most of the viewers would be well outside the New York area. Finally, it would add to the Garden's reputation as the nation's premiere site of major sporting events.

Rosencrans and MSG formed a partnership to launch the project. To run it he hired a young public relations executive who, along with her husband, had worked for UA previously in the effort to win franchises in New Jersey and upstate New York.

Kay Koplovitz had become a devotee of satellite distribution when, as a college student bumming around Europe in the summer of 1966, she had happened to attend a lecture at the London School of Economics delivered by Arthur C. Clarke, the science fiction novelist who had been the first to propose the idea of geostationary satellites.

He talked about how satellite communications would fundamentally change the world, knock down walls, make democracies out of dictatorships. "He was very compelling in talking about his ideas. I just knew that these dreams would come true," Koplovitz later recalled.

"Listening to him had a profound effect on my life. I just thought 'You've got to do this.'"

She ditched plans to enter medicine and instead wrote her college thesis on communications satellites, something

Children's Block:
MSG unveiled a block of children's programming in 1978 entitled Calliope. Pictured (l-r): Mark Smith, Calliope producer; Peggy Charren, Action for Children's Television president; MSG's Kay Koplovitz, and Gene Francis, host of Calliope.

nobody knew much about. "I just never understood why people thought I was from outer space," she recalled about her enormous enthusiasm for a technology most people thought was science fiction.

After a stint as a producer for a TV station in Madison, Wis., she and her husband Bill looked for a way into the cable business. Bill went to interview with Rosencrans about a franchising job and called his wife to say, "You won't believe this guy. He is so down-to-earth, such a straight shooter and such a visionary."

As Kay Koplovitz remembered years later, "That wasn't the type of person we were used to dealing with in the television business."

The only bad news, Bill told his wife, was that they would have to move to New Jersey.

The Koplivitz team won a slew of new franchises for UA. But the state of New Jersey then put a mortorium on

131

the issuance of new franchises and Koplovitz left the company to set up her own public relations agency in Washington, D.C. Among her clients was Home Box Office. She produced many of its early launch events, including the first one via satellite in Vero Beach, Fla., where she was on hand to witness the dreams of Arthur C. Clarke come true less than 10 years after she had first heard them.

When Rosencrans decided to start a network of his own, he knew instantly who should run it.

"He called and said he wanted to do this, but wouldn't do it unless I agreed to run it," Koplovitz remembered. "He was enormously generous," she said, referring to his decision to take a chance on a young, inexperienced former TV producer to run a network. His offer wasn't just generous, it was smart too.

Koplovitz laid out a business plan that called for charging cable operators 10 cents a subscriber per month for the service. This, she figured, would just about cover the cost of programming, the transponder rental, marketing and overhead. Any profits would come from selling advertising.

This was the first time cable operators had been asked to pay for programming directly. HBO had agreed to split revenues with the operators. WTBS was available from Southern Satellite Systems for three cents a subcriber a month, a fee designed to cover the cost of the satellite transmission.

But the new MSG Network pioneered the concept that operators would pay a fee for the programming as well as the distribution based on their total number of subscribers, whether anyone watched or not.

Koplovitz also offered cable operators the chance to sell local advertising to insert into predesignated "avails" of 30 or 60 seconds each during the network programming. It was the first chance local systems had had to sell ads within a national network.

The operators loved it. Greg Liptak, by then running marketing for the United Cable Television system in Tulsa, Okla., recalled that any programming that could give the urban operator an edge over the local broadcasters was desperately needed.

"At the beginning we were welcomed by the industry," Koplovitz recalled, "because they knew they needed more programming to penetrate the urban markets."

Cox Cable was one of the first big cable companies to sign up. After a presentation to Henry Harris and his colleagues, Koplovitz came away with a commitment to launch on all the Cox systems.

When the network went on the bird in

Baseball Contract:
MSG's regional carriage of a N.Y. Yankees-Boston Red Sox game led to the signing of cable's first major pro sports TV rights contract. On hand for the April 1979 announcement were Baseball Commissioner Bowie Kuhn, UA-Columbia's Robert Rosencrans, Young & Rubicam's Bill Donnelly, sportscaster Jim Woods and MSG's Kay Koplovitz.

April, 1977, it was received by cable systems serving 750,000 subscribers. Even without advertisers, the network was able to cover its costs in the first year from the fees charged to cable operators.

But expansion and profits would depend on the ability to crack Madison Ave. "We set out with missionary zeal to bring in advertisers," Koplovitz recalled.

Without any ratings numbers, the pitch was based on the concept that viewers across the country would certainly want to see the huge events at Madison Square Garden. The first break came when agency Young & Rubicam, anxious to experiment with this new medium, committed its client, Gillette, to spend $100,000 to advertise on the new network.

MSG was the first cable programming service to achieve a dual revenue stream — part from advertising and part from subscriber fees. This financial model, replicated by dozens of ad supported networks in the coming decades, would give cable networks a huge leg up in their battle with broadcast networks for viewers and advertising.

Using this model, a cable network could begin to purchase programming and market its service before selling a single commercial slot. The revenue stream from cable operators could be leveraged to fund the pre-launch expenses and then used to pay for the first months of a service's operation.

What Rosencrans understood was that by adding new services, cable operators in turn could increase the rates they charged their existing subscribers and add new customers, boosting their own revenue stream more than the cost of the additional programming.

Koplovitz then began to seek programming from other venues besides MSG. She cut a deal with New York Yankees owner George Steinbrenner to carry the games of the Bronx Bombers on her new network. The first game she televised was a hotly contested battle between the Yankees and their arch rivals, the Boston Red Sox. Koplovitz came into the office the next morning elated. Then the phone rang.

On the line was Major League Baseball commissioner Bowie Kuhn, who told her that the deal "flagrantly disregarded League rules" because each team had the exclusive rights to sell TV rights to their games in their own markets. The Red Sox were livid that their game with the Yankees had been imported into New England without the permission of the team.

"He told me he was going to court that afternoon to seek a restraining order to prevent the telecast of the Yankee games on MSG. I put him on hold and called my husband (an attorney) and asked him whether Kuhn had a case. He told me that Steinbrenner would be the one in trouble, not us, but that it was certainly possible that Kuhn could block the deal.

"So I got back on the phone and asked him if he would be willing to talk to me about a league deal. 'I'll see you in my office tomorrow at 9 a.m.,'" Kuhn replied. The result was the first deal between Major League Baseball and a cable network, giving baseball a couple of cents per sub per month in added revenue and requiring MSG Network to black out games in the home cities of the teams that were playing.

The blackout provision would continue for decades, forcing MSG to put up a second feed to provide alternative programming to the systems where the games were blacked out and forcing cable operators to install equipment capable of blocking those games.

Still, it was the beginning of an enormously lucrative, if at times contentious, partnership between cable and major league sports, giving the leagues a new revenue source and cable hundreds of hours of highly desirable programming that was already being produced, but not being televised except in local markets.

Koplovitz followed with deals for NBA, NHL and North American Soccer League events.

Rosencrans and Koplovitz had always planned that their network would have a broad array of programming, like the broadcast networks themselves. Almost from the start they had been looking for ways to expand beyond the sports schedule. In 1978 Koplovitz unveiled a block of children's programming to run in the mornings. She called it Calliope.

"At the time there was a huge public debate over the disintegration in values in children's programming," she recalled. "The broadcasters were being accused of commercializing children's programming. We found that there was a huge block of programming available for children, mostly short films, of wonderful quality and at a very low price."

Calliope remained part of the network schedule for 15 years. Koplovitz followed by creating a segment on health, aimed primarily at women viewers, called Alive and Well, sponsored by Bristol Myers. And she provided time for a couple of other programmers as well. One wanted to offer public affairs programming and the other had a notion to develop fare geared to minorities. The first was Brian Lamb and the second Bob Johnson.

By the end of the decade the network had expanded its programming mix well beyond sports and changed its name to USA Network. In 1981 UA and MSG sold their interests to a joint venture of Time Inc., MCA and Paramount Pictures.

The launches of USA, WTBS and HBO were soon followed by dozens of new networks. This onslaught of new, cable-exclusive programming fundamentally altered the nature of the cable business. Hastening that change were the activities of two newcomers to the business, both with deep roots in program production and distribution: Warner Communications and Viacom International.

Warner had entered the business in 1972 when it purchased TV Communications, the cable company that had been started by Alfred Stern and had built a two-way interactive system in Akron, Ohio. Warner also bought out Continental Telephone's systems, including the two-way plant in Reston, Va.

Warner was headed by Steve Ross, a flamboyant deal-making genius who followed his instincts rather than any set business plan. Ross had started out running a funeral parlor in New Jersey, expanded into parking lots and in the

QUBE Man: *Warner Communications hired Gus Hauser, a veteran of GTE and Western Union, to run the company's cable operations that included the two-way interactive system, QUBE, in Columbus, Ohio. Hauser founded his own MSO after leaving Warner in the 1980s.*

late 1960s acquired the foundering Warner Bros. studio.

In 1973 Ross reached outside the cable industry for an executive to run his systems. He hired Gustave M. Hauser, who had been executive vice president of Western Union International, and prior to that for 10 years the CEO of General Telephone & Electronics International.

Hauser had already compiled a distinguished career when he decided to "take a flyer" on cable as he later recalled his decision to join Warner. A graduate of Western Reserve University, Hauser had earned degrees from both Harvard Law School (where he met his wife, Rita, also a Harvard Law graduate) and from the University of Paris Law School. As a lawyer in the Pentagon in the late 1950s he had written the first treatise for the Department of Defense on legal issues in outer space after the launch of the Sputnik satellite by the Soviet Union.

Ross ran a decentralized business, allowing his lieutenants wide leeway to operate as they saw fit and backing them with financing for any idea that caught his imagination.

When he bought his collection of cable systems, Hauser said, Ross "hadn't the faintest idea of what to do with it."

The one thing that was clear was that Ross and Hauser weren't interested in running an old fashioned reception service. "We shared a vision that cable could become a big business," Hauser said of his first meetings with Ross. But they realized together that there was one central problem: "Cable had no product."

The retransmission of broadcast television signals, which had been the mainstay of cable in the 1950s and 60s in rural and small town America, didn't work in big cities where broadcast reception was generally very good without cable. The FCC continued to limit the number of distant signals a cable system could import. Cities were insisting that any of the cable systems built needed to offer dozens of channels, including those devoted to public access and government.

The first thing Hauser did when he came to Warner in late 1973 was to bring in a group of people from outside the industry to reorganize the cable operations and stop the financial bleeding.

"We brought in people from ITT and other companies accustomed to running trucks," Hauser recalled. They instituted a computerized system to monitor the installation and repair schedules for cable systems, saving the systems thousands of dollars a month in operating costs.

Hauser cut staff, laying off as many as 50% of the employees in some divisions.

And he raised rates. In most Warner systems rates had been unchanged for

Answer Man:
John Dean answers viewer questions in 1978 on Columbus Live, a call-in program that featured QUBE's two-way interactive technology. Susan Goldwater was the host of the program that allowed QUBE viewers to instantly respond to on-air questions by hitting their remote control.

135

20 years. Again Hauser brought in a group from outside, a "swat team" as he called it, to tour the country and persuade city councils to allow Warner to raise rates in its systems.

In many communities Hauser himself made the pitch. "I remember in Palm Springs, Calif., everybody in town turned out for the meeting. I made the presentation myself. Afterwards the head of the community group opposing the increase took me aside. 'Don't worry,' he said, 'rates aren't really the issue. This just gives us a chance to make a fuss.'"

But the most fundamental problem Hauser faced was how to create a product that people really wanted in communities such as Akron where off-air broadcast signals were so readily available.

Ross, who owned a movie studio, had always dreamed of a way to sell movies into the home. So did the head of his movie distribution arm, Ed Bleier, who had come to Warner after at stint at ABC where he did a strategic study of cable TV. Bleier, as he later recalled, was sitting on a library of films that "were collecting dust."

Phenomenon:
While QUBE never made a profit, it was a public relations success. Phil Donahue came to Columbus to demonstrate QUBE's network technology with Vivien Horner.

After looking at the Akron system, Bleier realized that this could provide an additional outlet for Warner Bros. films. "Bingo, there's pay TV," he remembers thinking.

At first they attempted to run pay-per-view movies in Akron, where there was a two-way system. But the equipment installed by Jerrold proved to be highly unreliable, and Hauser decided to junk the entire system.

Columbus, where Warner owned a state-of-the-art 36-channel system, seemed to be the next best test market, Hauser recalled, "because people weren't buying cable. We just couldn't get any penetration. When we did get somebody to sign on we couldn't keep them more than a month."

The other attraction of Columbus was that the city council was willing to allow Warner to experiment. "They weren't interested in regulation. You could do anything."

Hauser created another swat team with such people as Mike Dann of CBS, Vivien Horner of the Children's Television Workshop, and radio executive John Lack.

Now he needed the equipment to pursue his two-way interactive dreams. To build the hardware needed to provide interactivity, Hauser recruited Pioneer Electronics, the giant Japanese manufacturer, to enter the cable business.

Pioneer and Warner had developed a relationship because Warner produced records and Pioneer produced stereos.

Hauser and Ross decided to call the system QUBE (standing for nothing in particular).

Among the first programming efforts at QUBE was a children's channel, developed by Horner, the CTW veteran with a Ph.D. and a national reputation as an expert in child development. "We bought all the children's programming we could get," Hauser recalled. "It was easy to do; nobody wanted it."

The channel started with no advertising, and even after it became Nickelodeon, it continued to limit advertising and to place restrictions on what could be advertised. By 1979 it was being delivered nationally via satellite. QUBE also launched pay-per-view and a monthly movie package called Star Channel, later renamed The Movie Channel.

The QUBE systems devoured programming. "We would go around and buy anything that flickered," Hauser recalled. "Everything was fair game."

One day John Lack came in with the news that some of the record companies were experimenting with a new format: short-form videos to match the music.

Warner started putting the clips — those few that were available — on a segment of QUBE called "Pop Clips."

Because the QUBE technology could monitor how many homes were watching any given channel at any given time, it was easy to measure the response to "Pop Clips." It was huge. Not only did it draw a sizable audience of teenagers, but they in turn proved to be ready buyers for the music shown on Pop Clips. Warner quickly determined that music videos could be a hit for everybody — cable operators, advertisers, music companies, record stores and viewers.

They called the new service, MTV: Music Television. One minute after midnight on Aug. 1, 1981, MTV went on the satellite.

In a few months, Lack brought a hot young radio programming executive, Bob Pittman, who originally came on board to work on the Movie Channel,

Video Vision: *MTV hired radio executive Robert Pittman, who helped turn the music video network into a cultural cornerstone.*

over to head MTV. Within a year, American youth was rallying to the theme, "I Want My MTV."

QUBE was a hotbed for innovations in programming and cable operations. It attracted enormous nationwide attention (its launch made the news on all three broadcast networks). And it proved an enormous plus for Warner during the franchise battles of the 1980s as city council members around the country were shipped to Columbus to see the cable system of the future up and running.

Unfortunately, QUBE never made any money. "It was," Bleier said later, "the most successful failure known to mankind. While QUBE didn't work (financially), it made everything else work."

Viacom, the other new entry into the cable business in the mid-1970s, was born by government intervention in the communications business. The FCC forced broadcast networks to divest their ownership of cable and syndication businesses. CBS, which had extensive operations in both areas, spun off those businesses into a company headed by Ralph Baruch, who had been president of CBS's syndication business.

Viacom inherited about 30,000 subscribers from CBS, most of them in clusters in the San Francisco, Seattle and Long Island areas. As the executives of the new company sat down to plan their strategic objectives, Baruch recalled, "It was obvious that within the next two to three years costs were going to outpace revenues, which would have been an unsupportable condition. It seemed to us the only thing on the horizon was pay television."

But Viacom was dissatisfied with Home Box Office, which it had launched in several of its systems. The churn was high, and Viacom, which had experience in TV production and syndication as well as cable operations, felt it could do a better job of creating a pay service.

Viacom started with a pay-per-view experiment in the system in Suffolk County, Long Island, using a device developed by a company called Transworld, a subsidiary of Columbia Pictures. The system didn't work very well, and Viacom shifted its focus to a monthly pay television service. To run it the company hired a 33-year-old executive who had worked with Michael Eisner and Barry Diller at ABC TV and developed that network's strategy of two-hour made-for-TV films. Jeffrey Reiss had been born in Brooklyn and studied biology at Washington University in St. Louis.

He worked in the theater in New York, producing several off-Broadway plays one of which won an OBIE award. In 1971 he became involved with a startup company that had developed one of the first versions of the videocassette recorder.

At Viacom Reiss took a close look at HBO and found what he believed was the reason for its high churn rate: "The product was overpromised and oversold."

When customers first looked at the HBO schedule, they were overwhelmed by its choice, as many as 80 or 90 movies a month plus sporting and other special events. But after a few months they began to discover that many of the films were repeated. Pretty soon many of them would be saying, Reiss recalled, " 'Gee, there's nothing on this month that I haven't already seen.' "

Reiss designed a service that would offer only 12 movies a month, but would change the entire schedule every month so that there would always be something fresh each month.

He also brought to the mix some of the programming and marketing expertise he had honed at ABC. He developed interstitial programming to go between the films so that the movies could begin on the hour (he adopted the word interstitial from his days in biology. The word refers to the space between the cells of a living creature). He hired Jules Haimovitz, another ABC veteran, to do the programming.

The service, called Showtime, launched in Dublin, Calif., outside San

Touch Now:
That was the command QUBE viewers followed to vote on everyhting from public opinion polls to which music videos played.

Francisco, in June, 1976. It was a success, drawing 35% penetration in the first marketing sweep. But signing up customers wasn't the goal. Keeping them was.

Reiss and Haimovitz monitored the subscriber counts closely. After a couple of months they began to decline. The small Showtime crew set up a war room, taking over the Viacom board of directors room, to continue to watch the movement. After a few more months the levels began to stabilize and then to increase, eventually going past the penetration levels that had been achieved in the first push.

They charted the growth, decline and more growth in a pattern that became known as the Haimovitz curve (or Haimovitz hump as Reiss called it because it resembled the humps of a camel.)

Like HBO, Showtime relied heavily on its guide to promote upcoming films and to retain customers.

The second guide was to carry on its cover a picture of Gene Wilder, star of the film Young Frankenstein. It was set to go to print when the studio called and said the movie couldn't be used because its producer, Mel Brooks, had never heard of Showtime. Reiss picked up the phone to call Brooks and was told he had the flu.

"So I sent down to the delicatessen and had a large order of chicken soup with matzo balls sent to Brooks' home with a note asking for a chance to speak to him." The director called the next day. "Who are you?" Brooks bellowed into the phone at the startled Reiss. "Do you know who I am? Do you think Jewish jokes are funny? How was I to know you weren't some nut? How was I to know the soup wasn't poisoned?"

Showtime Vets:
These four men (l-r) - Jules Haimovitz, VP; Terrence Elkes, EVP, Viacom International; John Sie, VP-marketing/affiliates, and Jeff Reiss, president - were among the key Showtime executives in the late 1970s.

Brooks in real life was just like the characters he portrayed on the screen.

Reiss sent him over some copies of the guide and the next day had his approval.

It wasn't so easy when it came to getting Viacom to approve satellite distribution. Reiss recalled that the brass at Viacom was skeptical about his ability to make satellite distribution pay, and by the time they had the go-ahead, it was 1978, and a huge amount of territory had been lost to HBO.

To make up the ground, Reiss made a radical hire, bringing on board not somebody from an operating company or a programmer, but John Sie, head of the Jerrold cable division. Again some of the Viacom brass were doubtful.

"I was just so impressed with his analytical ability and his communication skills," Reiss recalled. "He had one of the best pitches I have ever seen, and he had passion. Above all he had passion."

But the two of them would have to wait until 1978 before they could offer cable operators a satellite-delivered pay service. And HBO in the meantime was signing up subscribers and affiliates.

DISTANT SIGNALS: HOW CABLE TV CHANGED THE WORLD OF TELECOM

The launch of satellite programming, and the new wave of cable customers it produced, particularly in the more populous areas of the country, forced major changes in the way cable companies were run.

As long as cable remained a small-town phenomenon, with the primary goal of retransmitting broadcast signals, most systems could be run by a couple of people: an engineer who kept the system running and an office manager to answer the phones, send out the bills and deposit the checks.

But as cable companies grew, they required people with more complex skills to run them. During the late 1970s many of the largest cable companies began to bring in new talent, often people who had no experience in cable but who could bring skills from other businesses to the industry.

After wresting control of TelePrompTer from Irving Kahn, Jack Kent Cooke installed as company president Russell Karp, a Yale Law School graduate who had held top positions with Hollywood studios Screen Gems and Columbia Pictures. Bill Bresnan remained as senior vice president of TelePrompTer and president of the cable division. The new lineup also included two under-30 marketing executives, Marc Nathanson, from Cypress Communications, and Jeffrey Marcus, from Sammons, to head up a major effort to boost subscriber levels.

At American Television & Communications Corp. CEO Monroe Rifkin recalled, "We had been promoting from within. I decided to bring in better talent to the industry." In short order he had visited many of the major business schools recruiting such people as David Van Valkenburg from Harvard Business School, John Rigsby from Harvard, Mike Kruger from Stanford, and Larry Howe from Dartmouth. A headhunting firm brought him a rising young marketing executive and Dartmouth graduate who had worked at Procter & Gamble, Trygve Myhren.

One of the first in this new wave of well-educated talent was Joe Collins, a 1972 graduate of the Harvard Business School who had done his thesis on cable television and sent out letters to the heads of all the major MSOs when he graduated. Rifkin was the only one to reply.

ATC assigned Collins to be marketing manger at the system in Orlando, Fla., which Rifkin later recalled was a place where the company "learned an ugly lesson." The Orlando market had a full complement of broadcast television stations, and ATC's expensive two-way system was achieving very little penetration. When Collins arrived, it had only 500 customers, among the 2,500 homes passed.

Collins recalled that he felt this was actually pretty good since the only programming the system offered that could not be received over the air was an educational station from Tampa and a local origination channel.

The only marketing effort the system was making when Collins arrived was the door-to-door selling that had been the mainstay of all systems up to then. So he tried some of the tricks he had learned at Harvard — point-of-purchase displays in the local stores, direct mail, telephone solicitation. With the marketing blitz and the addition of a couple of distant signals from Miami and Tampa when the FCC allowed it, Orlando, where Collins had been promoted to general manger, was able to increase its penetration to the upper 30% range from the 20% he found when he arrived. In 1973 he received a special marketing award from the National Cable Television Association.

After moving to ATC's Denver headquarters in 1974 Collins managed the installation of satellite dishes and the conversion to pay TV for ATC systems in the late 1970s. He recalled the headaches as systems went from coupon billing systems to computers. "In Reading, Pa., there was a man who had trained his dog to bring the coupon book and the payment to the office every month," Collins said. "He was heartbroken when we made the switch and the bill had to be sent someplace else."

Continental Cablevision faced similar challenges of growth and broought in new talent to help. Within a few months in 1974 Amos Hostetter and Irv Grousbeck had hired ex-broadcast news producer Jim Robbins, psychology professor Robert Clasen, Harvard College history major and assistant director of the Bedford Stuyvesant project Barry Lemieux, and banker Tim Neher. Each would wind up being president of a major cable operating company. So would Ray Joslin, who had been one of the first employees of Continental. Joslin was serving as regional sales representative for the Jones & Laughlin Steel Co. in 1966 and living in Findlay, Ohio, when he met Grousbeck, his next door neighbor, and signed on to work for Continental.

In those days, Joslin later recalled, the system managers did everything, including bailing the construction workers out of jail on Monday mornings so they could go back to work and build a few more miles of plant.

Clasen recalled an early experience in Upper Sandusky, Ohio. He was seeking a rate increase to $7.25 from $5.90 per month (almost 23%) and some of the city council members were threatening to oppose it. So Clasen, using his psychology background, took out ads in the local paper with a picture of a huge nickel, with a slice cut out, and a message stating that the increase would amount to less than nickel a day. The local paper ran a story on the matter with the headline, "Cable company asks for only a nickel increase," and the council approved the rate hike.

Robbins, a Harvard Business School graduate who had worked at WBZ-TV news in Boston, and had started in cable at Adams Russell in Massachusetts, was assigned to run the Continental systems outside Dayton, Ohio. His big breakthrough came when General Instrument introduced the first remote control, a long wire that attached the channel changer to the TV set.

He launched a marketing blitz to sell

Launching Pay:
HBO's Frank Biondi (center) joined Rogers Cablesystems' Robert Clasen (right) and the mayor of Syracuse, N.Y., to launch Cinemax in that city in the 1980s.

the remotes for $1 per month extra, reaping huge profit margins for the system which bought the devices from Jerrold for less than $5 each.

Robbins also leaped into the political fray. He was infuriated by the FCC rules that required systems to black out shows carried on distant signals when local stations owned the rights to air the same shows. The local Dayton broadcasters had purchased the rights to syndicated shows that they were warehousing, simply retaining the rights to air the shows without actually running them. The objective was to force blackouts in the distant signals carried on the cable system.

The system was required to black out these programs. So Robbins created a campaign against this so called "syndicated exclusivity" and used ads on his own system to urge customers to send in postcards (preprinted by the system on red paper) to the Federal Communications Commission and the Congress to protest the rule.

When Robbins himself traveled to Washington to lobby on the issue, he recalled, one FCC staffer opened the drawer of his desk "and it was full of red cards."

But nowhere was the impact of fresh new management felt more acutely than at Tele-Communications Inc. John Malone was a sailor, and when he arrived at TCI, even though it was far from the ocean, the waves were beginning to wash over the gunwales and the squalls on the horizon were dark and ominous. As he put it 25 years later, "There weren't any more deck planks to throw into the fire."

In 1972 TCI chairman Bob Magness had gone on an acquisition binge, convinced that the FCC's decision that year to relax the distant signal rules indicated a new era of deregulation in Washington. He had financed his purchases with so-called bridge loans, short-term financings that he expected to be able to repay by selling additional TCI stock after the acquisitions had been completed.

When the cable stocks crashed in the wake of Irving Kahn's conviction and interest rates began to skyrocket, TCI was caught in a squeeze. Magness' bridge loans had become pier loans, and the banks weren't happy. Magness needed help. When he brought Malone on board as TCI president Magness told friends, "I just hired the smartest son-of-a-bitch I ever saw."

Eye For Talent:
After convincing Dr. John Malone (top) to join TCI, Bob Magness told associates "I just hired the smartest son-of-a-bitch I ever saw."

Malone knew full well how bad the situation had become. TCI had been a major customer when Malone ran Jerrold, and the cable company had made full use of the financing program Malone had set up. When Malone moved to Denver, TCI had $132 million in debt, and only $18 million a year in gross revenues. Its stock dropped as low as $.87 a share. Its market capitalization was down to $3 million.

Malone cut everybody's pay, including his own, by 30%. He visited the communities where TCI had pledged to upgrade the systems and warned them "there wouldn't be any picnics this year, and probably not next year either," he later recalled.

He and his colleagues slept two or three to a room at Holiday Inns. They paid their employees in stock (a payment scheme that would make some of the clerical help millionaires a decade later).

Malone situated his office so he could hear who was coming to the reception area and make a quick and quiet exit when creditors arrived. And he found a way around the rules he himself had set up when he was at Jerrold.

John Sie, who became head of the Jerrold cable division when Malone left, recalled that TCI's precarious economic conditions and tardy payments forced Jerrold to put the company on credit watch, denying it additional products until it brought its payments up to date. But Sie noticed that Jerrold was still shipping out product to TCI systems. When he looked into it he found that Malone had set up many of the TCI systems as separate companies which were then able to apply for credit with Jerrold on their own with clean bills of health. It was the only way Malone could keep his systems operating while he fought his biggest battle: with the banks.

TCI's debt had been financed by a group of seven banks. The loan agreements imposed a repayment schedule as well as a variety of covenants that restricted what TCI could do (limiting such items as debt-to-cash-flow ratios, purchases of additional equipment, and even stating, in a Catch 22, that TCI could not spend more on launching pay TV than it was earning from pay TV). The agreement stated that if TCI failed to make an interest payment any bank could call its loan. If TCI violated the covenants, a majority of the banks could call the loans.

Two of the banks had been on the short end of a deal that TCI shareholder George Hatch had done to purchase a motion picture company called Republic Pictures. Hatch had guaranteed his loan would be repaid by TCI without informing Magness. But Magness stood

Hollywood Debate:
TCI's John Malone and Viacom's Ralph Baruch testify before Congress as MPAA's Jack Valenti (far right) looks on. All three were regular guests on Capitol Hill in the 1970s and 1980s.

by his friend even as the banks filed suit charging criminal fraud. It was a messy situation.

Both banks involved in the Republic fracas were ready to call their loans to TCI at the drop of a pin. As Malone recalled later, "There were a whole lot of covenants, and we weren't meeting any of them."

The two banks that had been burned on the Republic deal were all for calling the TCI loans. But the other five had been willing so far to hold back as long as TCI made its interest payments.

The standoff among the banks came to a head when TCI needed them to grant a waiver of the covenants so that the company auditors could give TCI a clean bill of health for its annual report to shareholders. Without a waiver of the covenants, the auditors would not have been able to sign off on the TCI financials and the company would have collapsed.

That was the scene facing the new TCI president, all of 33 years old, when he met with the bankers for the first time to seek the waiver. Magness, Malone later recalled, was so angry at the banks and his blood pressure was so high he left before the meeting even started.

Malone outlined the situation and asked the somber-faced bankers for help. "They asked me, 'What is your plan to repay these loans?' And I answered 'I don't know. Heck you guys made these loans before I got here. What were you thinking? How did you expect us to repay them? Maybe there's a memo in your files someplace that will tell me how you expected to get your money back.'"

After a few hours the bankers adjourned to discuss what they would do. When they came out, the lead banker went down to Malone's office where the CEO was waiting. The bankers, Malone was told, were prepared to grant TCI the waiver... if TCI agreed to pay a higher interest rate.

Malone didn't miss a beat. "We have this thing screwed down as much as possible," he said. "If you raise the interest rate, we won't be able to continue to operate."

He reached into his pocket and took out the keys to the office. Sliding them across the table he said, "If you guys think you can run this thing better, you run it." And he walked out.

The next day the bankers called and told Malone he would have his waivers. But the relationship between TCI and its bankers remained one of thinly disguised hostility. Malone did get a few breaks in the next few years, some of them almost by providence.

One of the companies Magness had acquired was Athena Cable. The deal allowed TCI to take all of the Athena assets. Among piles of paper that TCI personnel found when they assumed control was a group of warrants for the purchase of stock in Resorts International, a company which had recently won the rights to build a casino in Atlantic City. The warrants were worth millions.

But, as Malone later recalled, "We were in the woods" until he was able to strike a deal with a group of insurance companies for a $76 million, 15-year line of credit at a bearable interest rate of 7 5/8%.

Malone called a meeting with the bankers, all of whom expected more bad news and were already grumbling over their coffee when Malone opened the meeting.

"From time to time," Malone said,

144

"some of you have expressed a desire to reduce your exposure to TCI. I am pleased to announced we are prepared to allow you to redeem your loans."

Malone then introduced a representative of Chase Manhattan Bank, the agent for the insurance companies, who said he was ready to write out a check then and there to any bank wanting to get out. And the TCI president once again got up and started to leave a room full of open-mouthed bankers. As he reached the door he paused.

"And by the way," he said over his shoulder. "From now on TCI is a prime rate borrower."

He later heard that there was silence for several minutes in the room after he left, until one of the bankers exploded: "TCI isn't a prime rate borrower and never will be."

Two of the banks took the offer and bailed out. The other five stuck with TCI. As Malone later recalled it, "The weight of 18 months was off us and we could stop biting our fingernails down to the bone every day. After that, we never looked back."

With the banks off his back, Malone could begin to look at the possibility of launching pay television in his systems. By this time TCI was one of the few cable companies that had not yet launched pay service.

Malone had the luxury of choosing between HBO and the new upstart Showtime, and he made the most of it, driving a hard bargain in simultaneous negotiations. HBO president Nick Nicholas flew out for the showdown and tore up his company's rate card to meet the TCI demands for a multiyear contract. Showtime matched his pricing but could not match HBO's delivery system. (The dealmaking was typi-

Lines In The Streets:
The arrival of HBO and Showtime brought lines in the street and the corresponding apt description "truck chasers" for those hungry for an alternative to broadcast channels.

cal of Malone who would many times play off one supplier against another to get the best deal for TCI.)

At the end of the day HBO had agreed to finance the TCI earth stations and provide help with marketing, a nearly no-lose situation for the operator. But the final call was made because HBO was already up on the satellite and Showtime wasn't.

"I remember he took me out to the parking lot at TCI and put his arm around my shoulder and pointed to TCI's earth station," Showtime's Reiss recalled. "He told me 'I can receive HBO's signal in my parking lot today. I have an obligation to my subscribers to give them something I can deliver on. Showtime says it will go up on satellite, but it might and it might not.' " TCI went with HBO.

As the years went on, Malone again adopted some of the characteristics of another mentor, this time Magness. Like many easterners who move to the West, Malone reveled in the freedom of the region. He shed the tie and business suit (except when the bankers came to town) and adopted the feet-on-the-desk, shoot-from-the-hip, part cowboy, part outlaw characteristics of the old West. And when it came to a model, Magness, part Cherokee Indian and raised in Oklahoma, was the genuine article.

The parking lot visit with Malone and other similar experiences with cable operators convinced Reiss that Showtime would never catch up with HBO under the game as it was being played. Cable operators were signing up with one or the other, and most had gone with HBO.

There were only two avenues that would allow Showtime to catch up. One would be to convince a major HBO customer to switch to Showtime. The other would be to convince MSOs that they should carry both services.

The MSO came in the form of TelePrompTer, whose original HBO contract ran out in 1979. Russell Karp, TelePrompTer's chairman, had been one of the first to sign up for HBO when it launched on satellite. But he had also become convinced that his company should share in the ownership of a premium service he was delivering to some 300,000 TPT subscribers.

Viacom offered him a half ownership of Showtime in return for an agreement to switchout HBO for Showtime in all the TPT pay homes. It was a huge undertaking, involving major marketing efforts to prepare consumers for the change. Showtime used satellite-delivered training sessions to teach TPT customer service representatives and managers how to deal with customer complaints.

But the big breakthrough for Showtime came in the unlikely town of Thibodaux, La.

Until 1979 most operators, and the pay services as well, had assumed that every system should offer subscribers only one pay service. After all, they reasoned, who would want to pay to watch more than one network?

But Sie and Reiss had noticed that in a few newbuilds where there was enormous channel capacity, both services could be offered and both could do very well. Womecto Cable TV then decided to invite both HBO and Showtime to launch at the same time in the Wometco system in Thibodaux. It was to be the mother of all battles between the competing pay services, and each side assumed it would be a battle to the death, with the loser being force to play second fiddle forever.

John Sie sent in a new hire, Sara Levinson, to run the Showtime effort, and HBO assigned one of its hot young marketing guns, John Billock, who had just joined the company from Colgate Palmolive, to manage its side of the launch. "I remember Jerry Levin told me 'Go out there and come back a winner,'" Billock recalled, "and I figured 'All right, here we go.'" The town was plastered with posters, newspaper ads, door hangers and other marketing materials in the weeks prior to the launch.

"I literally lived down there for eight weeks," Billock recalled. "I wrote copy for the ads and took the slicks to the

Program Chase:
In the early 1980s Showtime landed the rights to 'The Paper Chase,' the first network series to eventually find a home on cable.

146

local papers. I took the radio spots over to the local radio station. We both overmarketed, but we had a great time."

Just prior to the launch both Sie and Billock were at the system headend and Billock managed to spill his soft drink, a Mr. Pibb, into the character generator that ran a Showtime alphanumeric ad on the system. It was a "very sticky, unfortunate accident," Billock later recalled. Sie wasn't so sure. But the battle, they both later agreed, was all in good spirits.

What happened next amazed everybody. There wasn't one winner and one loser. Everybody won: HBO, Showtime and the cable system. "We found 15%-20% of people were taking both services," Sie later recalled.

Billock remembered that the total penetration for HBO after six weeks was about 38%, almost double what might have been expected had HBO launched alone.

Launching HBO and Showtime together had proven more than twice as profitable as launching either one separately. The sum was worth more than the parts.

The consumer had decided that the two services, rather than being competitive, were complementary. Showtime's schedule of all fresh movies combined with HBO's bigger inventory and original fare was an attractive combination. Once again the consumer had proven to have an appetite for more choice than anyone had expected.

"That was the defining event for the entire category," Billock later said. "The marketplace had spoken."

At the Western Show in Anaheim that year Showtime rented a huge ballroom and unveiled the results of the "Showdown." Sie worked up a frenzy, signing up new affiliates, many of them already

Political Ally:
Tim Wirth (D-Colo.) was an advocate for cable throughout his tenure in Congress. He later ran the charitable foundation Ted Turner funded to help the United Nations.

carrying HBO. "It was like there were 100 John Sies," recalled HBO affiliate rep Bill Grumbles. He was tireless.

Paul Kagan trumpeted the numbers in his newsletters. Within six moths HBO had launched Cinemax as a flanker brand to be able to offer two pay services to systems. The era of multipay had begun.

As HBO and Showtime battled in Thibodaux and Malone arm-wrestled his bankers, the ice at last began to break in Washington.

For the entire decade it had been difficult to get much of anything done in Washington, cable related or not. The Democratic Congress tangled with a Republican President and later the Watergate scandal and impeachment hearings consumed the entire political agenda at both ends of Pennsylvania Avenue.

Then President Nixon resigned, to be followed by the more moderate Gerald Ford. The Democratic landslide in the congressional elections of 1974 brought to town a new generation of lawmakers, Democrats from marginal or normally Republican districts. These new congressmen were not as rigidly ideological as some of their colleagues and were more inclined to look for new ways to do things.

Among them was a freshman from the suburbs of Denver, Timothy Wirth, who had won by the skin of his teeth with the financial support of a group of Denver cable operators. He sought and received a post on the House Telecommunications Subcommittee.

In 1976 another unconventional Democrat, former Georgia governor Jimmy Carter, was elected President, and for the first time in eight years both the Congress and the White House were controlled by a single party, the Democrats. But these Democrats were a different breed from those who had imposed such stringent regulations on the cable industry a decade before.

Carter chose as his FCC chairman Charles Ferris, a card-playing, wisecracking but very, very savvy politician who had served as Secretary to the Majority in the Senate, the top staff position for the Senate Democrats.

Ferris, like Carter, was liberal when it came to such issues as civil rights. But he also had an instinctive distrust of media monopolies and a bent toward allowing competition to replace regulation as a way to rein in the power of the broadcast networks and promote program diversity.

This philosophy was reflected elsewhere in the Carter Administration which championed deregulation of the trucking and airline industries.

The turnaround in cable's fortunes began in 1976, before Carter took office, with passage of a copyright law governing the broadcast signals that cable operators carried.

For years the cable industry had argued it had a right to pick up broadcast signals off the airwaves for free and then charge customers for the right to watch over the cable system. For years the broadcast stations and the program producers had argued that cable operators could not do this without payment of copyright fees to the stations and the producers.

In 1968 the Supreme Court had ruled the issue should be decided by the Congress, and for nearly a decade the various parties and the key members of Congress wrestled with various different solutions to the issue. The battle involved not only the copyright issue. Also at stake were prospects for sweeping legislation to regulate the entire cable industry, which had been proposed in various forms during the Nixon Administration. The National Association of Broadcasters and the organizations representing copyright holders effectively blocked any congressional consideration of broader cable legislation until the copyright issue had been resolved.

The ultimate outcome was a law that set up a special agency of government, the Copyright Royalty Tribunal, which

Pole Rates:
Continental chairman Amos Hostetter testifies on the 1978 pole attachment legislation that would set up payment structures for cable operators to string wire on utility poles.

was assigned the task of setting copyright payments by cable system operators. The fees would then be put into a fund which would be distributed to copyright holders.

In return for agreeing to this payment plan, the cable operator won what was called the compulsory license. Under this concept broadcast stations had no right to prevent cable systems from carrying their signals.

Passage of the Copyright Act settled one of the most contentious and volatile issues the cable industry faced in its first quarter century of life. The law would not completely resolve the matter. The various interests battled back and forth in proceedings at the CRT for the rest of the century over what was a fair payment. But the legislation did at least set up the format and arena in which copyright disputes could be settled. And it resolved an uncertainty that had caused concern among the financial institutions backing cable and among cable investors. (It also gave a boost to the career of Tim Boggs, legislative aide to the House sponsor of the bill, Rep. Bob Kastenmeier. Boggs would become the Washington representative of Warner Communications and later Time Warner.)

But the bill came at a cost. As it had several times before when faced with legislative issues, the cable industry split over the issue of copyright. Large MSOs wanted the matter resolved in order to move forward with wiring of the major cities. Smaller operators feared that the imposition of copyright fees would cut into their profits.

The dispute became so contentious that a group of the smaller operators split off from the NCTA, forming their own group, The Community Antenna Television Association (CATA), and hiring the law firm of Rick Brown and Steve Effros to handle their Washington lobbying.

Effros, a feisty former FCC staffer, was successful in negotiating an exemption from the standard copyright payment for operators with less than 1,000 subscribers. Instead those systems were required to pay only $27 every six months into the copyright fund.

"We just made ourselves such a pain in the ass," Effros later recalled, "that (Motion Picture Association of America president Jack) Valenti, just threw up his hands. I told him that he wasn't losing any money anyway since the real copyright money was in the big systems. So he just gave in to get us off his back."

Two years after the copyright bill was signed into law, the cable industry achieved another legislative victory. For 20 years cable operators had battled with utilities over the use of poles on which cable wire had to be strung. It was like the situation in the Middle East. Battles would flare up from time to time and then an uneasy truce would ensue, followed by another period of conflict.

Advocate: Steve Effros helped write the FCC's 1972 cable rules before heading into private law practice. A split between big and small in the cable industry over copyright issues led to formation of the Community Antenna Television Association which Effros headed.

The issue had been partly resolved in the 1960s when the NCTA reached an agreement with AT&T for a uniform, nationwide rate policy for Bell poles. But the deal didn't cover poles owned by independent phone companies or those owned by power companies.

In the mid-1970s as cable's cash registers began to swell with new revenue from pay TV, the pole owners saw a chance to add to their own revenue streams. Most had the simple objective of taking in more fees for the pole rights. But some phone companies also hoped to force cable systems into so-called leaseback agreements in which the phone company would build the cable plant and then lease it back to the cable operator to manage.

"What was happening," recalled Harold Farrow, an attorney for the California Cable Television Association, "was that telephone companies in an effort to encourage, shall we say, the cable operators to use facilities owned by the telephone company, would delay or deny pole attachment agreements

Knock, Knock:
Once HBO, WTCG, MSG and the Family Channel went on the satellite in the late 1970s, cable's door-to-door salesmen had more to sell than better reception.

and tell the cable operator, 'Look, you don't want a pole attachment agreement. You will get that with the leaseback service I'll provide.' "

"But the leaseback process was a pretty negative and anti-competitive process. The cable operator had to put up in advance all the capital to build a plant. He was limited to use of the plant for only 12 channels in one direction. Everything else was reserved for the phone company. He was then charged a rather exorbitant annual fee for the maintenance of the plant. He couldn't touch the plant...He would own the little piece of wire at the end of the plant where it was in the customer's house. Basically everything else was under the control of the telephone company."

During the mid-1970s, for example, Pacific Gas & Electric (Calif.) doubled its rates to $5 per pole, Dayton (Ohio) Power announced an increase of 100% to $8 per pole per year, Carolina Telephone & Telegraph (N.C.) shut down service to three ATC systems which had refused to go along with a 50% rate hike, and Toledo (Ohio) Edison asked for a percentage of the cable systems' gross revenues.

The cable industry sought relief from the FCC, but was rejected when the commission decided it had no authority to regulate the activities of utilities. So the industry turned to Congress. There it found a sympathetic ear, partly because the phone companies had one of the worst public images of any American industry in the mid-1970s (reflected when Lily Tomlin played the phone operator, Ernestine, in the popular Laugh In TV series, answering customer complaints with the response: "We don't have to. We're the phone company.")

Rep. Wirth sponsored legislation to grant the FCC authority over pole rates and attached it to a measure that would also have allowed the FCC to impose fines on cable systems found in violation of FCC rules.

Once again the industry split over the plan. CATA opposed the bill because its members feared that smaller operators would not be able to keep up with the avalanche of paperwork the FCC was requiring and would be therefore subject to fines. The larger MSOs and NCTA were more inclined to support the bill to get some relief from the escalating pole rates.

The bill nearly passed in the 1976 session of Congress, but House Speaker Carl Albert blocked its consideration at the last minute because a group of cable operators from his home state of Oklahoma asked him to kill it.

The opposition from the smaller systems was fierce. At one point CATA president Kyle Moore wrote a letter to FCC chairman Richard Wiley stating, "I have an illegal cable system ... Come and get me."

In the end the legislation passed, and the FCC was granted the right to set the fees paid by cable systems for the right to use the poles owned by utilities and phone companies. Another huge uncertainty that had hampered the growth of cable had been resolved.

The National Cable Television Association in the late 1970s also found some stability in staffing after years of a revolving door, particularly at the top. The organization had three different presidents from 1965-1975. Two of them had known little about cable when they took the job, requiring a breaking-in period just to learn the basics of an increasingly complex business.

The public policy issues were difficult enough, but even tougher in some

Stabilizing Force: *Bob Schmidt became NCTA president 1975, giving the industry a stabilizing presence in Washington, D.C. Schmidt is pictured here testifying on Capitol Hill with Burt Harris.*

respects was the job of dealing with the NCTA board itself. When David Foster stepped down as president in 1975 after just two years at the helm, NCTA board member Amos Hostetter commented that he was quitting while he was ahead.

"Considering the fact that the board is made up of self-made individuals with strong opinions on everything, the presidency is a transitory job — for anyone," Hostetter told *Television Digest*.

The shuffle in the president's office also produced several periods in which the organization was headed by a lame duck or by nobody at all while the NCTA board conducted a search for yet another successor.

The revolving door slowed with the selection in 1975 of Robert Schmidt as the new NCTA president. Schmidt had been a staff member at the Democratic National Committee in the early 1960s and had good contacts with the Democrats on the Hill (most cable operators were conservative Republicans). He also had served for a decade as the public affairs vice president for ITT and understood some of the issues facing the cable industry. Finally, he was young, energetic and personable and a good politician able to get along with those who held opposing views.

Schmidt launched an aggressive campaign to position cable as the scrappy underdog taking on the giant telephone and broadcast monopolies. The strategy paid off handsomely in the next decade.

Even more important, Schmidt hired as his deputy a savvy executive from the trade association representing grocers, Thomas Wheeler. When Schmidt left the NCTA in 1979 (the first president ever to leave without being fired or forced out) Wheeler stepped in. Thereafter, each NCTA president would be succeeded by his number two, eliminating the interregnums and break-in periods for new presidents that had plagued the association for its first two decades.

As cable began to resolve some of its financial and political troubles in the late 1970s, it also had new technological developments to provide excitement. The 1975 NCTA convention saw the display of the first feed-forward amplifiers, able to boost the cable signal without the distortion that had plagued earlier amplifiers.

But some of the technologies proved to be ahead of the time. This was true of fiber optics, which captured the industry's attention for a while in the mid-1970s. Fiber technology had come a long way since the technique had first been used to send television signals in the previous decade. Manufacturing of fiber had been refined and the manufacture of lasers, used to generate the light beams that were sent down the fiber strands, had progressed as well. Fiber was being used by telephone companies and other telecommunications providers both in the U.S. and abroad.

In 1977 Irving Kahn, the irrepressible former chairman of TelePrompTer who had been an early advocate of using satellites to deliver cable services, arrived at the NCTA convention in Chicago with a new joint venture and a new product.

His joint venture partner was Times Wire, the largest supplier of coaxial cable to the industry. Kahn, who moved to New Jersey after spending 20 months in jail for a conviction on federal bribery charges, had run into a group of engineers for Bell Labs who had perfected a new system for producing fiber strands. Kahn took his invention to Times Wire and persuaded its parent company, Insilco, to spin off Times Wire into a joint venture that Kahn and Insilco would own, called Times Fiber.

"The plan was simple," Kahn later recalled. "If you were going to try to sell the stock of Times Wire, you would maybe get a multiple of five or six to one, and nobody wanted to be in a commodity-type business. On the other hand, if you changed the name to Times Fiber and became a high-tech company, Wall Street perceived you as a multiple of maybe 30 to 1. If we could develop the fiber we would phase out the wire, increase the fiber, use our marketing strength and let the profits from the wire fund the development of the fiber."

At the NCTA, Kahn demonstrated his product, which he said could send TV signals down a 10-mile stretch of cable with only 13 repeaters. The same length of coaxial cable would require 30 amplifiers, each of which degraded the signal.

By then end of the year, Times Fiber had installed its fiberoptic cable in the TelePrompTer system in Lompoc, Calif., sending 12 channels down 8.4 kilometers of fiber supertrunk at a cost of about $10,000 per mile for construction.

Because fiber could carry a signal so much farther than coax without amplification, it could produce much more reliable pictures at the TV set and required a lot less maintenance.

Unfortunately, Kahn was ahead of his time. When Times Fiber began to manufacture fiber in large quantities, it ran into trouble. "The truth is that for many, many millions of dollars and virtually no results, we were never able to manufacture a good piece of fiber that was competitive," Kahn later said. The lasers also proved to be expensive and unreliable.

But the plan was a success on Wall Street where Times Fiber, riding a wave of new construction in the 1980s, proved to be a go-go stock, earning a small fortune for Kahn.

While fiber proved to be ahead of its time in the late 1970s, the glitch was scarcely noticed in the industry-wide euphoria over new pay and basic services, deregulation and the prospect of opening up the major cities to cable.

It was a 180-degree turn for an industry that only five years before had appeared on the brink of disaster. The change had not gone unnoticed in the boardrooms of some of America's biggest companies, including those with such household names as Westinghouse, Times Mirror, American Express and General Electric. Together with those companies that had built the business from the beginning, the group would launch the telecommunications equivalent of the Normandy Invasion, bringing cable into the nation's biggest cities.

Marketing Networks

When Ted Turner put his broadcast station up on the satellite, he hired a small Atlanta advertising agency to help him market it to cable systems. The agency, since renamed CommunicationTrends, Inc., developed the first trade advertising for WTCG. It grew along with the cable industry and went on to handle trade and consumer marketing for a wide range of clients, including Turner, USA Networks, HBO, American Movie Classics, Black Entertainment Television and Bravo, among others.

CTI's CEO Toni Dwyer was no stranger to cable: She helped put herself through college going door to door selling cable subscriptions for systems in Ohio. "I made a lot of money knocking on doors," she remembered.

After a stint at several New York and Atlanta advertising agencies, Dwyer, then Toni Augustine, created one of the first national advertising campaigns for cable operators. The campaign was called "We bring it home to you" and it included radio and print ads, door hangers and other materials that could be customized by local systems anywhere in the country.

It was used by 360 systems to raise awareness of what cable offered.

When Dwyer signed on to help Turner launch WTCG on the satellite she found she had an unusual client. "Ted personally went to every little system," she recalled, "in every Podunk town. The cable operators were enamored with him and wanted to be associated with him. He was so ebullient and enthusiastic. There was no way you could stay around him and not get

Toni Dwyer in 1975

Above, two early trade ads Toni Dwyer created for WTCG; right, a 1984 campaign urged cable operators to support price increases to guarantee quality programming.

excited about cable."

Dwyer developed kits for the systems that agreed to launch WTCG, including press releases, photos, biographies of Turner, and ad slicks for newspapers.

Turner's first trade ads debuted in 1976. They emphasized growth, boasting that cable operators would be able to sign up thousands of new subscribers by offering the exciting new programming carried by WTCG.

Turner, Dwyer recalled, was deeply involved in all the creative decisions about the trade ads for WTCG, setting all the strategic direction and editing the copy and changing the creative work himself.

At an NCTA convention in the late 1970s, Dwyer's exhibit was next to one set up by the new Madison Square Garden Network. She met the new network's president, Kay Koplovitz, and the two hit it off, with Dwyer signing up to handle MSG's marketing to cable operators. This included developing the overall marketing plans and creating ad campaigns for MSG and its sister service, Calliope.

From then on, Dwyer was deeply involved in the launch and growth of many of the largest cable networks: TNT, CNN, Sci Fi Channel, American Movie Classics, Romance Classics, Cartoon Network and others. Through Dwyer's hard work and that of her husband Art Dwyer, president of CTI, the agency remained the premiere cable marketing company throughout the rest of the century.

Above, Madison Square Garden's Calliope service was an early client of CTI's. Above right, some of TBS's and CNN's trade advertising was showcased in a 1984 CTI campaign. Right, TBS promoted 'The New Leave It To Beaver' in 1986.

Growth

The 1970s ended with a bang. The noise came from an explosion that destroyed a small, navigational motor aboard the Satcom III satellite shortly after it was launched on Dec. 7, 1979. With the motor gone, the satellite drifted off into space, never to be found. With it went the hopes of a dozen cable programming networks that had planned to use the bird to launch new services.

Satcom I, which had carried the satellite signal from HBO since 1976 and which had been the home to a couple of dozen other cable networks, was getting old and unreliable by the end of the decade. Satcom III was expected to provide a much more reliable distribution system for new networks and such existing programmers as Home Box Office and Warner Amex Satellite Entertainment Co., each of which had reserved five transponders on the new bird. The vulnerability of programmers to satellite outages had been demonstrated in September of that year when Satcom II, used for other telecommunications services, veered out of orbit temporarily.

RCA, which owned the Satcom satellites, then moved some of its Satcom II clients over to Satcom I, bumping five cable networks off the air for up to 15 hours while it fixed the trouble on Satcom II. The satellite company was able to do this because some of the cable networks were leasing less-expensive, "preemptible" transponders, and therefore could be bumped off the bird if a client with a "guaranteed" transponder experienced trouble.

The loss of Satcom III was most devastating to a half dozen new programming services that had been scheduled to launch using the new satellite. The most visible and the most vulnerable of them was Cable News Network, the brainchild of Ted Turner.

Turner had been thinking about news for a while, he later recalled, ever since HBO went on the satellite. Of the four major categories of television programming — general entertainment, sports, motion pictures and news — only news was unserved by a cable network by the end of the 1970s.

"There wasn't much news on the lo-

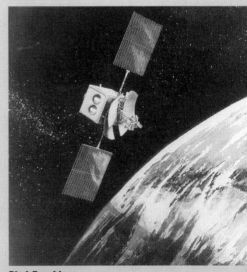

Bird Troubles:
Communications satellites held the key to growth for cable programmers, but the early birds weren't always reliable. Satcom II (above) experienced problems in September 1979, which caused RCA to move several cable networks to another bird. The problem got worse when Satcom III, launched on Dec. 7, 1979, fell into a useless orbit, causing headaches for a number of cable networks including CNN.

cal broadcast channels," Turner recalled, "and I didn't get home in time for the network news." So Turner figured there was a niche. Besides, his SuperStation WTBS (the name had been changed from WTCG) was doing well and he was ready to bet the ranch on something new.

Through Scientific Atlanta president Sid Topol, Turner had met Reese Schonfeld, the managing direcotr of the Independent Television News Association. ITNA, owned by a group of broadcasters, syndicated news to stations around the country for use in their local newscasts. Turner asked Schofeld to start up CNN.

The next task was to round up support among cable operators. Turner asked for a chance to meet with the NCTA board of directors at the Western Show in Anaheim in December 1978. As Terry McGuirk recalled "Ted took the floor, and I distributed contracts to everybody. He said he wanted to start a 24-hour cable news network and he needed 15 cents a sub a month with three million subscribers and wanted everybody in the room to sign the contract committing to carriage."

"They howled and laughed. After they calmed down, they asked if they could think about it for a little while."

Turner, Schonfeld, McGuirk and newly signed affiliate relations vice president Roy Mehlman worked all through the next year, pitching operators, signing up veteran newsman Dan Schorr to be an anchor, contracting for equipment and breaking ground on a new studio. Turner had to pledge not only his company's earnings, but his personal wealth as well to raise the money to get the network under way.

Then Satcom III disappeared.

"Ted was seeing stars," McGuirk recalled. "We were dead."

But McGuirk remembered that there was a clause in the transponder contract between WTCG and RCA that said something to the effect that if an extra transponder ever became available RCA would be required to notify Turner about it before leasing it to anybody else. The wording was vague enough so that the Turner folks felt they could make some kind of case that RCA owed them first crack at any available transponder on Satcom I. It was a thin reed, but the only hope they had. Turner asked for a meeting with the satellite provider.

"Ted," McGuirk later said, "went into his 'Rasputin the Mad Monk' routine."

He yelled and screamed and waved his hands around and promised he would sue RCA, break it up into little pieces and make sure "none of you guys will ever have jobs again."

When one of the RCA lawyers attempted to challenge him, Turner grabbed the fellow by the collar and backed him up against the wall. It was a full-blown performance.

Then Turner unveiled his contract clauses. "They just went white," McGuirk remembered. None of the RCA lawyers had read the old WTCG

Old Hands: Roy Mehlman (left), long-time Turner Broadcasting and Group W executive, joins Betsy and Bob Magness outside The Fort Restaurant in Denver.

contract or knew about the clause. And they were petrified that Turner would sue them. He, after all, had nothing to lose and was clearly not in a mood to be mollified. Turner did file a suit, seeking $34.5 million in damages.

RCA decided to compromise. It agreed to allocate to CNN a Satcom I transponder it had been reserving for its own message traffic. The deal was only good for six months, but it would do.

"It isn't everything we wanted," Turner said at the time. "But we were frozen in terror two weeks ago. Now we're breathing a lot easier."

But the crusade to find subscribers was tougher than they had thought. Many operators were grateful to Turner for launching WTCG on the satellite and wanted to support him. But they weren't sure he could pull off launching a 24-hour news channel, and they didn't want to promise one to their subscribers only to have it go dark after a couple of months. And nobody could quite grasp the concept. Why would anyone want 24 hours of news, they weondered, failing to understand that the concept was that news would be available 24 hours a day so anyone could watch a half-hour or hour any time. It would be like a water faucet, always ready to turn on when you want. That was a radical notion in the TV business in 1980.

The cable operators' biggest concern was that the enterprise would look second-rate next to the glitzy, expensive broadcast network newscasts and that cable would suffer a black eye. It was a risk few of them were willing to take before CNN launched.

Operators owned by companies with broadcast interests also felt heat from their station manager colleagues. At Storer, McGuirk recalled, senior cable

24-Hour News:
The launch of CNN on June 1, 1980, was perhaps Ted Turner's most audacious move, as he single-handedly took on the news organizations at the three broadcast networks.

executives "Jim Hall and Jim Faircloth got their necks chopped off" by the broadcasters in the company when they committed to carry CNN.

When CNN launched in June of 1980, it was carried on systems serving only 1.7 million subscribers, well below the five million Turner had been telling advertisers, and the three million he had told the NCTA board he would need to get the enterprise off the ground. (The cable industry had about 15 million subscribers at the time.) But Turner didn't hesitate. He announced the launch of the network, declaring that the last story it would cover would be the end of the world.

It would seem like the end of the world more than once in the next few years. CNN's initial budget had pegged the first year's loss at $12 million, McGuirk recalled. "Then, two weeks after we launched, it turned out to be $18 million. Then, three months into it we were going to lose $32 million. It was like a runaway train."

Turner was in debt up to his neck, borrowing $200 million to keep the network on the air. He cut every expense he could find. "We were like the Viet Cong," he recalled with a grin. "Living off a handful of rice a day."

Scientific Atlanta's Sid Topol recalled when Turner decided to buy an earth station to transmit CNN. "He called me up and said he wanted to buy an uplink. I said 'Great, send me the specifications and we'll do a design.' And he said, 'No, Sid, I want to buy a dish today. Send your man over right now.'" So Topol sent one of his salesmen over to see Turner who asked how much the entire package would cost. Told it would be about $250,000, Turner wrote out a check and handed it to the startled salesman. On the way out Turner cautioned him "Just don't deposit it until next Monday."

CNN survived. By the end of 1981 it would reach break-even with some 10 million subscribers. As the industry began to build the major urban areas, its prospects for growth were even brighter.

Of the five new services that had planned to launch on the ill-fated Satcom III, only CNN made it.

By taking on the broadcast networks in an area where they had seemed invincible — national and international news — CNN gave the cable industry a legitimacy it had not had before, particularly with more upscale, better-educated viewers and with political and civic leaders.

CNN wasn't the only network to boost cable's image as a provider of information as well as entertainment.

The story of C-SPAN is in some ways even more improbable than the story of CNN, HBO or USA. It was the creation of a trade magazine reporter with no business or TV production experience, financed as a non-profit venture by a group of cutthroat capitalists, and given to the people of America in what has to be the largest public service gift ever bestowed to any country by any industry in history.

Brian Lamb grew up in Lafayette, Ind., about as quintessentially a middle America community as can be found. He became interested in broadcasting when he took a course in television from a popular high school teacher, Bill Fraser. Fraser, Lamb later recalled, "taught me how to interview, taught me how to listen in an interview."

After graduating from Purdue University in 1963, Lamb joined the Navy and served in the Pentagon as public affairs officer, witnessing the coverage of the anti-war demonstrations in Washington in the late 1960s.

CNN Logo:
Its familiar shape has punctuated news events from the Challenger disaster to the massacre at Tianenman Square to the fall of the Berlin Wall and is arguably as recognizable to the world's population as the Coca Cola logo. But CNN's well-known label isn't a corporate icon created by committee after a couple hundred thousand dollars was spent researching colors and shapes. Its saga is much humbler: What's now Communication Trends Inc. created the CNN logo nearly 20 years ago for just $2,800. Graphic artist Guy Bost crafted the now-famous, rounded CNN tag for the launch of the news network in 1980.

He worked for Colorado Sen. Peter Dominick for a couple of years and then moved over to the Office of Telecommunications Policy in the Nixon White House where he got a behind-the-scenes look at how government policy is formulated and received an education in the media, including cable TV.

Lamb found he was irritated continually by the lack of information available to the public about what was happening in Washington. "I kept seeing things with my own eyes and realizing that I wasn't getting this on the nightly news," Lamb later recalled. "I wanted to do something about that."

He took a job in cable trade publishing, first with the Searle brothers and then with *CableVision* magazine where he found an extraordinary supporter in publisher Bob Titsch.

Titsch offered to pay Lamb a full-time salary while Lamb spent part-time working on his idea of a public affairs cable network. Titsch also agreed to help find backers for the project. He called 15 different CEOs at cable operating companies and persuaded each to contribute $1,000 to the project, enough to enable Lamb to buy a video camera and tape machine.

Lamb began to comb the corridors of the House and Senate office buildings, seeking out interviews with members of Congress and then sending tapes of those interviews to the cable systems in that member's district.

It was pretty rudimentary stuff. In the spring of 1977, a group of cable operators calling itself the Cable Satellite Access Entity met in Washington to listen to ideas for new networks to put on the satellite.

Lamb was among 30 invited to pitch the group. After he made his presenta-

Straight Shooter:
Frustrated by how little the public saw of Washington, D.C., policymakers, Brian Lamb set out to open the halls of Congress with C-SPAN and later C-SPAN II.

tion he was approached by Bob Rosencrans, the man who had bought the first satellite dishes to receive HBO and had started what became USA Network.

"We're interested in making this happen," Rosencrans told him. Rosencrans offered Lamb a check for $25,000 in startup money and said he would raise additional funds from his fellow cable operators.

It proved tough going. A month or so later Rosencrans called Lamb and reported sadly, "I can't get a penny out of these guys."

But Lamb kept on plugging. On Oct. 27, 1977, he had scheduled an interview with Rep. Lionel Van Deerlin (D-Calif.), chairman of the House Telecommunications Subcommittee. After the interview Lamb noticed a small TV monitor in Van Deerlin's office. It was used to receive the signal from a security camera that was trained on the House chamber. Van Deerlin could use it to keep an eye on what was happening on the House floor.

Lamb blurted out to Van Deerlin an idea: Put the House of Representatives on a satellite-delivered channel that could be picked up by some 200 cable systems with earth stations around the country.

"Can you write me a speech on that?" Van Deerlin asked. Lamb went back to his office, only to get a call later that afternoon from Van Deerlin. "Brian," he said, "I'm in the cloakroom of the House of Representatives. We're going to vote on this in an hour. I need everything you can get me on those things you mentioned."

Lamb scraped together what he could, and Van Deerlin took the floor of the House to promise his colleagues that if they authorized television in the House, cable would carry their deliberations full time, live and uninterrupted to viewers (and voters) all across the country. By a vote of 325-80 the House adopted a resolution authorizing the Speaker to begin television transmissions of the House proceedings.

With the prospect of televising the House, Rosencrans went back to the industry and found a much more favorable response. Lamb made a pitch at the NCTA board meeting in Boca Raton, Fla., and by the end of the year had signed up 22 cable companies

House Call:
Then House Speaker Jim Wright (D-Tex.) joined cable pioneers Ed Allen, Robert Rosencrans and John Saeman, to commemorate the launch of C-SPAN.

pledging a total of $425,000 to launch the Cable Satellite Public Affairs Network. Rosencrans was chairman of the board and offered to provide time on the satellite transponder used by MSG Network to get the network started. C-SPAN would run during the day and MSG in the evening.

The big hurdle remaining was who would control the cameras in the House. Lamb was hoping that the broadcast networks would not be given the franchise. He figured that if the broadcasters controlled the process they would charge big bucks for anybody wanting to take the feed. The broadcast networks, he reasoned, didn't know how to do anything on the cheap.

Together with NCTA president Bob Schmidt, Lamb paid a call on House Speaker Thomas P. O'Neill (D-Mass.) and his top aide, Gary Hymel. Lamb made the pitch not to allow the broadcast networks to control the televising of the House.

The Speaker decided that the House would maintain control of the cameras and offer the feed for free to anyone who wanted to take it. The decision provided C-SPAN with thousands of hours of programming at no cost.

Sid Topol cut him a deal for S-A equipment and Lamb built an uplink outside Washington in time to launch his service on May 19, 1979.

He was dying to know if anybody was watching. But with no ratings and no direct response advertising (indeed, no advertising of any kind), it was impossible to figure out what the audience might be, or even if there was one.

When the House was not in session C-SPAN needed to find other public affairs events to televise. The first was the Republican Governors' conference in

Austin, Texas, Thanksgiving weekend in 1979.

It was a watershed moment.

"The last event was a speech by Henry Kissinger," Lamb recalled. "I thought it would be interesting to know if we had any viewers. I called down to engineering and told them to put up a message on the screen asking viewers to call our number if they had a comment or a question. We only had three lines, and they all lit up at once and didn't stop ringing for the entire weekend. We were running around like crazy trying to answer them all."

The experience led Lamb to launch the first nationally televised call-in shows. They would transform the network once again, from one that delivered huge amounts of unedited public affairs programming into a kind of electronic town hall, where voters could actually comment on the proceedings.

Ever on the lookout for penny-pinching ideas, Lamb had struck a deal with CloseUp, a non-profit organization that brought students to Washington to meet with government leaders. CloseUp had agreed to buy $170,000 worth of cameras and equipment, and C-SPAN agreed to televise its events.

It would lead to what Lamb later called the most surprising moment he ever had on camera and another measure of the growing importance of the network.

In 1981 the CloseUp group had just met with newly elected President Ronald Reagan, and Lamb was doing a show with some of the students who had talked with the president. He opened the program to calls from viewers.

Midway through the show there came a call identified to Lamb as being from Washington, D.C. Lamb was inclined to put it on hold while taking calls from viewers who were calling long distance. But his staff put up a sign that read "TAKE THIS CALL NOW."

When he answered a voice said, "Mr. Lamb, hold one moment for the President." And a startled Lamb was speaking, live and on air, with the President of the United States who had been watching the show and wanted to clarify a point.

It was one of the few times Lamb showed any emotion on camera. His trademark is a deadpan expression even when listening to the most partisan and passionate political diatribes by guests or callers. He manages to hide whatever political views he has. His interviewing style, learned in high school in Indiana, is the antithesis of the broadcast networks where the interviewer is often more important and better known than the person being interviewed. He listens intently, allows his guests to finish their answers and almost never offers a comment himself.

This careful attention to non-partisanship was a key to the success of C-SPAN, which earned a reputation across the political spectrum as a genuinely even-handed institution.

C-SPAN and CNN broke the stranglehold the three broadcast networks had maintained over the delivery of news for decades, a lock that had forced all news to be compressed into a 24-minute show and delivered once a day at a fixed time.

On CNN viewers could watch the news any time of night and day including in-depth reports on issues the broadcast networks barely touched. On C-SPAN they could witness the entire debate in Congress, the full speech of a political candidate or the entire proceedings of a congressional committee without commentary or interruption. And neither network, at least at the outset, had the kind of media star/anchor who on broadcast television was often treated as more important than the newsmakers.

If information is power, CNN and C-SPAN did more to empower the voters of America than any development since the American Revolution. And together they made it imperative for those interested in public affairs to watch cable, including most public officials, candidates for office, newspaper editors, TV news directors, and citizen activists. Cable wasn't just entertainment any more.

CNN and C-SPAN were only indications of the kind of programming diversity that cable could provide.

By the end of the decade, just four years after HBO had first launched on the satellite, some 27 programming services were being carried by Satcom I.

Pay programmers included HBO, Showtime, The Movie Channel, Home Theater Network and Galavision.

SuperStation WTBS had been joined by Chicago broadcast station WGN, delivered by United Video of Tulsa, Okla.; New York City independent WOR, distributed by Eastern Microwave; and Oakland, Calif., station KTVU, uplinked by Warner.

Sports was covered by ESPN, Madison Square Garden Network, and Total Communications Service (offering Canadian football games).

Religious programmers included Christian Broadcasting Network, PTL (for People that Love) and Trinity Broadcasting Network.

In the news and information arena, C-SPAN and CNN were complemented by the Appalachian Community Services Network, an educational service that was an outgrowth of a federal program to provide educational programming for people living in that mountain region. UPI and Reuters offered alpha-numeric networks with the latest news bulletins.

Children's programming was offered by Warner's Nickelodeon and UA-Columbia's Calliope service.

Southern Satellite Systems, distributor of WTBS, had launched a smorgasbord network, called Satellite Programming Network, that carried a variety of programming from different producers ranging from religious shows to fishing programs. Programmers leased time on the network.

This programming bonanza was attracting more and more Americans to cable. By the end of the 1970s, 14.5 million homes, or 20% of U.S. households, were receiving their signals through a cable. After years of barely scraping by, cable operating companies finally had some financial breathing room. Some were even reporting profits. Cable revenue in 1978 exceeded $1.5 billion, a jump of 25% in single year.

Nowhere was the change more apparent than at TelePrompTer, the nation's largest cable operator with some 1.2 million subscribers, of which some 355,000 were also subscribing to Showtime, in which TPT had acquired a 50% interest.

TelePrompTer had risen from a loss of $9 million on revenues of $81 million in 1974 to post net income after taxes of $14.2 million on revenue of $146 million in 1978. Operating profits in the cable division soared 22% in 1978 to $9.2 million.

In three years the company chopped $50 million off its total debt load, reducing its debt-to-equity ratio to less than 1-to-1. By restructuring its debt in 1978 it was able to shave $4.3 million a year off its interest payments and achieve a long-term credit structure that assured the company of the funds it would need for expansion and construction well into the next decade.

Wall Street also began to catch on to the game. Between November of 1979 and April of 1981 the average price of a share of cable company stock more than doubled.

The hard-pressed hardware suppliers were beginning to see daylight. General Instrument posted sales of $95 million in 1979, up 34% from the previous year, and profits of $9.6 million, up 129%. Scientific Atlanta, building earth stations as fast as it could, went from $250,000 in profits on $16 million in sales in 1972 to $5 million in profit on $94 million in sales in 1979. C-COR, building much of the equipment for Warner Cable's interactive systems, saw its revenues and profits soar.

All of this success was bound to attract attention, and the late 1970s and early 1980s saw the entry into, or expansion of existing cable businesses, by a group of companies with well-known names: General Electric, Westinghouse, Times Mirror, The New York Times, Hearst Publishing, Time Inc., Newhouse and American Express.

Time Inc., which had traded its cable systems for a 9% stake in American Television & Communications Corp. in 1972, gradually upped its stake first to 12%, 15% and then 27%. In 1978 it purchased the entire company in a tax-free exchange of stock. The new company combined ATC's 750,000 subscribers with Time Inc.'s 100,000 Manhattan Cable TV subscribers. It also completed the transformation of Time Inc. from a company where the primary business had been print to one where the bulk of the revenues came from electronic media.

Times Mirror, owner of the *Los Angeles Times* and a group of broadcast television stations, purchased Communication Properties Inc. in January 1979.

The New York Times in 1980 bought systems serving some 55,000 subscribers from Irving Kahn.

Hearst formed a cable division headed by Continental Cablevision's Raymond Joslin to acquire and manage cable systems. Joslin had been one of the first employees of Continental Cablevision. He met Irv Grousbeck when the two were neighbors in Findlay, Ohio, where Grousbeck was looking for franchises and Joslin was handling sales for the Jones & Laughlin steel company.

Joslin ran systems for Continental in Ohio and later in California. At Hearst, he later recalled, his mission was threefold: to build a cable operating company with some 500,000 subscribers, primarily through acquisitions; to translate the Hearst magazines into television programming; and to develop an

Cable Man:
Hearst Corp. tapped Continental Cablevision executive Ray Joslin to run its cable division.

electronic publishing strategy for the company.

Hearst CEO Frank Bennack was a friend of ABC chairman Leonard Goldenson, who by 1980 was also seeking ways to get his company into the cable business. The two suggested that their cable deputies, Joslin for Hearst and Herb Granath for ABC, get together. ABC had been working on a plan for a cultural network, to be called the Alpha Repertory Television Service (ARTS). Hearst, meanwhile, had been doing some preliminary work on a service, to be called Daytime, that would make use of some of the same material covered by the Hearst magazines. The two companies joined to form Hearst-ABC Video Enterprises. By the middle of the decade ARTS had become the A&E Network and Daytime had merged with Cable Health Network, started for Viacom by Jeff Reiss and Dr. Art Ulene, to become Lifetime.

By the mid-1980s, Joslin had achieved his goal to create an MSO with over 500,000 subscribers.

Healthy TV:
Terrence Elkes joins Jeff Reiss and Dr. Art Ulene at the launch of the Cable Health Network, which later merged with Daytime to form Lifetime.

Other media companies also were building their cable portfolios. In February 1980, Newhouse, through its subsidiary Metrovision, started by ex-Cox Cable CEO Henry Harris, purchased systems from Daniels Properties serving more than 116,000 subscribers for a price that *TV Digest* estimated at between $80 million and $100 million.

Getty Oil Corp., in March 1979, paid $10 million for 85% of the Entertainment & Sports Programming Network, a fledgling network started to carry games of Connecticut colleges.

In late 1981 Tribune Co., publishers of *The Chicago Tribune* and owners of a string of broadcast television stations, formed an alliance with United Cable Television to bid for franchises in Sacramento, Calif., and Montgomery County, Md.

Dow Jones & Co., publishers of *The Wall Street Journal*, purchased a 24.5% stake in Continental Cablevision in October 1981 for $80 million. The move came after Dow Jones and its partner Knight Ridder had lost an attempt to purchase UA Columbia Cablevision and its 450,000 subscribers.

The joint venture lost out to a veteran of the cable business in Canada, Rogers Cablesystems Inc. Rogers bid $152 million for a 51% share of UA Columbia.

Rogers, chaired by Ted Rogers, the 48-year-old son of a broadcast television pioneer, was the largest cable operator in Canada, with some 1.2 million subscribers. But Canada had imposed severe limits on the industry, prohibiting pay television, for example. Rogers in 1980 posted profits of $1.4 million on revenue of $70 million from 1.2 million subscribers while UA Columbia posted profits of $4.8 million on $55 million in revenues from only

400,000 subscribers. UA's profits were more than triple Rogers' with only about one third of the customers.

Rogers and other Canadian companies had been casting their glances south for several years. Limited by the government in terms of national programming, the Canadian operators had become masters at local programming. When they entered the franchising battles in the U.S., the Canadians were able to point to innovative and long-standing local programming ventures at their home systems that could not be matched by U.S. cable companies.

By the time of the UA purchase, Rogers was already operating systems in such U.S. cities as Syracuse, N.Y., Orange County, Calif., and Portland, Ore.

Its successful entry into the franchising fray ignited a debate over whether U.S. cable companies should be owned and operated by companies based outside the country. Various members of Congress introduced legislation to ban foreign ownership of cable systems, and in such cities as Minneapolis, the notion of a non-American owning the cable system cost Rogers the franchise. But overall the company did very well in the U.S., hiring to run the American operations former Continental Cablevision system manager Bob Clasen.

American Express Co., in October 1979, purchased a 50% interest in Warner Cable for $175 million. At the Warner-American Express Co. joint venture, the duties were split into two parts. The programming section, Warner Amex Satellite Entertainment Co. (WASEC), was put under the management of former CBS executive Jack Schneider. The cable operating division was left to Gus Hauser to run. It was called Warner Amex Cable Communications.

Sports Beat:
Bill Rasmussen (top) hired former NBC executive Chet Simmons (bottom) to launch and run ESPN, one of cable's key programming networks.

American Express brought more to the table than its huge financial resources. Steve Ross, who had been tainted by allegations of wrongdoing in connection with a theater operation in Westchester County, N.Y., was looking for a partner that would be a "real white shoe with an unimpeachable reputation," as Hauser later recalled. Such an alliance

Second Generation:
Ted Rogers took his father's Canadian television company and expanded into cable in both the Canadian and U.S. markets.

Cable Connection:
Although GE's merger with Cox Cable fell through, GE sent Robert Wright (right) to run the MSO in the early 1980s. At left is Art Dwyer, Wright's marketing chief at Cox at the time.

would prove particularly valuable in the struggle for big city franchises.

American Express, on the other hand, was looking for a way to diversify and figured its entry into cable TV would give it a head start in the era of electronic commerce that was fast arriving.

Also in October, 1979, General Electric Co., signed to purchase the broadcasting and cable properties owned by Cox Communications.

As the new corporate entities came into the business, they brought with them a new management style. Although the Cox-GE merger was derailed when, after months of regulatory delays, Cox insisted on a sweeter price, GE did send a new management team in to run the company for the period between the announcement of the deal and its collapse. Heading the team was an executive from GE Capital Corp. and a protégé of Jack Welch, Bob Wright.

Wright brought with him another GE executive, Arthur Dwyer, a Massachusetts native with degrees in journalism and business from the University of Missouri.

When the Cox-GE deal fell apart, Dwyer stayed and brought to Cox some of the management techniques that were transforming GE.

"When I got there every system was doing its own purchasing," Dwyer recalled. "All the control was at the system level. We had 100 different deals for trucks."

All the programming decisions were made at the local level as well. In doing a deal with a programmer, Cox could not guarantee how many subscribers it would deliver. And the programmers had to deal with each Cox system manager individually to gain carriage.

"I remember one day getting a call at 7 a.m. from Ted Turner telling me that San Diego had just taken TBS off the air," Dwyer recalled. "That cut it."

Cox consolidated all its purchasing at corporate headquarters, assuming control of the major contracts for equipment and programming. The result gave the company far greater clout with its suppliers, gave the suppliers a chance to make a single sale rather than dozens and dramatically reduced costs for Cox.

It was only one example of the kind of corporate management techniques that were to be instituted in the cable companies as they matured beyond their entrepreneurial phases.

In October, 1980, Westinghouse Electric Co. announced the biggest deal in cable history: the purchase of TelePrompTer Corp., for $646 million in cash and the assumption of $200 million in debt, or about $650 each for TelePrompTer's 1.3 million subscribers. Westinghouse put the cable operations under the management of its Group W Broadcasting subsidiary headed by Daniel Ritchie. Group W already owned eight cable systems with about 50,000 subscribers.

The entry of so many huge corporations into the cable business guaranteed the industry the funding it would need to wire the nation's major cities and to upgrade the thousands of older, small-town systems, many of which had only 12-channel capacity. TelePrompTer alone had estimated its need for capital at $600 million in 1979 to meet its newbuild and upgrade obligations.

The cable industry also got another boost in the late 1970s and early 1980s from the folks in Washington, D.C. The thaw in the hostile relations between cable and the government had begun in 1972 when the FCC, at the prompting of Cable Bureau chief Sol Schildhause and with the political support of the Nixon White House, had relaxed the 1968 ban on distant signal importation, allowing limited use of distant signals in most cable systems. This is what made it possible for Ted Turner to launch WTBS.

The FCC action was followed by a series of legislative and judicial victories for cable, including passage of the copyright and pole attachment legislation. The court victories included the Midwest Video case, which struck down an FCC requirement that all cable systems provide original programming. The court also ruled against the FCC regulations limiting the types of films that could be aired by pay television networks.

But in 1980 the FCC provided the industry with an even more sweeping victory.

President Jimmy Carter had installed as his FCC chairman a dyed-in-the-wool Democrat, former Secretary to the Senate Charles Ferris. But Ferris had proven to be a very different breed from the Democrats who had dominated the Commission under Presidents Kennedy and Johnson and who continued to be influential through much of the Nixon Administration. Ferris was more inclined to allow competition to resolve the issues between cable and broadcast than to attempt to dictate the relationship from the FCC headquarters on M Street.

Since the 1960s the FCC had been regulating cable on the assumption that without regulation — particularly limits on the importation of distant signals — cable would prove economically harmful to local broadcast stations. But there had never been any definitive studies to determine if this premise was correct. In 1979 Ferris ordered the FCC staff to conduct such an inquiry to find out if this fundamental underpinning of the FCC regulatory structure had any basis in fact.

The study was completed in 1980 and concluded that there was no rea-

Carter Man: *FCC chairman Charles Ferris handed the cable industry a victory in the late 1970s by pushing through new rules that relaxed restrictions on importing distant signals.*

son to believe that the importing of distant signals would harm broadcast stations in anything more than the most minimal way.

Another main reason for the FCC regulations, the desire to protect copyright holders, had been made moot by the passage of the Copyright Act, the study found. Both the broadcasting and cable industries were healthy and growing and the FCC did not need to protect one from the other, the report found.

The study provided the basis for an FCC vote on whether to eliminate the syndicated exclusivity and distant signal limitation rules. The former had required that cable systems black out TV shows carried on distant broadcast signals if they duplicated programming aired by local stations. The latter rule limited the number of distant broadcast signals any cable system could import into its community.

Broadcasters lobbied hard for the regulations, but the commission voted 4-3 to support chairman Ferris' contention that "these rules lack the slightest hint of justification."

While the vote was a huge victory for cable, the industry had in fact moved far beyond the stage where it depended on the importation of distant broadcast signals for its livelihood. Most cable systems were carrying as many satellite-delivered, cable-exclusive networks as they could and had little room to add distant broadcast signals, which carried copyright as well as distribution fees.

But the FCC decision was the signal that the era of regulation was finally at an end and that a new breed of Democrats would be prepared to lift regulation even more in the coming decade.

The hardware suppliers meanwhile were busy concocting new equipment that would significantly expand the programming and other services the cable systems could offer. Motorola in 1979 introduced a 53-channel amplifier priced at $30, the same as the older 35-channel model. Jerrold unveiled its 400 MHz system at the 1979 Western Show.

Then manufacturers began to roll out addressable converters.

The concept behind addressability was that when a subscriber wanted to change services or order a pay-per-view event the cable operator could make the change by sending an electronic signal to the converter in the subscriber's home rather than by having to send out a company employee to reconfigure the converter on site.

Dallas-based equipment manufacturer TOCOM offered a 55-channel addressable converter at the NCTA convention in 1980 that would sell for $150-$200.

Scientific-Atlanta, which had begun making antennas and then expanded down the cable to the headend and distribution system, unveiled its entry into the addressable converter market. C-COR offered an addressable converter in the fall of 1981.

Veteran cable engineer Archer Taylor said the technology had the potential to be as revolutionary as the satellite delivery of signals. Taylor told *TV Digest* that his only reservation was "whether the equipment can work as well as proposed when produced in mass quantities." It was a legitimate concern.

The addressable systems enabled cable operators to offer all kinds of new services. Among them were such exotic services as electronic reading of gas and electric meters in customers' homes, banking and shopping at home and, most exciting in the early 1980s, home security.

With new capital, plenty of attractive programming, exciting new services, a favorable regulatory climate and new technology, cable was ready by the late 1970s to launch its version of the Normandy Invasion: an effort to franchise and build the nation's largest urban markets.

In the space of six years, from 1978-1983, the following major cities awarded cable franchises: Boston, Chicago, Cincinnati, Detroit, Dallas, Denver, Houston, Indianapolis, Kansas City, Miami, Milwaukee, Minneapolis, New Orleans, Omaha, Pittsburgh, the outer boroughs of New York City, Tampa, and Washington D.C. Thousands of other communities, particularly the lucrative suburbs of the major urban centers, awarded franchises at the same time. And countless other towns with existing cable systems watched the process, anxious to upgrade to the most advanced systems as soon as their franchises expired.

Franchising in itself became a big business. Expenses just to bid for a franchise in a major city could exceed $1 million. Major cable operating companies hired teams of franchisers — many of them with good political credentials — to set up camp in a city and work it until the award was made.

A city typically would appoint an employee to oversee the process and then hire a consulting group, such as the Cable Television Information Center, to advise it on the franchising process. CTIC and other similar consultants had been involved in the franchising process in other cities and were adept at ensuring their clients received the most advantageous bids possible from each bidder.

The city would then issue a request for proposals inviting companies to bid. The RFP would outline the city require-

Trenching:
The late 1970s and early 1980s saw intense cable construction in major cities like Milwaukee, as MSOs wired large urban areas.

ments covering such issues as the expected number of channels, the number that would be devoted to government and public access, schools and other purposes. It might also suggest a level of acceptable rates and ask for an upfront payment of franchise fees (which had been capped at 5% by the FCC and which were supposed to be used entirely to pay for city regulation of cable). The RFP might also suggest that the bidders involve local citizens in planning the system.

In most cities the bidding companies would form alliances with politically influential local citizens, offering them a share in the system for little or no money down. Many of these locals stood the chance to make millions of dollars if the franchise were awarded to the company they backed. These individuals became known as "rent-a-citizens." Because many of them were related or connected to members of the city councils that made the franchising decision, their involvement was depicted in local press reports as little short of bribery.

In fact, however, most cable franchis-

Cable Construction:
Local government and cable officials in Dallas inaugurate system construction in the early 1980s.

ing personnel were very careful not to cross the line into anything that was illegal. The memory of Irving Kahn going to jail in the Johnstown, Pa., franchise case was still fresh. And most of the politicians who ran the franchising operations understood that if they were ever caught trying to do something illegal they would ruin their chances of winning franchises in other cities. Besides, most of them had plenty of more sophisticated ways of influencing the process aside from the old-fashioned cash in an envelope.

Cable companies used the press to play up their own strengths and point out the weaknesses of the opposing bids. In Denver, for example, the franchising process pitted three bidders: United Cable Television Corp., which operated in the suburbs of the city; an alliance of ATC and Daniels & Associates; and TelePrompTer Corp. The bidding was especially vicious because Daniels, United and ATC were all headquartered in Denver. The CEOs all knew each other (Bill Daniels and United CEO Gene Schneider had built the Casper, Wyo., system together in 1953). The contest took on personal as well as business overtones.

The Daniels/ATC bid had included the promise that the system would carry the fledgling Playboy Channel, sure to draw fire from the city's more puritanical crowd. One of the two rivals ginned up press accounts about the controversy, which was neatly defused when Bill Daniels appeared before the city council to announce that yes, his system would carry Playboy, he was damned proud of it, nobody who didn't subscribe would receive it, there would be security to ensure no children could watch and if anybody didn't like it they could just vote against him, but he wasn't going to back down.

It was a classic Daniels performance, and the attempt to damage his bid backfired.

Then, just before the franchise was awarded, the Daniels/ATC team turned the press on to another story. United Cable had said in its bid that it could get a jumpstart on construction of the Denver system because it had already wired the suburbs and could simply extend its wires into Denver.

But on the weekend before the council vote, the Daniels team took the press on a tour of the city limits, pointing out that in many areas the United wires were nowhere close to the Denver borders. United had no chance to reply to the negative stories. Daniels/ATC won the franchise.

Without a doubt the most successful franchising operation in the nation was that of Warner Cable, later Warner Amex. To run the franchising operation Warner Cable president Gus Hauser and his boss, Steve Ross, hired Richard Aurelio, an ideally qualified person for the job.

Aurelio, the son of Italian immigrants, grew up in Rhode Island and Massachusetts. His father was a policeman and involved in local politics for as long as the younger Aurelio could remember.

Aurelio graduated with a degree in journalism from Boston University and after a stint in the Army in World War II and Korea went to work as a reporter for *Newsday* on Long Island. He covered the police beat, local and state politics and in 1960 became the administrative assistant to New York Senator Jacob Javits, running two statewide campaigns for the Senator.

In 1968 he left Javits to work for John Lindsay, then the Republican mayor of New York City. Aurelio engineered Lindsay's upset bid for reelection in which he won the office as a candidate of the Liberal Party after losing the race for the Republican nomination. Aurelio served as New York City deputy mayor under Lindsay and was also manager of Lindsay's unsuccessful campaign for the Democratic presidential nomination in 1972, an experience that took him into politics around the country.

Aurelio's background in politics in the nation's largest city as well as the nationwide contacts he had made through Sen. Javits and the Lindsay presidential campaign gave him a great base to run the Warner franchising operation, which was, above anything else, a political adventure. He also looked the part of the local pol: gravely voice, fair drinker, back slapper, workaholic.

With the blessing of Hauser and Ross, he set up a nationwide operation, masterminded by a 30-person "boiler room" in New York. At any given time Warner would have as many as 100 franchise applications pending in 30 different states. The company bid for most of the major cities: Pittsburgh, Dallas, Houston, Milwaukee, Boston. By the end Aurelio's record was 82 wins and 30 losses.

At times the very size of the franchising operation could be a problem for Warner. Aurelio recalled he needed to study hard to remember which cities had been promised which services and the names and backgrounds of the hundreds of local politicians with whom he was dealing. On one occasion the boiler room messed up and put the numbers for the Dallas system into the bid for Omaha and vice versa. Because a company could not change its bid once it had been submitted, Warner was forced to find justifications when the consultants asked about the cockeyed numbers. But none of the consultants or city officials ever figured out what had happened. Warner lost in Omaha but won in Dallas.

Warner made full use of its well-publicized QUBE operation in Columbus, shuttling officials from cities with franchises up for grabs into the Ohio city to view the cable system of the future. It also made the most of its Hollywood connections, bringing movie premieres and closed-circuit boxing matches and other entertainment events into the towns

Pol: Richard Aurelio used his extensive political experience to help snare cable franchises for Warner Amex Cable Communications throughout the 1970s and 1980s.

Changing Hands:
Warner Amex's Drew Lewis, and TCI's John Sie and John Malone share a lighter moment at a Pittsburgh council meeting where TCI's purchase of the Warner's system was discussed.

where Warner was seeking a franchise.

Typically the company would hire a local coordinator to run the franchise and bring in support from Warner consultants including lawyers, public relations experts and political operatives. Above all they worked like dogs: Saturdays, Sundays, holidays, nights.

In Pittsburgh, which Aurelio remembered as the first big "Super Bowl" of the urban franchise era, American Television & Communications offered free basic service and locked up support from many of the city's leading businessmen and opinion makers. But Warner ran a grass-roots effort, giving away 20% ownership of the system to a coalition of 17 community groups that included the National Association for the Advancement of Colored People, the United Negro College Fund and the Pittsburgh Urban League. Competing against TelePrompTer, ATC and TCI, Warner won on a vote of 8-1.

Warner didn't bid in every city. It passed on New Orleans and St. Louis, because, Aurelio later recalled, the politics weren't right. And it pulled out of Tucson, Ariz., when the city insisted that the franchise winner allow the city to buy the system at any time for its "book value."

But in most areas Warner forged ahead, offering more and more services at lower and lower prices. In Pittsburgh it had proposed a 60-channel system with basic service priced at $5.35 per month for 13 channels. By the time it came to bid for Dallas a year later, Warner won with a bid of a 100-plus-channel, dual-cable, interactive system with 24 channels for $2.95 a month, 48 channels for $7.50 and 80 channels for $9.95. The system also promised to provide 52 channels for an institutional network linking the city's schools and government agencies. Hauser pledged it would be the most advanced cable system in the world.

As Warner-Amex racked up more and more wins and took on greater and greater obligations for construction, the American Express management began to get cold feet. They voiced ever louder concern that Warner-Amex had overbid and would be financially overextended. At the urging of American Express, Hauser and Aurelio adopted a more "realistic" approach when they bid for Boston, only to lose the franchise to Charles Dolan's Cablevision Systems Corp., which offered a basic service for free. (In Boston the franchising decision was made by the mayor, Kevin White. Just before the decision was to be made, Cablevision sent out to all Boston voters a brochure promoting its system. It looked like a piece of campaign literature. On the cover was a photo of the Mayor with the quote, "'Boston will have the finest cable system in the

World' — Kevin White.")

The TelePrompTer franchising operation was similar to Warner's. To run it, Bill Bresnan and Russell Karp hired Pennsylvania's former Secretary of Commerce, Norval Reece. Reece had been a top aide to Pennsylvania Governor Milton Shapp, the founder of Jerrold Electronics.

Starting in 1979, TelePrompTer developed a strategy designed to double the number of homes it passed in five years, with particular attention to those areas where TelePrompTer already had systems. It achieved its goal in three.

Reece found he had to wage a continual battle within the company against the experienced operators who didn't believe that dual-cable, hundred-plus channel systems would ever be able to make money. "They felt I was giving away the store everywhere," Reece recalled. The company adopted a standard that if the system could not achieve a 15% internal rate of return on the investment it would not bid. And TPT did pass on some contests.

The engineers, Reece recalled, were particularly suspicious of dual-cable systems, fearing interference and problems with electronics. Many of the dual systems, he later guessed, were built with only one active cable. The other "shadow" cable was strung but never activated. By the time some of these dual-cable systems needed the extra capacity of a second cable, fiber optics had come along.

Sometimes the local politics could get surprisingly complex. In St. Paul, TelePrompTer hired as its local coordinator the well-respected former president of the city council. After a couple of months the fellow, Bob Sylvester, simply disappeared. His wife couldn't find him, he didn't answer his phone or show up at his office. Reece figured he would never come back.

Then, a couple of months later he showed up — transformed by the magic of modern surgery into Kimberley Simpson. TelePrompTer wasn't quite sure if the former he, now a she, would continue on the franchising operation.

When they held a meeting of the local partners a few days later there was a knock on the door and there stood Kimberley. After a couple of minutes of silence TelePrompTer president Bill Bresnan got up, escorted Ms. Simpson to the table and held the chair for her as she sat down.

TelePrompTer lost the franchise.

Not every cable company entered the franchising frenzy to the same degree as Warner Amex, TelePrompTer and ATC.

Continental Cablevision, Comcast, United Cable and other MSOs whose pockets were not quite so deep picked their targets selectively and concentrated on a fewer number of bids.

And Alan Gerry at Cablevision Industries had success with his tactic of focusing only on smaller communities where he could make the argument that his company would give them special attention that bigger MSOs, preoccupied with building the big cities, could not.

For each franchise application Gerry would produce a 10- to 12- minute video with a section about Cablevision Industries and a part showing a town where Cablevision operated, which was similar to the community considering a franchise. The video would include interviews with the mayor, with customers and with leading citizens, all praising Cablevision.

And Gerry himself would spend weeks in the town. "They wanted to meet the man," Gerry later recalled, "not some suit." By playing up his down-home, small-town roots, Gerry was able to beat some of the bigger companies. "The biggest mistake they made," he later recalled, "was to tell the towns how big they were. We were able to convince people that even though we were smaller, we had a commitment they (the larger companies) didn't have."

Tele-Communications Inc., meanwhile, tried a few franchising forays — most notably in Pittsburgh. But it largely remained on the sidelines as franchising battles in such communities as Dallas, Cincinnati, Minneapolis and Denver upped the ante to levels which TCI president John Malone felt were unacceptable.

"We just didn't want to join the liar's club," Malone would later recall about his decision to forego competition for many of the larger urban franchises.

Instead TCI embarked on a strategy of buying up smaller systems, many of them classic, independent operations. The owners had in most cases long since paid off the debt of building the system and had used up the investment tax credits and other tax benefits that came with building a system.

Malone had earned his degree in industrial management where a key tenet of the business was economies of scale: once a factory was built the more widgets it churned out, the cheaper each one would be. He adapted this philosophy when building his cable company.

Typically, he later recalled, TCI was able to structure a deal so that it paid 10% down and then paid the rest out on an installment plan. (All the deals were done for cash since TCI's stock was still worth so little).

"If we could improve the cash flow of the system enough, we could generate enough additional cash to pay the installments," he later said.

TCI had plenty of ways to improve cash flow. Because of its size it could command lower prices for equipment and programming than smaller operators. And as it found systems in adjacent areas it could consolidate operations, combining several systems under a single manager and engineer. Many tiny TCI systems were left to run on automatic with a technician checking in at the deserted headend every few weeks just to make sure things hadn't collapsed. If there was a problem, the subscribers would let the manager know.

And TCI could upgrade the system to add such services as pay television, boosting revenue and cash flow.

Finally, when TCI purchased the systems it would trigger the depreciation allowance, allowing the company to write off the cost of the system on its taxes.

The tax breaks, economies of scale and the tight-fisted management led by chief operating officer JC Sparkman enabled TCI to squeeze enough fresh cash flow out of the newly acquired systems to cover the installment payments. And the additional subscribers provided additional leverage to borrow more money to pay the down payments on even more systems.

TCI had little competition in this endeavor. Most of the other major cable companies were so focused on winning and building the big cities that they paid little attention as TCI gobbled up a system a week during the late 1970s and early 1980s. While Westinghouse was paying $650 a subscriber for the TelePrompTer systems, TCI was buying smaller classic systems for $300 or $400 a sub.

The growth was phenomenal. TCI went from just under 600,000 subscribers in 1978 to more than 1.6 million subscribers by the start of 1982. Revenue during the same period increased to $180 million from $63 million.

But TCI never during this time reported any profits. "The scheme of this company is to go contrary to what they teach in business school which is to earn money, pay taxes, pay dividends and then have your shareholders pay 70% taxes on them," Malone told *Business Week* in 1981. "We think that's stupid."

To ensure the company was not vulnerable to a hostile bid, he created a new class B Stock which had 10 votes per share versus the single vote per class A share. Management was able to exchange class A shares for class B, ensuring that insiders retained control of the company. When a deal proved too large to fund alone Malone brought in partners, including publishing/broadcasting companies Taft and Knight-Ridder, and formed joint ventures to buy systems.

To help him out in managing so many deals, Malone brought on board a group of new employees. Among them in 1981 was a recent graduate of Harvard Business School who had offered to work for TCI for free for six months just to prove his worth.

Peter Barton was born in Washington, D.C., and graduated from Columbia University.

After college he traveled out west, working as a ski instructor and later earned a few bucks as a professional freestyle skier, jumping off cliffs and doing acrobatics while wearing Hexcel skis. He even took a job for a few months as a craps dealer in a Las Vegas casino.

He came back east and got involved in politics, founding his own political consulting firm, Partisan Artisan. He worked on Hugh Carey's successful campaign for governor of New York and stayed on to serve as a top aide to Carey, including a stint as deputy secretary of agriculture where he became an expert on potato diseases and helped write the Farm Winery Act of 1975.

Figuring he would never make any money in politics, he applied to Harvard Business School. As he approached graduation in 1982, Barton decided to look for work in places he wanted to live. San Francisco, Boston and Denver were at the top of the list. He decided he wanted to work for entrepreneurial companies headed by "somebody who was smarter than I am" and be involved in major transactions.

He wrote to dozens of companies, offering to work for free for three months and then to leave if the deal didn't work out with no financial obli-

TCI Triumverate:
John Malone, Peter Barton and John Sie formed a formidable management team at TCI and Liberty Media in the 1980s and 1990s.

gation by the employer. His resume listed his work as a daredevil skier, craps dealer and his Harvard Business School degree. He got a call from, among others, John Malone, who started the conversation by telling Barton, "I hate guys from Harvard."

"That's okay," Barton replied, because he didn't want to work for any company with a lot of Harvard MBAs in it anyway.

For the job interview, Malone invited Barton to sit in on a discussion about a complex deal for cable systems in which the sellers were proposing some tricky and maybe not entirely above-board schemes. When the meeting broke up Malone asked him what he thought. Barton said it seemed to him that what was being discussed was illegal. Malone laughed and agreed, and the two hit it off.

The interview with TCI chairman and founder Bob Magness followed. The two had lunch and talked about horse diseases and the difference between raising milk cows and beef cattle (Magness was a major rancher, and Barton was drawing on his experience as New York's deputy secretary of agriculture). Magness downed first one and then a second double martini. Barton, despite being a little woozy from the altitude, tried to keep up.

They went back to TCI's spartan, threadbare office. "Well, old Pete," Magness said, coining the nickname he would call Barton for the rest of his life, "Let's talk about you." And the chairman of TCI leaned back, propped his cowboy boots up on his desk and hauled out a huge cigar. Barton, allergic to smoke and a little under the weather with the drinks and the lack of oxygen, was horrified. What happened next horrified him even more.

Magness proceeded to take his trademark bite out of the cigar, chew and swallow it while listing to Barton's story.

"I thought to myself 'This is the toughest man I have ever met.' He had me," Barton later recalled.

A few days later Barton got a call from Malone inviting him to come to work for TCI. No mention was made of title, duties or salary and nothing in writing ever appeared. When Barton was driving out to Denver with all his belongings in the back of his car, he suddenly got a sinking feeling that maybe he hadn't heard things right.

He pulled into a truck stop and called Malone. "I got the impression he had forgotten he had hired me," Barton later recalled. "I had a little moment of disquiet at that Phillips 66 in Iowa City." But Malone then decided to remember and told him to come ahead.

When he got to Denver, Malone was about to leave for his annual three-month summer vacation in Maine (he drives there every year since his wife does not like to fly). Barton had no place to sit, so he took Malone's office and learned the ropes from Sparkman and Donne Fisher, the company's chief financial officer.

Barton cut his teeth on deals for smaller systems. But it wouldn't be long before the winners of some of the major urban franchises would get cold feet and the TCI gang would be buying up some much bigger properties.

While Barton was adapting to the TCI style of Wild West capitalism, some of the entrepreneurs who had started the cable business were chafing under the rules of much bigger, East Coast style corporate structures.

Monroe Rifkin, who had founded and served as CEO for ATC, spent two years working for Time Inc., but left thereafter to start his own cable operating company, Rifkin & Associates. Rifkin later expressed admiration for the executives of Time Inc. and said they had treated him very well, allowing him latitude to run the company and providing such amenities as a corporate jet. But he missed the old days and the old ways.

"Suddenly everything was meetings," he later recalled. "I was never very big on meetings. We were an entrepreneurial company. After a while it was starting to wear a little heavy."

Bill Bresnan had the same experience under the new regime at Group W and left in 1984, like Rifkin forming his own cable operating company with his own name on the door. So did Ben Conroy, one of the founders of Communications Properties Inc., the company sold to Times Mirror. After the sale Conroy formed his own cable company, Conroy Management Systems, and began to build systems in small towns in Texas. And Robert Hughes, another key executive of CPI, founded Prime Cable.

Rosencrans also found life difficult under a new ownership structure in which Rogers had a 51% interest and United Artists Theaters 49%.

"Once you own your own business it's not easy to be subordinate and get told, 'You just take care of management, we'll take care of ownership.' I could never get that through my head," he recalled.

Rogers and United Artists Theaters found they could also not get along and divided up the assets. UA divided the cable systems into two groups with roughly equal value and Rogers got to pick which half it wanted. Then UA formed a partnership with General Electric, which had been shut out of the Cox deal.

But the new owners insisted on a much more centralized operation than Rosencrans had run. And they resisted his efforts to allow company employees to share in the success of the firm, through a stock ownership plan for example. (Top UA-Columbia employees such as Marvin Jones had been given a financial stake in the company and were able to walk away with hundreds of thousands, or even millions of dollars when it was sold.)

In the end Rosencrans, one of the first to take a cable company public, the founder of USA Network and C-SPAN, the first cable operator to use satellite transmissions of programming and by anyone's account one of the classiest, most honest and most progressive men in the cable industry, was summarily fired by United Artists.

"It turned out to be the best thing that ever happened for me," Rosencrans later recalled. Together with his longtime partner, Ken Gunter, and other veterans of UA-Columbia, Rosencrans formed his own cable operating company which by the mid-1990s had well over a quarter of a million subscribers.

Moving On: *Monroe Rifkin was one of several cable presidents who left the big corporation life behind to launch their own MSOs.*

They called it Columbia International.

Another MSO chief who experienced difficulty operating in a new corporate culture was Warner Amex Cable president Gus Hauser.

Hauser found himself caught between the riverboat gambler style of Warner chairman Steve Ross and the Byzantine corporate politics of the much more conservative American Express.

American Express had entered the business with almost no experience in

The CAB: Courting Madison Ave.

As advertising-supported cable networks grew in number, they realized the need for a united front to tell cable's story to the big advertising agencies and advertisers collectively known as Madison Ave.

Cable operators also started to see that selling local advertising on national cable networks could provide systems with an additional source of revenue, one which would not require any increases in the bills cable customers were paying.

Broadcasters had an organization to pitch TV advertising, the TV Bureau of Advertising, and radio had the Radio Advertising Bureau, which had launched an enormously successful campaign built around the slogan "Radio is Red Hot," to revive interest in advertising on radio.

The idea to launch a similar organization for cable grew out of the Cable Television Administration & Marketing Society whose members were increasingly concerned about how to boost cable ad revenues. In 1980 a group of CTAM members launched a search, headed by former Cox Cable executive Leonard Reinsch, to find the right person to head up a Cabletelevision Advertising Bureau. The search settled on the number-two executive at the RAB, Robert Alter.

Alter, a native of Martha's Vineyard in Massachusetts, graduated from the University of Iowa in communications and after a stint in the Army in Korea went to work for the Radio Advertising Bureau in the late 1950s. He served as the group's executive vice president from 1958 until 1980.

Alter was impressed with the cable industry, still in its infancy stages when it came to advertising, and saw a chance to build the business from the ground up.

"It was an era when people were demanding more choice," he later recalled. "Consumer product manufacturers were getting into line extensions — Coke spawned Diet Coke and Sprite and other spinoffs. But when it came to what people consumed more of than anything else, TV, they had no choice."

Alter's first task was to raise the funding needed to launch the new organization. He held a meeting in Denver to pitch cable operators on the need for an advertising-focused organization. "I made the pitch on how advertising worked, shares, ratings. One of the guys asked, 'What's a rating, what's a share?' They had no concept at all how big a business advertising was."

TCI president John Malone noted that TCI's dues for the CAB would be more than the company earned from local advertising each year. But he signed up.

Ted Turner sponsored a reception in New York for the ad-supported networks, and Alter recalled he "screamed up and down and passed the hat."

One of the biggest problems the new organization faced was how to prove cable had an audience. A.C. Nielsen used diaries to measure audiences, and in a cable system with 24 or 35 or more channels family members often couldn't remember which

television or entertainment, Hauser later recalled, "They didn't understand anything about cable. Zero."

Within American Express the cable investment became a football in the power struggle between company chairman Jim Robinson and Sanford Weill, president of Amex subsidiary Shearson Lehman. Weill, Hauser later recalled, launched an effort to undermine confidence in Robinson's decision, including planting stories in the press about how channel they were watching. Also, in multiple-set homes the diary might be filled out by the head of the household, underreporting what the other family members were watching on other sets.

Alter set about to call on the major ad agencies and TV advertisers, asking them to try a schedule on cable just to learn about the new medium. "I told them to consider it R&D, to get in on the ground floor. I encouraged the agencies to appoint a cable expert. I invited them to be speakers at our meetings, to help us bake the cake. They understood this was going to happen and they needed to be prepared to answer questions from their clients."

Gradually he began to make progress. The agencies saw cable as a chance to experiment with ads targeted to selected audiences. Cable could also help give advertisers some leverage over the broadcasters.

Alter developed the notion of two universes. Cabled America, he said, consisted of only about 30% of all TV homes. But, according to Alter's research, those homes consumed 40% of the products sold in the country, and in some categories as much as 60% to 70%. Cable subscribers had higher incomes, better educations and more disposable income than non-cable homes.

The campaign against diaries began to pay off. A CAB-sponsored phone survey found that in diary homes cable was underreported by as much as 50%. Nielsen began to rely more and more on meters,

Radio Man:
The cable industry tapped Radio Advertising Bureau executive Robert Alter to launch and run the Cabletelevision Advertising Bureau in 1980.

which automatically registered what a TV set was tuned to, and to hook them up to the second and third sets in each home.

Alter arranged meetings between newspaper editors and the heads of major networks to encourage the papers to run cable program listings. He encouraged program guides. And he laid siege to *TV Guide* which gradually increased its listings of cable networks.

CAB held its first national conference in conjunction with CTAM in 1981. By the following year the organization was off on its own, drawing 1,000 attendees to a meeting at the Waldorf Astoria in New York.

From 1980 when the CAB was founded through 1997, total advertising revenue for cable television, both network and local, grew from $58 million to $8 billion per year. Local advertising revenue grew from $8 million in 1980 to $1.2 billion in 1994.

bad an investment cable TV had been for American Express.

By the end of 1982, Hauser recalled, he had had enough. He left and formed his own company, joining Rifkin, Conroy and Bresnan in the ranks of those former executives at major companies who started their own smaller MSOs with their own names attached.

Within the next 15 years nearly all of the big companies that entered the U.S. cable operating business in the early '80s would be gone, including American Express, Westinghouse, Hearst, Tribune, Times Mirror, General Electric, The New York Times, Rogers, Dow Jones. Time Inc. would remain and some others (Hearst, Times Mirror, GE) would continue to be heavily involved in cable programming.

Meanwhile, some of the entrepreneurs who sold out to the big companies, or took them on as partners — Bill Bresnan, Bob Hughes, Bob Rosencrans, Gus Hauser — would go on to form their own cable operating companies and many of them would continue to operate these with great success well into the '90s.

The big companies that found cable so attractive in the late '70s and early '80s simply didn't have the stomachs to take on the huge capital expenditures, the enormous risk of bidding for urban franchises and the huge dent in corporate profits that the cable business periodically required for those who wanted to remain in it for the long run.

This was particularly true for companies held by the public where the management was rewarded with bonuses tied to profits and where the companies were vulnerable to hostile takeovers. These companies found it difficult to bet the ranch over and over again on a new generation of cable television.

Most of the companies that remained the mainstay of the industry for the balance of the century were privately held or closely controlled: TCI, Continental, Jones, Cox, Comcast, Cablevision Systems, Cablevision Industries.

TCI president John Malone would later recall that when he first became involved in cable he felt it was not a game for any public company because it was such a capital-intensive business. In cable a successful strategy involved reporting minimal profits and making huge investments in technologies where there was often no track record and where the risk of failure was too high for corporate executives vulnerable to attack from shareholders or board members.

The premium television business continued to boom throughout the late 1970s and early '80s. From its beginnings in 1975 the satellite-delivered premium television business grew to serve more than nine million subscribers in just five years. By 1982 cable systems

Regional Flavor:
The Eastern Show was among the dozens of regional cable gatherings that emerged in the 1960s, 70s and 80s, as the industry and state association membership expanded.

had sold some 21 million pay TV subscriptions.

While Home Box Office, in particular, included more and more original programming in its schedule, the mainstay of all the premium services during this time were theatrical films, purchased from the major motion picture studios.

HBO, because it had a 70% share of the premium market, was the prime player in negotiations with Hollywood. The studios, although they earned huge incremental revenue for their films by licensing them to HBO, nevertheless grew to resent the power of this one pay service which could play one studio off against another in bidding for film rights. HBO's muscle in Hollywood increased when it began the practice of pre-buys. In a pre-buy, HBO would provide some of the funding for a film up front in return for a right to air the film exclusively on HBO for a period of time after its theatrical run.

To counteract the power of HBO, four major Hollywood studios in 1979 announced plans to form their own pay service, dubbed Premiere and backed by the financial power of Getty Oil, already the owner of ESPN. The four studios — 20th Century Fox, Paramount, Universal and Columbia — agreed to give all their films to Premiere on an exclusive basis for a period of nine months before allowing them to go to HBO, Showtime or The Movie Channel.

The plan by the studios to hold back films from the established pay services would have been devastating, particularly for Showtime and The Movie Channel, which relied almost entirely on Hollywood product for their programming. "It would have put us under," John Sie, then senior vice president of Showtime, later recalled.

Cable Hand: *Hollywood executives hoped Burt Harris' history in the cable business would help it sell its Premiere pay TV venture, but the Justice Department blocked the company from ever launching.*

The Premiere threat united the pay services which had been bitterly jousting for most of the previous three years. (Sie recalled a series of debates he had at various trade shows with representatives of HBO, often Tony Cox. At one of them, Sie found a statement by the HBO executive particularly egregious. "I come from China," Sie replied. "In China we spent a lot of time following the water buffalo around in the fields. That experience taught me something that can apply here. I can recognize bullshit when I see it.")

To run Premiere, the studios turned to Burt Harris, former chairman of Cypress Communications. After Cypress was acquired by Warner Communications, Harris had served for a time as vice chairman of Warner Cable, then left to form his own new MSO, Harris Cable.

"The motion picture companies did not have a good rapport with the cable industry," Harris recalled. It was an understatement.

Motion picture producers were also producers of television programming which they sold to broadcast TV stations. For the better part of three decades they

183

had been fighting to force cable systems to pay copyright fees for this programming. Passage of the copyright law did not lessen the prevailing view in Hollywood that cable systems were pirates. Motion Picture Association of America president Jack Valenti became a regular on the lecture tour, denouncing the cable industry as a parasite.

The extraordinary success of HBO and the other pay services added fuel to the fire, particularly as attendance at movie theaters began to decline.

As Harris recalled, "I had a good reputation in the cable industry and a name that they thought would get them further ahead very quickly with the cable operators."

Harris, and former Viacom president Chris Derrick, who was hired as

Women and Minorities on the Rise

With the explosive growth of cable in the late 1970s and early 1980s more women and minorities began to ascend to positions of authority in the business. Although racism and sexism were no less prevalent in the cable business than in any other American industries, the desperate need for qualified people to fill a rapidly expanding business created more opportunities for women and minorities in the cable business than might have been the case in other, more static industries.

The senior female executives in the cable business began to get together informally at the NCTA conventions in the mid-1970s, often in the hotel room of NCTA public relations official Lucille Larkin. Some of those involved in the early meetings were June Travis of ATC, Vivien Horner of Warner Cable, Barbara Ruger of *Cablevision* magazine and Carolyn Chambers of Chambers Communications.

As Gail Sermersheim, one of the original group recalled, "We talked about how few women there were in the business and how we needed to support each other and try to bring other women into the business."

Sermersheim had begun to work in the cable business in the early 1960s when her parents became partners in the cable system in Jasper, Ind. After getting a journalism degree at Indiana University, she went to work for Telesis, a multiple systems operator with operations in the Midwest. After 12 years running the marketing operations for Telesis, she asked Tony Cox for a job with Home Box Office in 1977. Cox offered her a choice between the HBO offices in Kansas City, Philadelphia, New York or Florida.

"It was June, and it was snowing outside," Sermersheim recalled. "I picked Florida." Twenty years later she was vice president for HBO for the southeastern region.

By 1979 the informal meetings in Lucille Larkin's room had grown to the point where Sermersheim and Larkin and a few others began to talk about setting up a formal organization. With support from a couple of dozen cable companies, Women In Cable incorporated and began to launch a series of programs designed to help women improve their skills, network and make a greater contribution to the cable industry.

During the following decade the organization launched its Accolades award program, a professional education course in conjunction with the University of Denver, a National Management Conference and a variety of other seminars, studies and programs. By 1996 the number of members

Premiere's president, didn't have much of a chance. Shortly after the announcement, the Justice Department filed suit to block the project on antitrust grounds. It argued that by joining together to deny product to HBO and other pay services, the movie studios were acting in restraint of trade. Arguments were heard before a federal district court judge in New York.

"Each of the companies had one law firm, and Premiere had its own," Harris recalled. "We went to court with about 50 lawyers." The legal army didn't help. The federal court granted the Justice Department's request to block the launch of Premiere, and the pay service lost on appeal. The company was disbanded.

But the wounds the studios felt after had grown to 3,000.

The National Association of Minorities in Cable had many of the same goals and a similar origin to WIC. NAMIC was the brainchild of Gracie Nettingham, who had worked for the United Church of Christ's television unit. The group became official in 1980 with Gayle Greer of ATC as its chair. Others involved in the founding included Gail Williams of HBO; Barry Washington, owner of the system in Newark, N.J. and Patrick Mellon of Telecable.

As Greer later recalled, the main purpose of the group when it started was to provide a framework for people of color in the industry to get together and to raise issues of importance to minorities, particularly with regard to hiring and promotion. But the agenda quickly expanded as the industry called on NAMIC's expertise to help expand cable subscriptions in the urban markets.

By the mid-1980s NAMIC was involved in a wide range of activities, including sponsorship of an annual breakfast at the NCTA convention and an annual seminar in New York on marketing cable in urban areas.

Greer, a native of Tulsa, Okla., received her undergraduate degree and a masters in organizational planning from the University of Houston. In 1978 she was running a community organization in Ft. Wayne, Ind., when the ATC franchising team came to town and asked for her support. She declined, but ATC later came back and asked her to join their nationwide franchising division. By the mid-1990s she had risen to be senior vice president of corporate affairs for Time Warner Cable.

New Guard:
Women In Cable was formed in the late 1970s to help women advance in the cable industry. The early founders included (front row): Kathryn Creech, June Travis and Vivien Horner, and (back row) Kay Koplovitz, Lucille Larkin, Gail Sermersheim and Barbara Ruger.

You Don't Have To Miss A Minute Of CNN, Headline News, Or SuperStation WTBS.

With a VideoCipher II descrambler, it's easy to subscribe to your favorite satellite programming networks. Once you install your descrambler, just call your nearest cable system or satellite dealer. They'll offer you a choice of programming packages at surprisingly affordable rates. And they can activate your service subscription. Act now to ensure uninterrupted viewing.

Cable Marketing: *Turner Broadcasting System used the media to cross-promote WTBS and its new CNN and Headline News services in the early 1980s.*

their losing battle over Premiere would remain tender for years. Soon Hollywood would have a powerful new avenue into the home, one which did not depend on HBO or any other middleman. It was the videocassette recorder, only one of several new technologies that would seriously challenge the cable industry's dominance of home-delivered video entertainment during the 1980s.

Deregulation

In the autumn of even-numbered years, the lights burn 24 hours a day in the U.S. Capitol building. For sponsors of legislation this is the moment of truth.

Each Congress lasts two years, and every new Congress starts with a clean slate and dozens of new members. Legislation that does not pass in one Congress must start the process all over in a new one, with hearings, markups, conferences, floor votes and amendments all subject to the actions of a new set of congressmen and very often a new set of committee and subcommittee chairmen.

Any member of Congress who hopes to get a new law enacted must do it within a two-year time frame or risk an entirely new set of circumstances with a new legislature.

Such was the case in 1984 with a bill to deregulate the cable industry.

The notion that the country needed a law to govern cable television had been kicking around Capitol Hill since the late 1950s when the industry was first subject to legislative hearings.

Various attempts to pass such a law had failed, most notably in 1960. Cable television involved a lot of powerful players — phone companies, broadcasters, Hollywood studios, city and state officials, and the cable industry itself. And within each of these groups there were several factions, often warring with each other.

That factor alone made it difficult to pass legislation governing cable because, under the American system of government, it is much easier to play defense, blocking a new law, than it is to play offense, getting a new law passed. With such powerful players, the political reality was that a compromise acceptable to everybody had to be crafted if legislation were to be passed. And the deal that had been crafted in 1984 was so fragile that it was unlikely it could be reconstructed if it were not passed in 1984.

Thus it was that in early October, 1984, the lights were burning bright late at night at the Capitol, including at the office of Rep. Timothy Wirth, a young Democrat from Colorado who had become chairman of the House Telecommunications Subcommittee.

Wirth was a different kind of Democrat. Elected in the Watergate landslide of 1974, he represented a marginal district which could just as easily have elected a Republican. And he had been elected with the support of several executives of cable television companies based in Colorado.

For the first 30 years of the cable industry's history, Democrats in Washington had most often taken the view that the best way to ensure diversity of programming was to encourage the growth of broadcast television stations. And to do this, they believed, the fed-

Point Man: *Colorado Democrat Tim Wirth led the charge in the House to pass the 1984 cable bill that substantially deregulated the industry.*

eral government needed to restrict the power of cable. Because the Democrats during all that time controlled the Congress and for much of it the White House and the FCC as well, their views were important.

But Wirth took a different view. In the eight years since the launch of satellite-delivered programming, a slew of new cable television networks had been born, including C-SPAN, Nickelodeon, The Disney Channel, Black Entertainment Television and A&E. Wirth believed cable, not broadcast, was most likely to provide the kind of programming diversity he favored, particularly public affairs programming and programming for children, minorities and fans of the arts.

And something else had happened that changed the political landscape. In 1980, Ronald Reagan, a devout opponent of government regulation, had been elected President, and the Republicans for the first time since the Truman administration took control of the Senate.

All of these factors provided a political landscape fertile for the passage of legislation that would lift many of the regulations governing cable television, particularly the regulations cities imposed on cable rates and franchise renewals.

Wirth and a group of other congressmen and Senators had worked hard for several years on the issue. The Senate had passed a bill in 1983, but it took another year before the House was able to do so.

Agreements between the cable industry and organizations representing the nation's cities had been forged and fallen apart three times in the previous two years. Both groups were split over the wisdom of supporting legislation.

But by the first week in October, the two sides had agreed to a compromise, and the bill appeared headed for passage. Then, just hours before Congress was slated to adjourn, another thorny issue arose.

At the urging of a group of African-American congressmen, led by Mickey Leland of Texas, Wirth had inserted into the House version of the bill a measure designed to increase the number of minority employees in cable companies. It specified that each cable system would employ a percentage of minorities and women to reflect their presence in the community the system served. Wirth had given his word to Leland that the provision would not be dropped from the bill.

But two conservative Republican members of the Senate — Orrin Hatch of Utah and Jesse Helms of North Carolina — objected to the provision on the grounds that it would set racial quotas for hiring, something they opposed. Both sides were locked in on a matter of principle. The stalemate seemed insurmountable.

As the clock ticked past 10 p.m. on Oct. 10 most of the lobbyists involved in the bill and most of the congressmen as well concluded it was dead. NCTA president James Mooney later recalled going home that night with little hope that the bill would pass. But the lights were still on in Tim Wirth's office.

There, the chief counsel of Wirth's subcommittee, Tom Rogers, had an idea. He went in to see Wirth to ask him to make one last effort.

The Cable Act of 1984, like so many developments in American culture and politics in the 1980s, had its genesis in California. The nation's most populous state had long been a proving ground for developments that affected the industry. A landmark California decision in the late 1950s had prevented the industry from being regulated as a public utility. One of the early experiments in pay television had been defeated in a public vote in 1960. Most of the major cable companies had at least one system in the state.

Most of all, California had an energetic, imaginative and forward-thinking state association led by a remarkable man named Walter Kaitz.

Born in Russia, Kaitz and his parents had fled the chaos of the Russian revolution and settled in South Boston. Kaitz won a scholarship reserved for newsboys to go to Harvard College. After service in World War II he earned a law degree and settled in California where he established a business as a lobbyist for real estate companies, for broadcasters and for public employee groups.

In the late 1950s he was approached by a group of California cable operators who asked him to help them lobby the legislature for a measure to legitimize the franchising process in the state. The result was a law that formally authorized California cities to grant cable television franchises. And out of that effort the California Cable Television Association was born.

Kaitz built the association from the ground up, and he did it as a family affair. His first wife, Idell, served as secretary-treasurer for the group, and their children all played roles as well, helping to register attendees at the California Cable Convention, the so-called Western Show, which came to rival the NCTA convention as the premiere cable industry gathering. Eventually Kaitz's son, Spencer, would succeed him as head of the CCTA.

"What made Walter so remarkable," recalled Ray Joslin, who ran the Continental Cable systems in California in the 1970s, "was that he treated everybody as an equal. And he had a great affinity for people who had been born on the other side of the tracks. He really organized us and brought us all together in a common effort."

Joslin recalled that when he first came to California, his boss, Continental president Amos Hostetter, was not popular with the California cable operators. Hostetter, as chairman of the NCTA, had worked out an agreement on pole attachment rates with AT&T which the California operators felt was too onerous.

But Kaitz, Joslin said, "went to great extremes to get to know me and to introduce me to other operators. He asked me to serve on a foundation that was designed to help minorities get employment in the cable industry." Joslin would eventually serve as chairman of CCTA.

As Spencer Kaitz later recalled it, by the end of the 1970s cable operators in California were faced with an increasing number of cities demanding upgraded cable service but refusing to grant rate increases.

"It wasn't every city," he later said, "and it wasn't even a majority." But some of the cases were pretty serious. In Los Angeles, during a period in which the city and 10 other neighboring municipalities each granted at least two rate increases, the city of Beverly Hills refused to permit a single increase. In San Jose, where Al Gilliland had built one of the first dual cable systems in the country at enormous expense, the city council refused for four years to approve a rate increase.

"We had talked for some time about how to deal with this," Spencer Kaitz recalled later, "and we approached a member of the legislature, Bruce Young from Cerritos, who suggested that we make an effort to enact a bill to end city regulation of cable rates."

What followed was a "knock-down battle" with the California League of Cities, which bitterly opposed the measure. Finally, Marc Nathanson, who had just recently left TelePrompTer to start Falcon Communications, suggested the association hire Monroe Price, a professor of law and friend of California Gov. Jerry Brown to work on the bill.

Price rewrote the measure from top to bottom. Instead of providing for automatic deregulation, the new measure allowed cable companies to choose to deregulate. If they chose that alternative, they would be required to contribute a portion of their gross revenues to a statewide fund for production of public service programming.

The idea attracted Brown's support, and the measure was enacted into law in 1979.

The new law did not usher in a rash of huge rate hikes. In fact, many cable systems chose not to deregulate. But they were able to use the law as leverage in negotiating with their local cities for rate increases. Their argument was that if they were denied reasonable rate increases they would opt for deregulation and would have to contribute to the state programming fund. Cities preferred to grant rate hikes and keep the money for programming at home.

As John Goddard, then head of California-based Viacom Cable, later said,

Kaitz

When Walter Kaitz died in 1979 a group of cable industry leaders on hand for his funeral gathered in the kitchen of the Kaitz family home. Among them were Ray Joslin of Hearst, Spencer Kaitz, Doug Dittrick and John Goddard of Viacom Cable and Amos Hostetter of Continental Cablevision.

The talk turned to how to create a memorial to this remarkable man, and Joslin suggested that they establish a foundation that would encourage the hiring and promotion of more members of minority groups in the cable business. The result was the Walter Kaitz Foundation, which maintains a program of recruitment, training and placement for minorities seeking entry into the business.

By the mid-1980s the foundation's annual fundraising dinner, held in New York each September, had become the premiere black-tie event in the business, attracting thousands of attendees at several hundred dollars a seat to honor a leading cable executive and to acknowledge the current class of Kaitz fellows.

By 1998 the foundation had placed nearly 500 Kaitz Fellows in management positions in the cable industry.

Helping Hand:
Rep. Lionel Van Deerlin (D-Calif.) joins Walter Kaitz and his son, Spencer.

the act was a cable television bill of rights.

Even more important, the California bill provided a model for other states — Massachusetts and New Jersey — to follow in adopting their own deregulation measures. And it provided a precedent for the U.S. Congress in fashioning national legislation. Perhaps most important, it effectively neutralized the huge California delegation to the Congress. With deregulation already in place in their home state, California lawmakers had little objection to passage of a similar bill for the nation.

When a new law is to be passed in the United States all the stars need to be aligned just right. With a new cast in the Congress and the FCC, a Republican President and Senate, and the example of California's deregulation law, many of the heavenly bodies in the legislative firmament were lining up just right. So were the economic factors and public perceptions that also play a role in the pressures Congress feels.

Cable by 1984 was no longer an economic darling. During the previous three years the news media had been full of articles about expensive financial disasters in programming, operations and technology. At the same time a series of new programming distribution systems were providing real competition to cable television for the first time in its history.

In the early 1980s all three of the broadcast networks attempted to get into the business of programming for cable television. They had watched from the sidelines as a slew of new programming services had launched in the late '70s and early '80s. Wall Street analysts had begun to lob negative comments at the big three for failing to enter the fray.

But the broadcasters didn't really know how to do cable programming. The culture of the broadcast networks encouraged huge expenditures on programming and relied entirely on advertising for revenue. This formula didn't work for cable.

Nevertheless first CBS then NBC and later ABC entered the fray. And like lemmings, they followed each other off the same financial cliff.

CBS was first. It launched its CBS Cable cultural service in October of 1981, treating cable operators the following spring to a lavish party in a huge tent located in the desert outside Las Vegas during the NCTA convention. The theme was "Desert Oasis," and the musicians, waiters and entertainers were all dressed in Middle Eastern costumes. It was one of the most expensive parties ever produced at a cable convention.

But when the party ended, thousands of cable operators had to stand in the rapidly cooling desert night, with the wind whipping up the sands, while an inadequate number of busses attempted to take them back to Las Vegas. It would be in the minds of many, the most

Black Rock's Black Eye:
CBS executives — Thomas Leahy, Gene Jankowski, Thomas Wyman, William Paley and Dick Cox — were all smiles at the launch of CBS Cable.

memorable entertainment event of any NCTA. It was also a metaphor for CBS Cable itself, a lavish beginning with a disastrous end.

CBS promised high-class, well-produced programming ranging from dramas to opera to concerts, more than half of it originally produced.

But within a year the network shut down the service after losing some $30 million. With five million subscribers, CBS simply was not able to draw the advertising revenue it had expected when it launched. And its progress in cable carriage was impeded by its decision to transmit on the Westar satellite, rather than Satcom I or III where most other cable programmers were berthed. This meant that to pick up CBS Cable many cable systems would have to install a second satellite dish.

The CBS enterprise may also have been a victim of the internal politics of the network. It was axed by CBS president Thomas Wyman within weeks after he took office.

The next broadcast network-backed service to bleed to death in the brave new world of cable was The Entertainment Channel, a joint venture of RCA (parent company of NBC Television Network) and Rockefeller Center Television and headed by former CBS president Arthur Taylor (CBS presidents came and went quickly in those days as CBS chairman William Paley installed a revolving door in the office of his number two and presumed successor).

Like CBS Cable, The Entertainment Channel was to be a high-brow service offering Broadway shows, foreign films, jazz concerts, adaptations of classic novels such as "Great Expectations" and "Gulliver's Travels," and programming produced by the British Broadcasting Co. Unlike CBS Cable, TEC was to be a premium network relying entirely on revenue from subscribers for its income. But The Entertainment Channel ran into a brick wall when it came to carriage on the cable systems.

Years of aggressive marketing by HBO, Showtime and The Movie Channel had ensured that most cable systems were already carrying what they regarded as a full complement of premium services. When The Entertainment Channel launched, it reached fewer than a dozen cable systems. It folded in nine months, after losing nearly $10 million, and merged with the ARTS network being run by Hearst/ABC Video Enterprises. The result was the Arts & Entertainment Channel, headed by Warner Amex Satellite alumnus Nick Davatzes.

A&E and Charles Dolan's Bravo, which also launched in the early 1980s, proved that cultural programming could work on cable. But both also proved that the best way to do it was to make use of inexpensive, already produced programming rather than spending millions to produce original programming themselves.

"The Entertainment Channel would spend $800,000 to produce a live broadcast of La Traviata while we would find a tape of the same opera, at La Scala, for $1,000," recalled Ray Joslin, head of Hearst's cable division.

In addition to merging their ARTS service with The Entertainment Channel, the Hearst ABC joint venture merged its Daytime service with Viacom/ Jeffrey Reiss' Cable Health Network. Each had launched in 1982 and each had found it difficult to gain carriage and advertising. The merged service, called Lifetime, was headed by former RKO Radio president Tom Burchill.

But by far the most widely publicized programming casualty of the early 1980s was an attempt by ABC Video Enterprises and Group W Cable to join forces for an assault on Ted Turner's CNN. The two big broadcasting companies called their new service Satellite News Channels.

The concept was to use the news gathering capabilities of ABC News, together with a group of local broadcast stations (some owned by Group W and some by ABC), to produce two 24-hour cable news channels. One would offer updated news every half hour while the second would combine longer features and in-depth reports. The two services would be offered free to all cable systems and the venture offered to pay an incentive of 50 cents per subscriber for any cable system that signed on.

The announcement of SNC in August, 1981, was a huge blow to Turner Broadcasting System, whose CNN was just beginning to see a chance for profitability. But it was exactly the kind of challenge that the company's boss, Ted Turner, loved. Here he was, in his favorite spot as the underdog — the lone Confederate fighting the huge Union Army, the Greeks at Thermopolae, David versus Goliath. The big broadcasters had thrown B'rer Ted into a briar patch.

Turner immediately counterattacked. At a hastily improvised press conference at the CTAM convention in Boston he announced he would launch his own second news channel, Headline News, six months before SNC was to start service. And he derided the broadcasters, calling SNC "a second-rate, horse-shit operation" and claiming ABC would never pass up the chance to report a scoop on its broadcast news in order to report it first on SNC, an allegation that the SNC staff later conceded was correct.

The press conference was more like a pep rally, with operators standing and cheering and offering to sign up on the spot.

But when the cooler heads back at MSO headquarters began to examine the situation, the SNC offer looked pretty attractive. Most of the cable operators, particularly John Malone at TCI, liked having two suppliers competing with each other. It gave the cable operators the chance to play off one against the other and to hold down costs.

Bringing the broadcast networks, particularly long-time cable foe ABC into the cable fold, would be a bonus, operators reasoned. And the lack of a per sub fee contrasted nicely with the 10 cents per sub per month CNN was charging. Finally, the offer of 50 cents per subscriber just to launch SNC would give the MSOs, always strapped for cash, a nice boost on the bottom line.

When SNC launched in June, 1982, it had some 2.3 million subscribers and 12 national advertisers.

Turner played on his long-standing

Counter-Attack:
Ted Turner won the war against SNC, the 24-hour news joint venture between ABC and Group W, but his sojourn into music videos was less successful. TBS' Cable Music Channel sold out to rival MTV less than two months after launch.

support for the cable industry, reminding operators in trade ads and posters (of Turner standing before a dish farm dressed in overalls) that "I was Cable before Cable was cool."

Still, SNC kept on coming. By December the situation at CNN headquarters resembled Valley Forge. The financial picture was that Turner could operate with two cable news networks for only about 90 more days, TBS executive vice president Terry McGuirk recalled.

Then the Turner folks got a piece of good news.

"We heard that ABC and Group W closed their offices for Christmas," recalled McGuirk. "They went on vacation for two weeks. They would be defenseless in the marketplace."

Like George Washington crossing the Delaware, Turner planned a Christmas offensive when his opponent was least expecting it. "We decided to offer one dollar a subscriber per year for the next three years to any operator willing to sign up before the end of 1982 and guarantee that they would take CNN and Headline News as the first two news services on each of their systems. We blitzed the entire industry, and everybody signed up. When the SNC folks came back from vacation they gave up almost immediately."

By November of 1983, Group W and ABC agreed to accept a payment of $25 million from Turner in return for withdrawing from the business, dropping all lawsuits (both sides had filed against each other) and handing over 7.5 million SNC subs to TBS.

The joint venture had lost in the area of $100 million in less than two years of operations.

The lesson was that the concept of cable networks supported by advertising revenue alone would not work. CBS Cable and SNC were both offered free to cable systems. But they were unable to sell enough advertising time to support their huge programming costs. Turner, in contrast, was able to defeat much larger opponents because he had two sources of revenue: per-subscriber fees and advertising.

Another lesson was that it was very difficult to take on an existing cable service. CBS Cable and The Entertainment Channel both were competing against existing services: ARTS and Bravo. And CNN had already been well established before SNC entered the cable news business. With limited channel capacity, cable operators saw little need to offer more than one news channel or more than one cultural service. Even if they did, they would tend to offer complementary services from a single provider and which they could buy at a discount rather than two competitive services that were likely to duplicate each other.

This was a lesson that Turner himself would learn the hard way. In October, 1984, he launched a rival to MTV, called Cable Music Channel. Inexplicably, Turner adopted the same strategy against MTV that SNC had attempted against CNN. He offered CMC free of charge, while MTV, with more than 24 million subscribers, was charging between 10 and 15 cents a sub per month. And MTV followed a strategy similar to Turner's crusade against SNC: it launched its own second channel, VH1.

Turner also shot himself in the foot with MTV's core audience. He announced that CMC would be a softer version of MTV and would not have the hard rock and sexual innuendo of the raucous network that teenagers had come to love. MTV, Turner said, was a "bad influence" on young people. And he announced all this in an interview with *Rolling Stone* magazine, the Bible of the hard-rock crowd.

It took Turner far less time than SNC to realize he had made a mistake. He shut down CMC in November, 1984, less than five weeks after it launched. At the ACE awards ceremony the following month he ran into Jack Schneider, chairman of MTV parent Warner Amex Satellite Entertainment Co. Turner dropped to his knees, pulled out a white handkerchief and began waving it.

The venture cost Turner only about $2 million. It earned him the gratitude of cable operators who had used the threat of CMC to pressure MTV to agreed to more favorable deals. And it gave him a look at a new employee, Scott Sassa, who would return to TBS and help build it into a major programming powerhouse in the late 1980s and early 1990s.

Premium television services, whose growth had been phenomenal in the late 1970s and early '80s, began to suffer some setbacks as well as the decade progressed.

In 1980 the Premiere pay service, backed by four Hollywood studios and Getty Oil, was killed when the Justice Department opposed it on antitrust grounds. But the end of Premiere didn't end the hostility some cable operators and most Hollywood studios felt towards Home Box Office, still by far the dominant player in pay television and increasingly using its clout to dictate prices to film producers.

Michael Fuchs, the vice president of programming for HBO, came in for special hatred in Hollywood, reflected

Wanted Man: Michael Fuchs may have been a wanted man among Hollywood studios but his forceful personality and programming vision thrust HBO to the top in the 1980s and 1990s.

in a series of press articles along the theme of an *Esquire* profile entitled "The Man Who Ate Hollywood." Fuchs, an attorney, had a background in politics serving as an advance man for Sen. Edmund S. Muskie (D-Me.) in the 1972 presidential primaries. He later worked for the William Morris talent agency in New York. His sometimes abrasive style was not what Hollywood moguls, not lacking in the ego department themselves, were used to. None of the studios, in fact, had ever been confronted with such an awesome power as HBO.

Cable operators as well feared HBO's power, particularly after it launched its Cinemax sister service and escalated the war with Showtime. In 1981, after the demise of Premiere, a group of cable operators (TCI, Cox, Times Mirror and Storer) launched their own premium service dubbed Spotlight.

But Spotlight didn't do very well. Most of the really big marketing efforts for pay television were coordinated and largely paid for by the big pay services, HBO and Showtime. Among the MSO's the marketing efforts ranged from spotty to non-existent. At TCI, in particular, the marketing budget was sparse. Nor could Spotlight get any carriage on the MSOs which owned HBO or Showtime or the Movie Channel: ATC, Group W Cable, Viacom Cable and Warner Amex Cable.

And the Spotlight partners even found it difficult to persuade their own systems to be very enthusiastic about the service.

As John Sie, then senior vice president of Showtime, later recalled, the MSOs found it "too hard to force the systems to carry in-house brands."

When the Spotlight partners decided to bail out there was a heated race for the subscribers. "I arranged a deal where (the MSOs) became phantom equity owners in Showtime," Sie recalled. "Under our arrangement, the more successful they became (in selling Showtime subscriptions), the lower their license fee would be. It was almost like they had equity in Showtime."

And the MSOs bit, selling Spotlight, with its 750,000 subscribers, to Showtime, which by that time had merged with The Movie Channel.

But while HBO would have liked to get the Spotlight subs for Cinemax, it was facing much more troubling issues by the end of 1984.

Pay television, which had been exploding for the better part of a decade, hit a brick wall in 1984. The total number of pay subs had jumped from 1.6 million in 1977 to 9 million in 1980 to 15 million in 1981 to 29.9 million in 1984. There it stalled. The category grew only 3 million units in 1983 and a heart-stopping 600,000 in 1984.

"It's not a disaster, but there is certainly not the euphoria that existed two years ago," Time Inc. executive vice president Nick Nicholas told *Business Week*. Wall Street read that as an understatement. When HBO reported its first-quarter numbers in 1984, the stock of Time Inc. dropped by almost 15% in a single day.

Most disturbing was the fact that there was no consensus about the cause of the slowdown, let alone a remedy.

Some blamed the duplication of product between the premium services. Customers who had been persuaded to take both HBO and Showtime, and even The Movie Channel as well, sometimes found that two or three services would be running the same movie in the same month, sometimes at the same hour of the same day.

Another obvious suspect in the slowdown of pay TV sales was the videocassette recorder. From its debut on

New Deal:
Viacom bought Group W's 50% share of Showtime to become sole owner of the pay network in the early 1980's. On hand for the announcement were Bill Bresnan, chairman/CEO of Group W Cable; Dan Ritchie, chairman/CEO Westinghouse Broadcasting Co.; Ralph Baruch, chairman/CEO, Viacom International; Terrence Elkes, president, Viacom International, and Mike Weinblatt, president, Showtime.

appliance store shelves in the mid-1970s, the VCR was in 20% of American homes by 1984. In a single week that year some 250,000 videocassette recorders were sold. The number of VCRs increased from 6 million in October 1984 to 20 million a year later and 26 million the year after that.

Paul Kagan Associates correctly predicted that soon Hollywood would be getting more revenue from videocassette rentals and sales than from any other source. And Hollywood, chafing at the bit to find a way to break HBO's power, opted to release its films on home video before licensing them to the pay services. Customers wanting to watch a hit movie at home would find that the quickest way to do it was to rent it and play it on the VCR. They would have to wait another six months to watch it on premium cable.

Eventually HBO would compensate by running more of its own original programming. But when it first confronted the problem, "our knee-jerk reaction was to pooh-pooh it," recalled Bill Grumbles, then an affiliate rep for HBO. "Then we tried to run ads against it, telling people: 'Don't drive to the video store when you can stay at home and see great movies.' But we found out people liked to go to the video store. And price didn't matter. For the price of a month of HBO a consumer could rent only about four movies at the video store. It didn't matter."

The screeching halt in subscriber growth in 1984 coincided with a huge backfire in the pre-buy strategy that HBO had been employing in Hollywood to lock up the exclusive rights to films. The scheme had been engineered by Frank Biondi, a former investment banker and Princeton graduate.

Biondi's greatest triumph in this

Program Strategy:
Frank Biondi expanded HBO's production and acquisition strategy of films in the early 1980s.

arena was the establishment of two new motion picture production companies. Tri Star Pictures was a joint venture of HBO, CBS and Columbia Pictures, and Silver Screen Partners was a public company that would raise millions to produce new films that could be run exclusively on HBO.

By 1984 Biondi, at the age of 39, had been named chairman and CEO of Home Box Office, leapfrogging over Fuchs, who had brought him into the company. But the slowdown in premium subscription sales and the huge costs that Biondi had incurred with his commitment to pre-buys combined to hit HBO like a two-by-four. The division, which had been a huge cash cow for Time Inc. since 1977, suffered three successive quarters of earnings declines in 1984. By Halloween Biondi was gone, replaced by Fuchs as chairman/CEO and ATC executive Joe Collins as president.

Programmers were not the only ones to suffer setbacks in the early 1980s. Cable operators who had competed in the franchising wars for urban markets had been forced to promise services they could not deliver or which were unprofitable. It was not long before they had to find a way out of the obligations they had undertaken.

The biggest winner in the franchising wars had been Warner Amex Cable Communications. With the lure of its interactive QUBE systems, Warner Amex had hooked franchises in such major urban areas as Dallas, Pittsburgh, Milwaukee and the outer boroughs of New York City.

The franchise wins obligated the company to huge expenditures to construct the systems. And the costs were proving greater than the company had anticipated. In Pittsburgh the cost to build the system had been estimated at $47 million when the franchise was awarded in 1979. By 1983 the price tag had hit $80 million and was still heading north. Selling subscriptions was also proving more difficult than anticipated. The Dallas system a year after it was launched was attracting only 20% of homes passed.

Warner Amex Cable was bleeding. It lost $40 million in 1982 and by 1983 would lose another $99 million. By the middle of 1983 it would amass $875 million in debt.

And the losses came at a difficult time for the parent companies. Warner Communications, never heavily capitalized, was suffering enormous losses from its Atari video game division (which fell $122 million in the red in the third quarter of 1983 alone). The stock tumbled. Warner chairman Steve Ross was at risk of losing control of his company to a hostile takeover. At American Express, meanwhile, chairman Jim Robinson was looking for ways to stem the losses to help him in his power struggle with anti-cable Shearson Lehman chairman Sanford Weill.

In January of 1983 American Express forced Warner to agree to replace Gus Hauser as chairman of Warner Amex. To run the company they brought in Drew Lewis, who had been serving as President Reagan's Secretary of Transportation and who prior to that had run a company that specialized in turning around troubled businesses.

Lewis didn't waste much time getting to work. He ended the franchising effort, demolished the QUBE system in Columbus, laid off half the headquarters staff and set off to visit Pittsburgh, Milwaukee, Dallas and other cities to deliver the news that Warner Amex would either demand a change in the franchise agreements or be forced to sell the systems. "We promised too much," Lewis bluntly informed the cities.

Lewis told the city council in Dallas that the cable system would cut back the number of channels to 47 from the 76 it had promised, build four rather than nine local origination studios, and cut substantially the number of local origination channels, pledged at 24. Lewis also proposed the city allow

Cable Convention:
Warner Amex's Drew Lewis, QUBE's Scott Kurnit and Rep. Tim Wirth (D-Colo.) meet on the set of a W-A system.

Warner to bypass installations of apartment buildings unless the apartment owners paid for the installations or guaranteed all residents would subscribe.

In New York Warner and five other franchise winners — Cablevision Systems, ATC, Cox, Queens Inner Unity, Vision Cable and Continental Cablevision — joined to petition the city for permission to scrap plans to build a dual-cable system and instead install a single 450 MHz plant with the promise to upgrade it to 550MHz when the technology proved feasible. Noting that the change would not mean fewer channels for customers, New York City franchise bureau director Morris Tarshis backed the plan which he said would save the cable companies $20 million in construction costs.

Within a year Lewis had renegotiated the franchise agreements in Dallas, Milwaukee, New York and a dozen other communities. In Pittsburgh he opted to sell the system to TCI, which owned systems in many of the surrounding suburbs.

TCI president John Malone had embarked on a new strategy earlier that year. Although TCI had remained on the sidelines in many of the urban franchise fights, Malone decided that a new sense of reality was coming into place in the cities and that it might now be possible to run urban systems that could make money. The first place the strategy was to be tested was in Pittsburgh. To implement it, Malone turned to his former colleague at Jerrold Electronics, John Sie.

Sie was at Showtime, but had become convinced that he would never get the top spot (when Showtime CEO Jeffrey Reiss departed, Showtime parent Viacom replaced him with a former broadcast network executive, Mike Weinblatt).

"So John (Malone) one day asked me if I would like to come to Denver and make some real money," Sie later recalled. Sie took a 50% cut in pay and moved to Colorado in February of 1984. There he joined a new team including Harvard Business School graduate Peter Barton and investment banker Stewart Blair, that Malone had assembled to take the company to the next level. Sie had no firm agreement with Malone on compensation or advancement. In fact, he had no specific duties. It was all done on a handshake and trust, Sie later said.

Sie's first assignment was to make sense out of the mess in Pittsburgh. Sie, a physicist by education, had held top positions at Jerrold and played a central role in the growth of Showtime. But he had never been involved in a franchising war and had no experience in politics.

The 73,000-subscriber system, Sie soon discovered, "was bleeding like a stuck pig."

Malone had agreed to buy the system from Warner Amex for $93 million. Sie's task was to make the deal work by wresting from the city an agreement to allow TCI to run a system considerably less elaborate than the one Warner had promised when it won the franchise in 1979.

The job, he later said, was "to convert a Cadillac into a Volkswagen."

Working seven days a week and 18 hours a day, Sie laid siege to Pittsburgh. He met with all the local officials, debated on the radio and television, testified before the city council and gave countless interviews to the press. He forged a working relationship with Brother Richard Emenecker, the priest who was head of the city office of telecommunications.

After seven months of negotiations, the city council voted unanimously to transfer the system to TCI. The deal allowed TCI to eliminate the 20% minority ownership provisions of the original franchise agreement, to build a system with 44 rather than 63 channels, to turn the public access programming division over to a non-profit agency, and to phase out the institutional network. TCI also won approval to junk the QUBE system, selling the interactive boxes back to Warner for $30 each. TCI did agree to add a surcharge to the cable bills to fund local programming.

"Our strategy," Sie later recalled, "was to demonstrate that less is more." It worked, largely because the city and the voters recognized that the elaborate system Warner had pledged to build was deep in the red and some cutbacks were necessary if the city was to have an economically viable system.

"We knew (the system) was in trouble," Pittsburgh's deputy city solicitor said at the time. The deal with TCI, he said, would give the city "a comparable system" to what Warner had built but with the assurance that it would be economically viable.

The sale of Pittsburgh and the cutbacks in Dallas and other cities generated enormous publicity about the risks of cable. "Warner Amex Chokes on Cable," said *Newsweek*. "Is Cable Losing its Luster?" asked *US News & World Report*. *Fortune* told its readers "Why Cable Is a Risky Investment."

But there were still cable operators who were willing and eager to prove the pundits wrong. One was a small venture that had been started by former TelePrompTer vice president of marketing and programming Marc Nathanson.

Nathanson was hooked on cable at an early age. His father, an advertising executive, was a close friend of cable pioneers Burt and Irving Harris. Marc's first go at the industry came at an early age. While an undergraduate at the University of Denver he had a chance to go to a cable convention in that city and decided he wanted to buy a system.

He paid a call on Daniels & Associates where he found the price quoted by Monroe Rifkin for a 2,000-subscriber system in Elk City, Iowa, a little too high.

After college, Nathanson went to work for Burt Harris at Cypress Communications. Harris made him manager of a small system in California where, Nathanson later recalled, "I learned that the engineer really runs the systems. The manager handles PR and government relations and some marketing. But the engineer runs the system."

After Cypress was sold to Warner, Nathanson left and went to work for Jack Kent Cooke at TelePrompTer. There, along with Jeff Marcus and 2,000 new sales personnel, he masterminded a marketing campaign that added more than 200,000 new subscribers in a single year, helping rescue the company from near bankruptcy.

Cooke Protegee: *Marc Nathanson learned the ropes of cable under Jack Kent Cooke at TelePrompTer before launching his own MSO.*

He also won an internal argument within the company, persuading Cooke not to start his own pay TV service, but instead to become the first big MSO to commit to a nationwide affiliation deal with HBO.

By the mid-1970s Nathanson had decided to pursue his college dream of owning systems himself. He went in to tell Cooke he was leaving.

"Marc," Cooke replied, "if you leave me I will make sure you never work in the cable industry again."

Nathanson replied, "Well, Mr. Cooke, I just want to do what you did, start my own company."

Cooke paused a minute and said, "Oh. Well, that's different. Let me know and I will invest."

Starting with a couple of systems and unbuilt franchises owned by his father and Burt Harris, and a million dollars in equity from his father-in-law, Nathanson in 1975 founded Falcon Communications.

The company's specialty was classic cable systems in geographic clusters. This was the kind of system that Nathanson had grown to know at Cypress and TelePrompTer. While most of the big MSOs were battling over the big cities, Falcon kept its nose to the grindstone, delivering service in small towns and cities in Southern California.

One company that paid very little attention to its smaller systems was Warner, which was pouring all its money and attention into the big urban newbuilds. Many of the smaller Warner systems had been owned by Cypress and Nathanson knew them well. And he knew what had happened to them after Warner took control.

In 1985 Warner Amex president Drew Lewis decided to unload 18 of these systems serving some 50,000 subscribers. Nathanson heard about the deal and called Warner Amex to ask if he could bid.

"You're too late," he was told. "We have already agreed to sell them." Nathanson asked if he could not at least make a bid. The answer was no.

So he contacted an old friend, Martin Payson, who had been a camp counselor at the summer camp Nathanson had attended as a boy and who was now conveniently a member of the board of Warner Communications. Payson advised him to decide what the systems were worth and send a letter making his bid to Drew Lewis with copies to Warner chairman Steve Ross and American Express president Jim Robinson.

The tactic blew the deal wide open, and with a bid of $50 million or $962 per subscriber, Falcon won the systems. It almost doubled the size of the company.

But with some simple techniques and a lot of tender loving care, Falcon transformed what had been an albatross around Warner's neck into a golden goose.

It didn't take rocket science to make the transformation. Falcon added channels, upgraded customer service and raised rates. It slashed corporate overhead. And it built line extensions. Many of the systems had not been upgraded since Warner bought them. The communities they served, meanwhile, had grown. By extending the cable lines to serve the newly built areas, Falcon was able to capture thousands of new customers at very little cost.

Most of all Falcon was a sales machine. At one California system the company bought in the early 1980s Falcon was able to take penetration from 20% to 48% in the first year, effectively cutting in half the $2,000 per subscriber price it had paid for the operation. There was nothing very fancy or revolutionary about Falcon's marketing: door-to-door, direct mail, telemarketing, tie-ins with local businesses. But it was done relentlessly. "This is a marketing company," Nathanson told *Forbes* magazine in 1985, "and I am a marketing man."

Falcon and TCI weren't the only companies to benefit from Warner Amex's troubles. Other MSOs found they could always point to Warner Amex when seeking to cut back on the promises made during the franchising wars. Companies such as ATC were given some shelter when they had to go in and renegotiate their own franchise agreements in such cities as Denver. After all, if the mighty Warner Amex was finding it too expensive to keep its promises, how could anybody fault ATC?

And ATC was finding some difficulties in meeting its franchise obligations. Like other MSOs which had won urban franchises, it discovered that the cost of building the systems was often greater than anticipated. This was particularly true in older urban areas where underground construction sometimes ran into unexpected sewer or gas pipes or, even worse, historical discoveries that could delay construction for months.

Cities' requirements that cables be laid underground increased the cost of wiring dramatically. One of the biggest headaches for ATC came in Rochester, N.Y., where sub-zero weather delayed construction and the city required that all homes be served. To meet the franchise requirement for universal service, ATC had to build 25 miles of underground plant just to reach four homes.

In Kansas City, the system had to cross the interstate highway system at five different locations. Each required a rerouting of the plant in order to use existing conduits over the roadways. And in such cities as Miami, the need to cross waterways and rivers added to the cost. In some urban areas the cost of wiring a mile of plant could zoom to $50,000 or more.

Make-ready was sometimes more expensive than imagined. In Denver the price of preparing telephone company poles for cable was 25% more than ATC had projected in its franchise proposal.

"The phone companies are asking for double and triple the make-ready charges of just a couple of years ago," Warner Amex senior vice president of engineering and construction Roosevelt Mikhail told *Cable TV Business Magazine* in 1984. "It's very obvious they are jacking up the prices on us and there's no way these high charges can be justified."

MSOs also sometimes underestimated the capacity of underground utility ducts to carry additional wires. "When the franchise bid is developed everybody just assumes that the cable operator will be able to use that duct system," Mikhail said. "But nobody really knows its condition because it hasn't been field tested. What often happens is that the ducts simply don't have the volume we need to run as many as six cables through. When you discover that the space isn't there you have no choice but to seek an alternative route — one which is invariably going to cost more money."

Just figuring out what to do with the dirt from digging up the city streets proved to be a big problem in such states as California, which had strict anti-dumping laws.

Construction contractors such as Burnup & Sims and Cable Services Co. became more and more adept at calculating costs. They also made use of new techniques to reduce expenses and speed construction. One new device that came into use was the rock saw, first used in constructing the Boston system. The rock saw cut a six-inch-wide trench through asphalt or concrete roads into which the cable could be laid. The cut could then be filled in and the entire process completed in a single day, far faster than the old technique of digging up the roadway with a backhoe or trencher.

Legal and financial problems also plagued some franchises. In Denver a lawsuit by the conservative Mountain States Legal Foundation challenged the ability of the city to award an exclusive franchise. The threat that the suit might succeed caused the bankers who had pledged to fund the construction of the system to seek to renegotiate the loans. Mile Hi Cablevision, a joint venture of ATC and Daniels & Associates, was forced to ask the city for permission to cut back on $30 million of capital commitments at least until the suit could be settled.

Cable operators also were plagued by the failure of some of the new equipment they had pledged to use in their systems. Suppliers, anxious to provide the latest state-of-the-art technology, sometimes rushed new devices into production before they were fully ready. And sometimes even when the new devices worked, the markets they were expected to serve failed to materialize.

"When you are on the cutting edge of technology you want to be just behind the blade." ATC president Trygve Myhren cautioned the cable industry in 1984.

One of the companies that suffered some cuts from the blade was Scientific-Atlanta, the high-flying firm that had started in cable by supplying antennas and then helped revolutionize the indus-

Mile Hi Man: *Trygve Myhren led the expansion of American Television & Communications through the 1980s before moving to the Providence Journal Co.*

try by building the earth stations used to demonstrate the first satellite delivery of cable programming.

S-A had grown like crazy in the decade following the launch of HBO on satellite. From a business that did $1 million in revenue a year in cable in 1972 it had grown to $300 million a year by 1980. It posted 40 quarters in a row of improvements in sales and earnings. And it had vastly expanded its product line down the cable plant. Starting with antennas, S-A had moved to supply headend and later distribution equipment.

S-A CEO Sid Topol was very taken with the model of IBM which supplied all the computer and data processing needs of its customers. He wanted to make S-A into such a one-stop shopping supplier for cable operators.

The only major gap in S-A's product line was converters, and there seemed no reason to believe the company couldn't find a way to make a better line of converters.

"At that time S-A could do no wrong," recalled Alex Best, later senior-VP, engineering, Cox Cable, "Everything we touched turned to gold."

Topol later recalled that he was approached by ATC chairman Monroe Rifkin who urged S-A to get into the converter business because, Topol said,

ATC was not fully satisfied with many of the high-channel capacity, interactive converters then on the market. This was true despite the fact that half a dozen major manufacturers, including Jerrold Electronics, Zenith, Oak, Tocom and Texscan, were making converters.

S-A had built a reputation as a company that produced high-quality, if sometimes more costly equipment. Its converter, the 6700, was, according to Best, the first 400 MHz, solid-state, 54-channel converter on the market.

It worked beautifully, in the prototype. When it came to producing the device in mass quantities, it was another story.

"We were used to producing 100 items a month," Topol recalled. "All of a sudden we were trying to produce 100,000 a month."

The high production schedule proved a major problem. The tuners on the converters were exceptionally delicate because of all the channels they carried. A small slip in the tuner would cause interference for several channels. S-A could build a reliable tuner in a few boxes, but not in thousands. Once they got into the field, after a few months, the tuners began to go awry.

Nor were the economics very good even if the device had worked. "We were producing them for $90 each and selling them for $70," Best recalled.

Not long after the first shipments of 6700s went into the field, the calls came about problems. Eventually S-A had to recall all of them, and Topol moved the manufacturing of the product from Georgia to a specially built plant designed just for this product by the Matsushita Corp. in Japan. The result was a much more reliable converter, the 8500. But it was also a major financial setback for S-A.

In 1982, the company reported a flat first quarter, the first time in a decade it had failed to post improved quarterly earnings. S-A stock, which had been selling at 36 times earnings in 1981 plunged to 12 times earnings by the following year.

The setbacks didn't dim Topol's enthusiasm for cable or his energetic salesmanship.

When he was ready to pitch the 8500 to TCI president John Malone the TCI folks suggested he join Malone, a big time sailor, on a cruise from Key Largo, Fla., to Ft. Lauderdale. Topol boarded the sailboat in the morning. TCI vice president Peter Barton prepared the

Disney Launch:
Walt Disney Co. entered the cable business on April 18, 1983, as Mickey Mouse flipped the switch on the Disney Channel. Exactly fifteen years later, on April 18, 1998, Disney launched Toon Disney.

brunch, including hash browns, bacon, sausage and eggs, all lathered in grease.

"I ate well, figuring it would be an easy trip," Topol later recalled. All the TCI folks on board were drinking and laughing. But then Malone headed out to sea to catch the Gulf Stream and its deeper swells. "We weren't out there 12 minutes and I wanted to die," Topol recalled. The irrepressible S-A chief spent the rest of the voyage doubled over in agony below decks. But when they landed, he won Malone over, proving if you can make the sail, you can make the sale.

The slower-than-expected pace of construction in the big cities and the cutbacks on some of the franchise promises hit the cable technology sector particularly hard. It had always been true that when the cable operators caught cold, the hardware suppliers got pneumonia. Never was this more true than in the period from 1982-1985.

And S-A wasn't the only company to suffer.

Times Fiber Corp. in 1981 had introduced the most revolutionary concept in cable system design since the start of the industry. It was called the mini hub.

Traditional tree-and-branch cable systems relied on a network of trunk cable, with amplifiers boosting the signal power every few hundred yards. A tap would connect the trunk cable to the drop cable which would run into the house. In the systems with more than 12 channels, the drop cable would be connected to a converter that would plug into the TV set.

The mini-hub concept relied on fiber-optic cable to link the headend to a series of hubs, each of which might serve a group of homes. By using fiber, the system eliminated the need for amplifiers. Each hub housed a computer that would direct the appropriate signals to the subscriber's home. In its initial design the mini hub computers also would receive and process signals from the subscribers' homes ordering pay-per-view and other services.

It was enormously efficient compared to the tree and branch system, but it was also far more expensive to install. When pay-per-view and other ancillary services proved less productive than cable operators had anticipated, Times Fiber introduced a scaled back version of the system, Mini Hub II, to offer only one-way service. Still the higher upfront costs dissuaded most cable operators from adopting the new architecture. By the middle of the next decade, mini-hub-type-switched network systems would become more viable. But in the mid 1980s the mini hub was just another technological development that proved to be ahead of what the market would support.

Home security was another example of a service that was technologically possible but a market failure. By 1983 Warner Amex Cable, the leader in home security, had signed up only 13,000 customers in five major cities for its service, and other cable operators were finding it difficult to operate a business that was very different from the core business of selling video services. Tocom, which had bet heavily on the home security business, was forced to sell to General Instrument after a disastrous 1983.

Oak Industries, which had invested heavily to get into the addressable converter market, suffered as the major urban franchises were built more slowly than planned. The company reported a loss of $166 million in 1983 compared with a profit of $4.1 million in 1982. RCA, long a supplier of cable equipment, announced it would exit the business. GTE Sylvania sold its cable equipment unit to Texscan.

Not only was the cable industry suffering from some internal illnesses, but it was also, for the first time in its history, facing some real competitors in the business of delivering multichannel television services to the consumer. In the early 1980s, four alternative delivery systems would rise up to challenge cable's de facto monopoly on the multichannel TV market: direct broadcast satellite (DBS) service, multichannel multipoint distribution (MMDS) systems, satellite master antenna systems (SMATV) and over-the-air subscription television (STV).

By 1982 there were some 150 single-channel microwave operations in the U.S. serving 750,000 subscribers. Although they provided some competition to cable, they were limited because customers had to have an antenna in the line of sight of the transmitter and because the service could deliver only a single channel of programming. With those drawbacks, microwave served largely to whet consumers' appetites for cable.

But in February of 1982, Microband Corp. asked the FCC for permission to construct a nationwide, five-channel MDS system that it said would cost no more than $35 million and could be built in two years. Low-cost reception equipment, developed in the early 1980s, allowed consumers to receive the service with an antenna that retailed for about $100.

The following year the FCC adopted new rules that allowed MDS operators to gain access to frequencies that had previously been reserved for educational institutions but which in many communities were unused. In effect the FCC was allowing MDS operators in major urban areas to offer up to 10 channels of service. The agency also authorized much stronger signals, giving the MDS systems the chance to reach customers as far as 50 miles away from the transmitter.

Subscription television, which had long been a dream of broadcasters and had been attempted at various times in the prior three decades, came into its own in the early 1980s. By 1983 there were some 27 STV operations serving some 1.5 million customers. The two largest operations, both based in Los Angeles, were ON TV with some 600,000 subscribers and SelecTV, with half a million. The two offered single-channel services with a mix of movies and local sporting events.

In 1982 the FCC lifted regulations that had confined STV operations to communities with at least five broadcast television signals, and the technology prepared to invade additional U.S. cities.

That same year the commission gave the green light to a group of would-be operators of direct broadcast satellite services. The first to launch, in 1984, was United Satellite Communications Inc., a joint venture of General Instrument

Lo Expansion:
Led by Cablevision Systems Corp. on Long Island, cable operators began expanding local news coverage in the 1980s. On the set of 'Cable Report' are Abbey Keningsberg and Pat Dolan.

Corp. and Prudential Insurance Co., which offered subscribers in the upper Midwest five channels of programming for $39.95 a month plus a $300 installation fee for the 2.5- to four-foot antennas needed to pick up the signal from the Canadian Anik satellite. Customers willing to shell out $995 to buy the dish could get a year's programming for free.

It was followed later that year by Satellite Television Corp., a unit of Comsat, which offered a five-channel service to the northeastern U.S.

Cable operators took the competition seriously. "There is no question that DBS will probably present the most serious competition the cable industry and particularly the non-urban cable industry has ever experienced. Be prepared," warned CATA president Steve Effros.

The most pesky competitors to cable in the early '80s were satellite master antenna television systems. These were mini-cable systems serving a large apartment building or a privately developed community. They were constructed just like a small cable system, with a dish and headend and trunk and drop cable. But because their wires did not cross city streets they did not need city franchises to operate and so were free from city regulation.

SMATV systems could not operate without permission of the apartment owner and would typically split revenue with the landlord in exchange for access to the premises. Often a franchised cable company would have difficulty gaining access to such apartments to offer service in competition to the SMATV, even if the apartment tenants wanted service.

SMATV received a boost in 1982 when the Supreme Court ruled that franchised cable operators had no automatic right to serve tenants without the permission of the landlord. New York landlord Jean Loretto had sued to challenge a New York State law that capped the amount a landlord could charge for access to tenants at $1 per apartment. The court rejected arguments by Group W Cable, seeking access to Loretto's apartments, that tenants could not be denied access to programming without violating their First Amendment rights. But the court ruled that the Constitutional ban on unjust taking of property was at stake and that "an owner suffers a special kind of injury when a stranger directly invades and occupies the owner's property."

One way the cable operators attempted to battle alternative distribution systems was to sign exclusive distribution contracts with programmers, forbidding cable networks from offering their services to SMATV or MDS or DBS systems that competed directly with franchised operators. Programmers, with cable as their primary market, generally went along, spawning a host of restraint-of-trade suits (except, of course, in those cases where the franchised cable operator itself built SMATV systems to gain revenue while the main cable system was still under construction.)

As ESPN vice president Andy Brilliant stated in 1984, "ESPN is in the business of distributing programming to the cable industry. We deal with the operator where there is a cable franchise."

That would remain the policy for most cable programmers until the alternative technologies began to gain enough customers, and clout in the courts and Congress, to put them on a footing more nearly equal to cable.

The combination of new, competitive technologies, cutbacks in the major urban franchises, setbacks in the technology arena, all combined to provide cable with the aura of an underdog by the time Congress came to consider the Cable Act of 1984.

At the same time, many of the industries that had been cable's nemesis during the previous three decades were otherwise occupied at the time, giving the industry a relatively free hand to work out a piece of legislation.

The National Association of Broadcasters was engaged in a bitter internal battle to determine who would head the organization and a fight for a broadcasting deregulation bill in the Congress at the same time. It regarded cable legislation as primarily a matter affecting the cities and the cable industry that would have relatively little impact on broadcasting. It was a miscalculation.

The Motion Picture Association of America was engaged with the cable industry in a bitter battle over copyright fees. After passage of the Copyright Act in 1976 the FCC had moved to lift many of the regulations on carriage of broadcast signals by cable systems. The issue of how cable systems would pay for the rights to carry those signals was left to a new federal agency, the Copyright Royalty Tribunal, to determine.

In 1982 the Tribunal had shocked the cable industry with a ruling that most systems would have to pay a fee equal to 3.75% of their entire gross revenues for each distant broadcast signal carried on the system.

The money would then be put into a fund that would be divided among the major copyright holders.

The ruling effectively increased copyright fees for some cable systems by as much as 500% and ignited another battle over the issue on Capitol Hill.

Cable systems began to drop distant broadcast signals, dealing blows to such companies as Turner Broadcasting System, whose WTBS Superstation was the most widely distributed distant signal, and to United Video, which was the common carrier for several distant broadcast signals, including WGN from Chicago and WPIX from New York. By 1984 the NCTA estimated distant broadcast signals had lost some 10 million viewers because cable systems had dropped them to avoid having to pay increased copyright fees.

Turner and United Video lobbied on the Hill vigorously to try to get Congress to pass legislation to reverse the CRT ruling, but to no avail as the MPAA and the broadcasters fought back in defense of copyright fees.

The effort was shortsighted on the part of the broadcasters. As cable operators dropped distant broadcast signals, they added cable exclusive programming, boosting the viewership for CNN, ESPN, MTV and the other cable networks which owed no copyright fees.

Embroiled in these other legislative battles, the MPAA and the NAB largely left the issue of cable legislation to the cities and the cable operators to work out.

There is no particular moment or incident in the years prior to 1984 that one can pinpoint as the beginning of the cable industry's effort to pass a comprehensive law governing the industry. Congress had been kicking around the idea for decades and one bill or another had been introduced in every Congress since the late 1950s.

But if one wanted to find a beginning for the process that culminated in passage of the bill in October, 1984, a leading candidate might be a meeting that took place in early 1975.

NCTA president Bob Schmidt and executive vice president Tom Wheeler were paying a courtesy call on Tim Wirth, the newly elected Democratic congressman from Colorado who came to Washington on the heels of Watergate and who had just been named to the House Telecommunications Subcommittee.

As they sat down together in the ornate Rayburn Room just off the House floor, Wirth began a tirade of criticism. He was a natural ally of the cable industry, he told Schmidt and Wheeler. Several of the largest cable operating companies were based in Colorado, and their executives had contributed to his campaign. And Wirth believed that lifting the regulations on cable would promote diversity of programming so that the viewers, not the broadcast networks, could make the decision on what kind of shows were seen.

But cable, he said, had done a miserable job of sending its message to the Hill and was widely regarded in Congress as unreliable, greedy and ineffective. "I'd like to be with you," Wheeler remembers Wirth telling them. "But you are totally incompetent."

Wheeler recalls that as he and Schmidt walked out of the Capitol that evening they looked at each other and vowed "Okay, buddy. We're going to show you you're wrong."

Schmidt and Wheeler set about to transform the NCTA. Before Schmidt had come aboard, the organization had been in disarray for the better part of 20 years, lurching from one president to another, unable to form a common policy and going for months at a time without any leadership at all.

Schmidt, young and energetic, began to lay down the law with the NCTA board, insisting that he be allowed to run the organization with direction from

Changing Hands: *Thomas Wheeler succeeded Robert Schmidt as NCTA president in 1979, and laid the groundwork for passage of the 1984 cable bill.*

the board of directors but without day-to-day interference. Wheeler recalled that a defining moment came when the board was invited to the White House sometime in the mid-1970s.

As the group gathered at NCTA headquarters, the members began to bicker about who would say what, what message they would want to convey to the President and the other officials who would be there. After a while Schmidt asked for quiet. Then the former USC football star began to give them a locker-room-style lecture.

"Okay you guys, how many of you have ever been to the White House before?" he asked. When nobody raised a hand, Schmidt continued, "Well I have, and I want to be invited back. So when we get there I will speak for this organization and nobody else is going to say anything unless I tell you to." It worked. As time went on the highly individualistic members of the NCTA board slowly began to grant more authority to Schmidt and Wheeler and to limit the board's involvement in the NCTA's day-to-day activities.

Wheeler was a master of marketing and positioning. He was naturally flamboyant, tall and good looking with dark hair and a dark mustache and a deep, resonant voice. He had been only 33 years old when he left the National Grocers Association to join the NCTA. He recognized early that the problem the cable industry had was that it approached every legislative and regulatory issue from the point of view of what this meant to the cable industry. Instead, Wheeler understood, the industry needed to make its case based on what this would mean for the consumer. The trick was to persuade the members of the federal government that regulation of the cable industry stifled the development of new, diverse programming services that would provide more choice for the consumer.

To drive that message home, Wheeler turned showman. He enlisted the cable programming services to host premieres of some of their original programming to which he would invite members of Congress, the FCC and key staffers. Warner's QUBE system in Columbus would tape every home football game at Ohio State University and ship the tapes to Washington where Wheeler, a native of Ohio, would host a party for the Ohio delegation to watch the game, not available otherwise in D.C.

When the pole attachment bill was up before Congress, the NCTA learned that one telephone company in Tennessee had cut down the cable it found on its poles and chopped it into foot-long pieces. Wheeler got hold of the cut-up coax and sent a piece to each member of Congress with a message stating that "this is what the telcos are doing to the cable companies."

And when the pole attachment bill passed, NCTA vice president Bob Johnson called a friend at C&P Telephone, purchased a telephone pole, had it sliced into inch-thick pieces and mounted a bronze plaque on each piece with a message of gratitude for each sponsor of the bill.

Johnson, a graduate of the University of Illinois and the Woodrow Wilson School of Diplomacy at Princeton University and a former press aide to D.C. Delegate Walter Fauntroy, was one of a cadre of energetic young lobbyists Wheeler and Schmidt brought on board at the NCTA.

Others included Bob Ross, an attorney for Southwestern Pacific Communications (a forerunner of Sprint), Brenda Fox, who had been assistant general counsel for the NAB, and Char Beales, who came to NCTA from a Washington broadcast station.

Wheeler promoted Kathryn Creech and eventually moved her into a spot running the Council for Cable Information, a Wheeler idea for an industry-wide advertising campaign to improve cable's public image. (The group died after only a partial effort when several large cable companies, led by TCI, refused to participate.)

Entrepreneur:
Former NCTA lobbyist Robert Johnson (right) with TCI Chairman Bob Magness whose company provided Johnson the backing to launch Black Entertainment Television in 1980.

The NCTA staff, which had been almost exclusively white and male, began to look more diverse.

After Wheeler succeeded Schmidt as NTCA president in 1979 he hired as the vice president of government relations James P. Mooney. Mooney had been the top aide to the House Democratic Whip, the third-highest-ranking position in the House. Mooney had been recommended to Wheeler by Tim Wirth. He was the first NCTA staffer ever hired who had had extensive experience on the Hill.

Wheeler also instituted a program of regular visits by cable company CEOs to the editorial boards of major newspapers such as the *New York Times* and *Wall Street Journal* to get the message about cable across to the opinion makers of the media.

The NCTA leaders also began to pay closer attention to local politics in the districts of key members of Congress. After his meeting with Wirth, Wheeler began to show up at the young congressman's campaign events, both in Washington and in Denver, on a regular basis, making sure Wirth knew he was there. NCTA urged its members to get to know their congressmen and worked with key state organizations to bring cable operators to Washington to lobby on a regular basis.

CATA, under the leadership of Steve Effros, also stepped up its Washington presence, with regular Cata-grams to members keeping them informed about D.C. events, and the APIL awards, which Effros named after a favorite saying of House Speaker Thomas P. "Tip" O'Neill: "All Politics Is Local."

In 1980 the cable industry's support of Tim Wirth paid off big time. Lionel Van Deerlin (D-Calif.), who had been chairman of the House Telecommuni-

Cable Pol:
Jim Mooney steered forces both inside and outside the cable industry to compromise during the final months of negotiations over the 1984 cable bill.

cations Subcommittee, lost his seat in the Reagan landslide. Wirth, only in his fourth term, was elevated to the post.

As the chief counsel of the subcommittee Wirth chose a young communications lawyer from New York, Tom Rogers. One of the first fights Wirth picked was over the funding for public television, making bitter enemies of the commercial broadcast networks in the meantime.

"In the drive to allow diversity of programming, cable became the white knight," Rogers recalled. "We recognized that we needed to do something to protect the emergence of cable as a pro-competitive force. There was a serious issue of whether cable could ever reach its full potential if you let every franchise regulate it in a different way."

Wirth had been particularly impressed with the creation of Nickelodeon, C-SPAN and Black Entertainment Television as examples of the kind of programming diversity cable could provide if left to its own resources.

And there was plenty of evidence that continued regulation by the cities would prevent cable from reaching its full potential.

Cities were constantly trying to deny franchise renewals to incumbent cable operators in an effort to strike a better deal with a new operator or to take over lucrative cable systems themselves.

The threat that a city could yank a franchise at any time or deny a renewal was proving to be a greater and greater hindrance to cable operators, and a major stumbling block in obtaining financing to upgrade systems. The same was true with regard to rate increases, which, except in California, had to be approved by each city.

It was a vicious circle. The threat that a city would deny a franchise renewal or a rate increase made it difficult to borrow money to upgrade the systems, which made it more likely that the city would deny the next request for a renewal or rate increase which then made it even more difficult to secure additional financing. And no city council member liked to be the one to champion a rate increase.

Even when cities did grant franchise renewals or rate increases, they often imposed new regulations at the same time.

All the political leaders were influenced by the franchising frenzy of the late '70s and early '80s in which major cities were able force cable operators to promise huge complex systems with a host of ancillary services. Every small town wanted just the same, if not better.

Sacramento included as a condition of its franchise the right to buy back the system after the franchise expired for the amount of capital that had been invested in it.

Fairfax, Va., insisted on the right to raise the franchise fee from 5% of gross revenues to 8% at any time and with no warning. Tucson wanted the right to buy the system at any time for book value. Boston at one point asked for the right to run its own pay television service, with all the revenue going to the city.

An NCTA study in 1982 found that a typical 35-channel system in a metropolitan area would have to spend 22% of its revenue to cover the costs imposed by local, state and federal authorities. This included franchise fees, copyright payments, expenses for local origination, government and educational channels, the cost of filling in forms and the odd other requirement such as planting trees or building a fire house (both of which were included in actual franchise agreements).

The report noted that competing distribution systems such as MDS, DBS and SMATV did not have such costs. "How can the cable industry compete if we have to give 22% of our revenues away for regulation and others do not?" the NCTA report asked.

In 1983 the U.S. Senate passed a bill that would have ended the ability of the cities to regulate rates charged by cable operators and placed the burden on the cities to prove that the incumbent cable operator had failed to provide decent service before a franchise renewal could be denied. The measure also would formally have recognized the rights of cities to grant franchises and increased the amount they could charge for franchise fees up to 5% from the ceiling of 3% the FCC had been imposing.

The Senate bill had been endorsed by the National League of Cities. But when its details became known the NLC backed off and insisted on major concessions in return for support for a measure in the House.

There the cities had much better strength since most big city mayors were Democrats and the House was controlled by the Democrats.

But Wirth was determined to fashion a bill that would pass. Slowly he began to put together the pieces.

To win the support of the Congressional Black Caucus, Wirth agreed to include a provision to require cable systems to hire minorities and women in the same proportion as they existed in the communities the cable systems served. He "crossed the aisle" to win support from deregulation-minded Republicans, and crafted a majority on his subcommittee for the bill.

And he wooed the powerful chairman of the Commerce Committee, which would have to approve the bill before it went to the House. John Dingell, an old-line Democrat from Michigan who had followed his father to Congress, was inclined to support the city position. But he recognized that the tide was moving in favor of legislation. He preferred to take credit for a new law on the books than to block legislation a majority of his own committee favored.

He urged the two parties to the deal — cable and the cities — to work out an agreement. As chairman of the committee, Dingell had a reputation as one who could intimidate witnesses and strong-arm lawmakers and lobbyists alike. He wasn't reluctant to do the same with regard to the cable bill.

In the late summer of 1983 Dingell agreed to appear in Colorado at a fundraiser for Denver congresswoman Pat Schroeder. Invited to hear him speak, and to make campaign contributions, were the leaders of some of the cable industry's largest companies.

Dingell had been on a hunting trip in Wyoming, and he appeared before the group dressed in work boots and a plaid flannel shirt. He looked like he hadn't shaved in a week.

Gatekeeper: *John Dingell (D-Mich.) wielded perhaps the most important gavel in the House of Representatives on communications issues over the past 20 years.*

Peering over his glasses, in deliberate, measured tones, he told the group that if the cable industry and the cities failed to reach an agreement on a bill, he was prepared to write his own and to send it to the floor of the House for a vote. And the bill he might draft, "I would observe to you," might not be one that the cable industry much liked.

He delivered the same message to the cities, and throughout 1983 and 1984 the two sides spent hours at the bargaining table attempting to work out a compromise.

Wheeler, Mooney and NCTA attorney Chuck Walsh led the cable side and New York City Council president Carol Bellamy, Pittsburgh Mayor Richard Caliguiri, and Tucson mayor Tim Voltz represented the cities.

By the spring of 1984, after several false starts, the two sides had at last crafted an agreement which they felt could win approval from their boards of directors. The measure watered down considerably what had been passed in the Senate. Instead of immediate rate deregulation, it allowed cities to continue to set rates for a period of five years (during which time cable operators were entitled to raise rates by 5% per year). Even after the five-year period, the compromise stated that the FCC could regulate rates in areas where there was no significant competition to cable. And the compromise also diluted the franchise renewal provisions of the Senate bill. Instead of giving the incumbent operator a presumption of renewal, it simply guaranteed that the operator would receive due process.

But the compromise would eventually lead to rate deregulation and would provide other benefits to the operators, even if not as many as the Senate bill had.

Wheeler took the measure to the NCTA board meeting in April and won a near-unanimous endorsement. It seemed a fitting end to Wheeler's tenure as NCTA president (he also announced in April that he would step down from his post in July to be succeeded by Mooney.)

Wheeler and Mooney unveiled the deal at the NCTA convention in Las Vegas the next month, and the measure was greeted by the trade press as a victory for the cable industry. That is the way most cable operators saw it as well.

Until it was read by Leonard Tow.

A graduate of Columbia University and a professor of economics, Tow had been the senior vice president at TelePrompTer from 1965 through 1973 under Irving Kahn. After leaving TelePrompTer, Tow founded his own MSO, Century Communications, with the financial backing of Sentry Insurance Company. By 1984 Century was serving almost 300,000 customers.

Well over six feet tall with wavy hair, Tow was an imposing presence. And like his former boss, Irving Kahn, he wasn't shy about expressing his opinions. In an industry of rugged individuals, Tow was more rugged and individual than almost anybody else. He wasn't even a member of the NCTA.

At bottom he believed that the government, city, state and federal, had no right to regulate the cable industry at all except to ensure public safety. He believed that cable had an absolute right to freedom under the First Amendment to the Constitution which states that "Congress shall make no law abridging the freedom of the press."

Cable, Tow reasoned, was simply an electronic publisher and entitled to the same freedom as the publisher of a newspaper. Cities wouldn't dream of trying to regulate what a newspaper could charge or how it should be distributed or what its content might be. Nor, he felt, should they be allowed to regulate cable.

When Tow got a copy of the compromise agreement on the cable bill during the NCTA convention in Las Vegas in 1984 he was outraged. The industry, he felt, was trading away its fundamental rights.

"I got a copy, and I was astounded at what I read. I was very, very, very upset," he told *Cable TV Business*.

Tow led a personal crusade to undo the agreement. He took his case to members of the NCTA board some of whom,

Point Man: *Pittsburgh Mayor Richard Caliguiri saw first hand the effect of franchise over commitments in his city. Caliguiri was part of the National League of Cities negotiating team for the 1984 cable bill.*

he later said, had not even read the agreement when they voted to endorse it. He took out full-page ads in the trade press denouncing the legislation, and he toured the country drumming up support among operators for his position. He filed suit against the NCTA and Wheeler personally, charging they had violated his First Amendment rights. He even charged they had violated the Racketeering and Corrupt Influences (RICO) law designed to help law enforcement agencies indict mob leaders.

Tow was making his case while the NCTA was in between leaders. Wheeler had announced his resignation but not yet left. The outgoing NCTA board, which had approved the compromise, was headed by two people who no longer owned cable systems (chairman Monroe Rifkin, who had resigned his post at ATC, and vice chairman Gus Hauser who had left Warner Amex).

In an unprecedented move at its April meeting on Captiva Island, Fla., in April, the board failed to approve Hauser's elevation to chairman and instead elected Western Communications president Ed Allen to the post. (The conspiracy to oust Hauser was hatched at a cocktail party the night before the NCTA meeting. The opponents of Hauser felt it would be bad for the industry to be headed by Hauser who, as head of Warner Amex Cable, had made many of the blue-sky promises to the cities that current Warner Amex CEO Drew Lewis was now trying to undo.)

Tow's effort picked up steam when the Supreme Court, in a unanimous decision, ruled in June that the state of Oklahoma had no right to ban liquor advertising on cable systems in the state. Cable, the court ruled, should be regulated by the FCC, not by local and state governments.

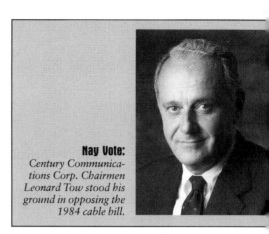

Nay Vote:
Century Communications Corp. Chairmen Leonard Tow stood his ground in opposing the 1984 cable bill.

The FCC the previous fall had adopted a ruling that would have preempted state and local regulation of rates for basic cable service (except tiers with broadcast stations) and allowed cable operators to move channels into upper tiers to avoid rate regulation.

After the Supreme Court ruling, more and more cable operators were inclined to take their chances in the courts and at the FCC than to support the legislation worked out with the cities. Wheeler and Mooney could feel the deal beginning to unravel. Mooney warned the cable industry "not to lose its nerve and fall into a condition of peevish grumpiness." Tow, and his attorneys, Harold Farrow and Sol Schildhause, were ejected from an NCTA board meeting when they attempted to make their case against the bill.

The California Cable Television Association voted to withdraw support for the compromise agreement and the New England Association and CATA quickly followed. By the end of July the NCTA itself had voted to change position and oppose the bill.

Mooney took on the task of explaining the situation to Dingell. "With Dingell," Mooney later recalled, "it was very important to listen very carefully.

215

He is very deliberate and formal and glowering. And what he calls you is very important. If he calls you 'My dear friend' you know you are in deep trouble. I made my pitch to him about the changed circumstances and our need to go back to redraft the bill. He wasn't happy. But I saw it as positive sign that at least he was calling me 'Jim' and not 'my dear friend.'"

Dingell agreed that the Supreme Court decision was a new element and promised Mooney that he would talk to the city representatives and ask them to come back to the table and to be reasonable.

But the city folks dug in their heels, insisting that they were prepared to push for a bill even without cable industry support. The NCTA wasn't about to back the measure that had been worked out before the Supreme Court decision. Dingell wouldn't allow the bill to go to the floor unless the two sides agreed. The measure seemed dead.

"What are the chances of 4103 becoming law and establishing a national regulatory policy for cable?" asked an article in the Aug. 13 issue of *Broadcasting*. "Between Slim and none. And Slim just left town."

Smooth Operator: Ed Allen's down-to-earth personality help restore cable's reputation in Congress in 1984.

But then the NCTA began to give a little. Ed Allen, an affable, avuncular man who had been in the cable business for 30 years, was an ideal choice to smooth the frayed tempers of the badly divided board.

"Ed Allen," Mooney later said, "is a great man. He is, in his folksy way, a visionary. He understood that cable needed to be more than a regulatory stepchild. And he had a way of dealing with people that was persuasive without being insistent. He didn't come across like a hard charger, but he was a leader. And he was persistent. Once he got hold of something he wouldn't let go."

Allen took the position that if Congress did not pass a cable bill in 1984, it would likely never pass one. Moreover, he felt, if the legislation was going to die, he wanted it to look like the cities killed it and that the cable industry had walked the last possible mile to reach a compromise. He recalled the 1960 experience where cable had switched positions and managed to kill a cable bill in the Senate, but in the process alienated powerful members of Congress who later found ways to punish the industry.

In August the NCTA board voted to allow Mooney to reenter the talks with a laundry list of requests but with some flexibility and with the intent that the industry would make the strongest possible effort to reach a compromise. The National League of Cities, prodded by Dingell, went back to the table as well in September.

At the end of September, with the deadline for Congress to adjourn less than two weeks away, the cities and the cable industry sat down again to work out a compromise. On Sept. 25 they emerged with a deal and the NCTA board voted, by 20-2, to support it.

Charles Dolan and Gene Schneider were the only 'nay' votes.

In the final set of negotiations Mooney had been able to win back much of what had been surrendered in the previous round. Rate deregulation was slated to take effect in two years after passage (a logical compromise between the five-year grace period in the previous compromise and the immediate deregulation in the Senate bill). In the meantime cable operators were allowed to raise all rates by up to 5% per year.

On franchise renewals, the compromise set up a specific timetable and process for renewal hearings. It also included a statement of purpose that the objective of the legislation was to prevent cities from unfairly denying franchise renewals to incumbent operators.

As the House session went into overtime, the staff members of the House and Senate committees scrambled to work out a version of the bill that could pass both chambers. By Oct. 5, Mooney was able to report to the NCTA board that all the communications issues had been worked out. But there was still a lot of politics in the way.

Roadblocks went up in the Senate, planted by Charles Dolan who had hired Charles Ferris, the former FCC chairman and former top Senate staffer, to kill the bill. Ferris worked hard to find opponents, and on the day before adjournment, Sen. Howard Metzenbaum, a firebrand liberal from Ohio, announced that he would filibuster the bill (effectively killing it) unless the franchise renewal process provided for automatic input from public interest groups.

As Mooney later recalled: "We were in the lobby outside the Senate and Metzenbaum came in with his Nader proposal." Mooney pointed out that reopening the entire renewal section of the bill would be impossible at this late date.

Eventually Metzenbaum indicated that he might be inclined to work something out if the NCTA would, as Mooney later recalled, "take a greater interest in some Ohio political races." The two worked out a deal that would state that the franchise renewal process would be open to input from all interested parties. Metzenbaum backed off.

But there was one more hurdle left to clear, as Wirth worked against the looming deadline for adjournment. The House version of the bill had included a section that would have required each cable system to hire minority employees in proportion to their presence in the community served by the system.

The provision had been inserted at the insistence of Rep. Mickey Leland, a Democrat of Texas, a member of the powerful Congressional Black Caucus, which was keeping a close eye on all legislation to determine how it might affect minority hiring practices. Wirth had given his word to Leland (and to Leland's administrative assistant Larry Irving) that the provision would remain in the bill.

But the Senate version had no such provision, and the conservative members of the Senate — led by Jesse Helms of North Carolina and Orrin Hatch of Utah — refused to allow a provision mandating what they viewed as "racial quotas" into the legislation. There seemed to be no way out.

By Wednesday night Oct. 10, Wirth returned to his office after yet another meeting with the Senate side. There were less than 24 hours left before Congress would adjourn. The Senate wouldn't give in. Wirth told his top aide, Tom Rogers, "There's no way. This won't happen."

As Rogers recalled, all the lobbyists had gone home, convinced the measure was dead. Mooney remembered leaving the Capitol late that night with the feeling that the bill "was in grave danger. I couldn't sleep." Time was running out.

"I went into a room with Howard Simons (co-counsel of the Telecommunications Subcommittee)," Rogers remembered. "We talked about how hard we had worked to get to this point. And we both knew that if the bill didn't pass now, it would never happen again. So I went in to see Tim and asked him to talk to Mickey and give us permission to negotiate" on the EEO provisions.

Wirth called Leland and told him that if the bill died there would be no

Cable Joins The Political Money Lobby

A key weapon in the cable industry's fights in the Congress was CablePac, a political action committee established in the late 1970s to raise funds that would be contributed to the campaign coffers of key members of Congress.

CablePac was started in 1976 by NCTA President Bob Schmidt who tapped his former law school classmate, Richard Loftus, to run the operation. Members of the industry had made political contributions before, but never in such an organized fashion.

The concept of political action committees had emerged from the post-Watergate campaign finance reforms. The new rules prohibited corporations from making campaign contributions and limited the amount any one individual could give to a candidate. But it allowed companies or industries to form PACs which could solicit funds from employees and then contribute those funds to candidates. It also allowed corporations to make contributions to enable the PAC to conduct its fundraising activities.

The PAC became a very powerful political tool for companies and industries seeking clout in Washington. But it required enormous energy to organize and execute.

Loftus had become involved in cable in the winter of 1961 when, just after he graduated from college, he was tending bar in upstate New York. A group of cable installers came in to get out of a blizzard. They got to talking about cable, and showed the young bartender how to climb a pole in the winter (putting his spikes in on the icy side so that his belt would be on the dry side and not slip).

After law school, Loftus went to work for Entron and eventually started his own cable operating company, building systems in Hoboken, N.J., Annapolis Md., and the Quantico Marine Corps base, among other sites.

Loftus went at CablePac like a crusade. He visited the chief executives of the major MSOs to enlist their support. It was a tough sell.

"I remember calling on (ATC Chairman) Monty Rifkin," Loftus said later. "He said, 'Let me get this straight. You want me to give you $3,000 so that you spend that money to solicit $30,000 from ATC employees to give to a bunch of politicians?!'

"I think you've got it," Loftus replied.

Eventually his efforts paid off with most of the major cable companies. The only exception was TCI, whose president, John Malone, was unwilling to contribute to a political process he regarded as flawed and irrational. TCI continued to refuse to participate in CablePac until the late 1980s when it set up its own political action committee.

But Loftus did more than raise money. He and NCTA vice president of public af-

progress on minority hiring, whereas if some compromise could be worked out it might be better than nothing.

"Mickey agreed," Rogers said, "and told Tim 'Just don't sell me out.'"

Wirth told Rogers: "See what you can do." Rogers phoned the counsel of the Senate Constitutional Law subcommittee, waking him around midnight, and asked him to come down to see if they could negotiate something each could support.

The two agreed to write a new provision that would eliminate any reference to numerical quotas, but would beef up the FCC's equal opportunity enforcement mechanism. Rogers took the deal to Leland and made the argument that the

fairs Lucille Larkin visited the editors of the top 50 newspapers in the country to brief them on cable. They launched an effort to educate cable company employees (from the CEO to the installers) on the public policy issues the industry faced. And they used the growing legions of CablePac contributors as shock troops in lobbying efforts, urging them to write letters or call their congressmen when a critical measure was up for a vote.

"For a member of Congress, a handwritten letter from a constituent is like a thermonuclear device," Loftus later said. "It has enormous impact."

Loftus would personally write "thank you" notes to each CablePac contributor and then keep the name on file, asking for help when a member of Congress from that district needed a little prodding on an issue.

Loftus launched an annual series of breakfasts at the NCTA conventions, inviting key government officials to speak (and dragging most cable executives out of bed earlier than they would have liked.)

He kept at the task for 13 years, without compensation and without even the gratification of giving away the money he raised. That task was left to the NCTA presidents and the CablePac board. The campaign contributions did not actually ever buy votes. But they did help NCTA officials gain access to members of Congress and helped raise funds to reelect the congressmen who were most likely to support the industry in critical votes.

In 1989, when Loftus stepped down from the CablePac chair, the organization raised almost $1 million dollars from over 12,000 contributors. The amount still lagged what the broadcasters, Hollywood studios and telephone companies were able to raise. But it was a long way from the $5,000 total that had been raised the year before Loftus took over.

Money, it is said, is the lifeblood of politics. CablePac had made the NCTA a major bloodbank, able to offer critical transfusions to its supporters on Capitol Hill.

Money Man:
Richard Loftus led cable's efforts to raise money to affect the political process in Washington, D.C.

Extra Mile:
Tom Rogers was chief counsel to the House Telecommunications Subcommittee in the early 1980s. In the final 24 hours before Congress adjourned in October 1984, Rogers devised a compromise on the cable bill's EEO provisions, clearing the way for passage.

compromise was "a major step forward," even if it wasn't everything Leland had wanted. Leland agreed.

If it hadn't been for Leland's willingness to compromise "to step up in a highly charged atmosphere and make a decision," as Rogers later described it, the Cable Act would never have passed.

The next day bleary-eyed members of the House and Senate both cleared the measure, just a few hours before the Congress adjourned.

The Cable Act of 1984, which had been cobbled together in the waning hours of the congressional session, changed the cable industry more fundamentally than any development except the satellite.

The legislation largely freed cable from regulation by the cities. It allowed systems to charge what the market would bear and to invest huge chunks of the new revenue from basic cable into the development of new programming networks, new technologies and new services. The bill gave cable operators enormous new clout with the financial community, which could make loans without having to worry about whether cities would permit rate increases or renew franchises.

But the Cable Act of 1984 also put the industry under the protection of the Congress rather than the Constitution or the courts, as people such as Leonard Tow had preferred. And what the Congress could grant, the Congress could take away, as the industry would discover less than a decade later.

Golden Age

Passage of the 1984 Cable Act ushered in the Golden Age for the cable industry. The new law ended regulation of rates and provided assurance of franchise renewals. And a series of legislative, judicial and business victories followed that propelled cable to the forefront of American business. All of a sudden this sleepy little industry became a darling of Wall Street and a financial and programming powerhouse generating billions of dollars of revenue and threatening to topple the broadcasters from the position of dominance they had held over television since the 1940s.

But before the fruits of the Cable Act had ripened, some of the biggest corporations that had entered the cable operating business at the beginning of the decade made their exits, demonstrating that just because a company has a well-known name doesn't mean it necessarily has an exquisite sense of business timing.

The departures of American Express, Westinghouse Electric Co., General Electric Co. and Dow Jones & Co. from cable system ownership enhanced the power of those companies that remained in the business for the long haul. These included the biggest MSOs such as Tele-Communications Inc., ATC and Warner. And the departure of some big names also spurred the growth of mid-sized companies such as Continental Cablevision, Century, Falcon and Comcast.

The story of Comcast in the last half of the 1980s illustrates how a mid-sized MSO parlayed rate deregulation, urban franchises, deft financing, a willingness to bet the farm over and over, and a deep belief in the long-term prospects of cable to enter the ranks of America's biggest companies. At the heart of the Comcast story were three men: Ralph Roberts, Dan Aaron and Julian Brodsky.

Julian Brodsky is a big man. He has big eyes, a big nose, a big voice and a big laugh. And he likes to do big deals. In 1984 he would do the biggest deal of his life, one that would double the size of his company and catapult it into the ranks of the nation's largest cable operating companies.

Twenty years earlier Brodsky's size almost cost him his first job in the cable business. He had applied for work with Ralph Roberts, a businessman who had just sold his men's accessories company and was in the process of buying his first cable system. When Brodsky came to see Roberts about a job, Roberts didn't know how to respond. There just didn't seem to be a way to squeeze a man of Brodsky's dimensions into the tiny office where Roberts was working.

"That's okay," Brodsky said. "I'll make do."

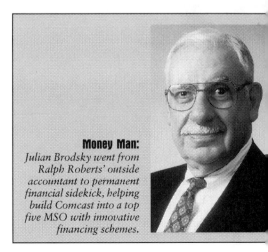

Money Man: *Julian Brodsky went from Ralph Roberts' outside accountant to permanent financial sidekick, helping build Comcast into a top five MSO with innovative financing schemes.*

He arrived for his first day on the job with a folding card table and chair which he set up in the hallway. Brodsky wasn't about to let this chance pass him by.

To join Roberts, Brodsky had left a Philadelphia accounting firm where he'd done tax work for several Pennsylvania cable companies and had helped Roberts sell his business. "I loved the cable business," Brodsky later recalled, "and I loved Ralph Roberts. When the chance came to put the two together I wasn't going to let it go by. I went to Ralph and said, 'You aren't doing this without me.'"

Roberts was born in New Rochelle, N.Y., and grew up in Philadelphia. He always wanted to be a businessman. After graduating from the Wharton School at the University of Pennsylvania and serving a stint in the Navy, he started a business making golf clubs. That folded when somebody forgot to heat the shafts of one shipment of drivers, causing them to bend in the shape of a pretzel when the golfer first teed off.

Roberts then worked for an ad agency in New York and later for the Muzak Corp. He founded his own men's accessories business, selling belts, suspenders and cufflinks. But two new developments – the Sansabelt pants and the button shirt sleeve – convinced him in the early 1960s that the time had come to get out. The sale left him with a handsome profit but no business to run.

"Then one day," Roberts later recalled, "in walks Dan Aaron." Aaron was a former Philadelphia newspaper reporter who had taken a job with Jerrold Electronics writing manuals and later moved to the cable operations side of the business. He was calling on Roberts in an effort to sell him one of Jerrold's cable systems in Tupelo, Miss.

Robert agreed to buy the system on one condition: that Aaron agree to join him to run it.

Roberts, Aaron and Brodsky made an unlikely looking trio. Brodsky was a huge bear of a man, Aaron a small wiry type and Roberts a dapper, courtly, fellow. Brodsky was always looking to expand and to find new ways to finance expansion. Aaron was more concerned with making the existing systems work and with not overloading the Comcast staff. And Roberts came down somewhere in between.

As Brodsky later recalled, "I had my foot on the gas, Dan had his foot on the brake and Ralph had his hands firmly on the steering wheel."

They structured each cable system as a separate entity so, according to Roberts, "If one went down the tubes, the whole company wouldn't be in jeopardy." Roberts, a child of the Depression, also insisted that the company always keep a large sum of cash on hand, just in case.

And in the early days he didn't put all his eggs in the cable basket. He became a major franchisee of Muzak and kept a small men's fragrance business. The men's accessories business, he re-

Jerrold Salesman:
Dan Aaron helped launch Comcast by selling Ralph Roberts the Tupelo, Miss., cable system. It marked the beginning of a lifelong business relationship.

called 30 years later, still seemed much more exciting than cable which, although profitable, seemed to him "dull as dishwater." (Years later, some Comcast employees would discover a stock of the cologne in a back closet when Comcast moved its headquarters.)

The first systems were financed by local Philadelphia banks where Roberts had done business, by Jerrold Electronics and by finance companies that would loan Comcast money at four to five points above prime.

In 1965 Brodsky discovered what he later called his "Holy Grail" of financing. It was a deal with Home Life Insurance Co., one of the first insurance companies to lend to cable, that provided 12-year loans at 6%-7% fixed interest.

Later Brodsky would make an even sweeter deal. To finance the acquisition of a system in Flint, Mich., Comcast arranged a 16-year loan from the John Hancock Insurance Co. with the stipulation that there be no principal payments, only interest, for the first 11 years.

For a cable operator, where the big construction costs come in the first three or four years of operation in a major market and profits often don't appear until 10 years after the system launches, this was a near-perfect financing vehicle.

Brodsky made use of limited partnerships and took the company public in 1972, just before the bottom dropped out of cable stocks. (Comcast opened trading at $7 a share in June of 1972 and had dropped to $5.50 by the following month. It would hit 75 cents a share at bottom, which Brodsky recalled as the days when the Comcast investor could swap a share for a bottle of beer.)

Brodsky also made use of new developments on the tax front. In the 1970s after a long search, he was finally

Steady Hand: With Dan Aaron and Julian Brodsky on board, Ralph Roberts had all he needed to transform Comcast from a small cable company to a dominant player in the telecommunications business.

able to find a law firm that would issue an opinion that the revenue from industrial revenue bonds would be tax-free to the investor. Comcast made use of seven such arrangements (saving 200 basis points in interest rates) until the law was repealed for entertainment companies (after Congress discovered that industrial revenue bonds were being used to finance, among other things, pornographic movie houses.)

In 1980 Congress, in an effort to spur capital spending, enacted a new law that allowed companies with excess tax write-offs to sell those write-offs to other companies that could use them. Through GE Capital Corp., Comcast was able to sell millions of dollars of prospective tax write-offs to such companies as Ford, IBM and GE itself. It would then use the money to construct or acquire cable systems that would generate even more tax losses that could be sold for even more cash.

The deal only lasted two years before Congress realized that it was losing billions in revenues. But Brodsky loved it while it existed. "You hit the ball where it is pitched," he recalled with a laugh.

By 1983 Comcast had become the

18th largest cable company in the country, with some 500,000 subscribers. But it still wasn't in the ranks of Group W, TCI or ATC.

Then Brodsky got a call from a friend at Merrill Lynch. Comcast was about to become a bit player in a much larger business drama of the 1980s involving such players as Kohlberg, Kravis & Roberts, Drexel Burnham Lambert, Shearson Lehman Bros., Merrill Lynch and other huge New York investment firms.

KKR in particular had mastered the technique of joining with the management of a company, borrowing huge sums of money and then buying out the shareholders to gain control of the enterprise. Drexel, under the leadership of Michael Milken, had pioneered the use of junk bonds – high yield instruments that were used to finance acquisitions. (Brodsky, although he never used junk bonds, admired Milken as a brilliant finance mind and adopted as his on-line names, "Highyield" and "Junquer.")

These were the days later depicted in the book and movie *Barbarians at the Gate*. The competition between the investment houses to do deals had become white hot by early 1985 when Storer Communications, one of the biggest MSOs with some 1.5 million subscribers, went into play.

Cable companies were logical targets for corporate raiders. Their big debt payments, huge construction costs and tax write-offs for depreciation depressed their earnings and with them the price of their shares. But they also generated huge internal pre-tax cash flow. And, after passage of the 1984 Cable Act that would deregulate basic cable rates in 1986, the Wall Street firms could see future increases in revenues.

Low stock prices, high cash flow and the prospect for big jumps in revenue made those companies where the majority control could be purchased on the open market tempting takeover targets. Many cable companies – Jones, Comcast, TCI, Cablevision Systems, Cablevision Industries, Cox – were structured so that control was in the hands of a single individual or a small group, usually including management.

But some others were vulnerable. Among them was Storer. Like many cable companies in the early 1980s Storer had won some big franchises and was busy upgrading other systems. Construction costs, depreciation and interest had depressed earnings and the company had posted a loss of $40 million in 1983 and $16 million in 1984.

The stock price by the middle of 1985 was $65 a share versus the $92 a share that Paul Kagan estimated the company would be worth if it were broken up and sold.

For the Wall Street sharks these numbers were blood in the water.

Coniston Partners opened the drama with a bid to buy out the company in partnership with Peter Storer, the son of the founder. But it wasn't long before KKR and others joined in the fray.

Dealmaker:
Stewart Blair was a John Malone aide at TCI who helped Julian Brodsky put together the cable MSO consortium that bought Group W Cable for $1.7 billion in 1985.

When KKR got in, Merrill Lynch was determined not to let this deal slip into the hands of its Wall Street rival. But it needed a partner who knew something about cable.

Comcast had begun a relationship with Merrill Lynch in the early 1980s when Brodsky, at Merrill Lynch's invitation, became one of the first cable CFOs to sell bonds in Europe to raise capital.

Merrill Lynch and Comcast figured Storer would sell for around $1.2 billion. It would be a huge leap for Comcast, a company that itself was worth only about $500 million. But Shearson Lehman, Comcast's investment banker, assured Brodsky the deal could be done. Merrill Lynch offered to raise $900 million, using its own balance sheet as collateral, and Comcast would have to raise the rest.

In the end the Merrill Lynch/Comcast team was outbid by KKR, which purchased Storer for $1.6 billion (and sold it three years later for $2.8 billion to a partnership of TCI, Knight Ridder and...Comcast).

Though the bid fell short, for Comcast it was like a trip to the majors for a AAA baseball player. It had given them a taste of what the big leagues were like and given other teams a chance to see them in action. It would not be long before they were back.

"We came in from no place," Brodsky recalled, and almost pulled off what was, at that time, the largest cable purchase in history.

Comcast's chutzpah was not lost on the other cable operators. A few weeks later, in July 1985, Brodsky was sharing a taxi out to the airport with Stewart Blair, the Scottish-born investment banker who had been hired as a top aide to TCI president John Malone.

Blair mentioned the rumors that Group W Cable, then the third largest MSO in the country, would come up for sale.

"Why don't we get together on this," Blair said. When Brodsky asked "Why us?" Blair replied, "Because you're the last guy I want to see coming down the alley the other way on this deal."

An excited Brodsky quickly called Ralph Roberts who reported that he also had an interesting conversation that day. His was with Nick Nicholas, executive vice president of Time Inc., asking if Comcast would like to join with Time in bidding for Group W's 1.9 million subscribers.

The Group W deal marked a major departure from previous cable system sales. Traditionally a seller would put his company up for sale and then select the highest bid from several that were submitted. Potential buyers would compete against one another.

But the Group W deal was different. Westinghouse Electric Co., Group W's parent, was facing some difficult times in 1985. Its manufacturing division was producing less than stellar results. Its investments in new technologies, includ-

Group W: *Dan Ritchie ran Group W Cable in the 1980s. He went on to become Chancellor of Denver University.*

ing a robotics company, had flopped. And the investment in cable had been a disappointment as well.

Group W had gotten into cable by buying TelePrompTer Corp. in 1981 for $764 million. But it had found the going tough. Many of the TelePrompTer Systems needed major upgrades and improvements in customer service. (When the Group W brass visited one TelePrompTer system the local newspaper ran an editorial asking "Who are you if you missed the last part of President Reagan's inauguration speech, the last quarter of the Super Bowl and the middle innings of the World Series? Answer: a TelePrompTer cable subscriber.")

Group W poured $800 million into building new franchises and upgrading existing systems. But the slow response of subscribers in the big cities hurt cash flow and earnings. And it stumbled badly when it attempted to get into the programming arena. Satellite News Channels, its effort to take on Ted Turner, was a failure. It sold off its interest in Showtime to Viacom after Westinghouse executives decided they didn't want their company to be associated with a network that offered R-rated films.

"I didn't realize it would be as difficult as it was," Group W chairman Dan Ritchie told *Business Week* in early 1985.

Like Storer, Westinghouse had found that the cable division was a huge drag on earnings. And like Storer, Westinghouse was vulnerable to corporate raiders. In 1985 the company, according to *Business Week*, had a market capitalization of $6.7 billion but a breakup value of more than $9 billion. It would be a huge deal for any hostile takeover artist to attempt, but not out of the question in the go-go '80s.

So Westinghouse chairman and CEO Douglas Danforth decided to dump the cable division. And he wanted to sell it in one piece, by the end of the year and to a buyer who would be certain to close the deal. To handle the transaction, Westinghouse hired a group of high profile New York investment bankers rather than a firm with more experience in cable.

Brodsky remembered the meetings to go over the deal. "We had no advisors, just a couple of lawyers, led by (TCI counsel) Jerome Kerns. On the other side were dozens of lawyers and investment bankers" representing Group W.

By placing so many strict conditions on the transaction, Westinghouse essentially invited the prospective buyers to get together to offer a single bid instead of bidding against each other. No single cable company could afford to buy all of Group W. Had Westinghouse sold its systems off in clusters or been willing to wait longer, it might have received a far better price than the $911 per subscriber it actually got. In fact, within six months of the Group W deal, a half dozen cable system transactions would be concluded at prices in excess of $1,400 per sub.

At $1.7 billion the selling price barely represented an increase over what Group W had invested in TelePrompTer: $764 million to buy the systems and another $800 million to build and upgrade them. Depending on how the accounting was done, some analysts speculated at the time, Group W may have been the only major company in history to have lost money by investing in cable systems.

But the deal gave Westinghouse what it wanted: a single sale by the end of the year to a buyer that was on firm financial footing. And it left the company in the cable programming business. As distributor of the Nashville Network and later Country Music Television, Group W would make millions in cable programming.

The deal also enabled TCI, Time and Comcast to add about 500,000 subscribers each at a bargain basement price (the three sold off additional pieces to Century Communications and to Daniels & Associates).

Westinghouse was not the only big name to exit the cable business in 1985 after less than a decade in the business.

American Express and Warner Communications had joined together in the mid-1970s to create cable operating and programming companies. But Amex quickly developed cold feet, especially when Warner Amex Cable began to win big city franchises that required huge sums to build. Warner's swashbuckling chairman, Steve Ross, was more bullish on cable, but by 1985 had his back to the wall after the failure of Warner's Atari video game system.

American Express figured that with Ross facing a possible hostile takeover from Australian media baron Rupert Murdoch, the time was ripe to make Ross an offer he couldn't refuse for a company that served 1.2 million cable customers and had interests in MTV and Showtime.

Amex offered Ross $850 million in cash and assumption of $500 million in debt or $1,125 per sub for the cable operating company. It then made a side deal with ATC and TCI to sell the two of them the operating company for exactly that amount. Amex also would purchase the programming services and sell them off.

With all the skids greased, Amex then triggered a put-call clause in the Warner Amex partnership agreement that allowed either partner to name a price and

Country Flavor:
Group W sold its cable operations in the 1980s but kept stakes in the Nashville Network. At Nashville Network's launch: Bob Boatman, executive producer; Ron Castell, SVP-marketing with Group W Satellite; Harlan Rosenzweig, EVP, Group W; David Hall, GM of Nashville Network, and Nashville Now host Ralph Emery.

the other partner to become a buyer or a seller at that price.

Amex thought they had Ross in a corner and that he would be forced to accept what was a lowball offer. But Ross had some cards of his own to play. He sent Warner Bros. president of pay TV Ed Bleier over to see Viacom president Terry Elkes. Bleier was a major seller of programming to HBO and Showtime and had an interest in seeing a good competition between the two.

"I told Terry that if Viacom didn't step up and buy Showtime and The Movie Channel my name at HBO would be 'take or leave it,'" Bleier said. Viacom agreed to buy Warner Amex's two-thirds of MTV and 19% of Showtime/TMC for $500 million in cash and $18 million in warrants.

The deal enabled Warner to turn the tables on Amex, choosing to become the buyer rather than the seller at the price Amex had set: $1,047 a subscriber. Within four years Warner would sell those same

Growth Rate:
Trygve Myhren led ATC's major growth in the 1980s, instituting multipay marketing campaigns and employee training programs.

subscribers to Time Inc. for a price Paul Kagan estimated at $2,295 a sub.

But, like Westinghouse, American Express got what it wanted: an exit from a business it had not fully understood and which drained its earnings and hammered its stock price.

Dow Jones' departure from the cable operating business was less spectacular. As an investor in Amos Hostetter's Continental Cablevision it had never really been a full-fledged cable operator. By the mid-1980s it was much more interested in developing its information delivery systems. Dow Jones had invested in a company called Telerate that delivered financial information to personal computers via telephone lines.

Anxious to delete the Continental debt from its books, Dow Jones sold back some of its interest in Continental to the company in 1985 for a price Paul Kagan Associates pegged at $989 per subscriber. Dow Jones had bought into Continental in 1981 for $757 per sub, so it made a tidy profit on its excursion into cable. But it could have done even better. By 1992 Continental would sell a portion of its company to Boston Ventures for a price Kagan estimated at $1,774 per sub.

General Electric's exit from cable operations was forced by its purchase of RCA Corp., parent of the NBC television network. FCC rules prevented GE from owning both cable systems and a broadcast network. It sold its systems in November, 1985 to United Artists for $985 a sub. But GE clearly understood where the future of television lay. To run NBC, GE tapped Bob Wright, a former president of Cox Cable. Wright quickly began to explore how NBC might enter the cable business.

Capital Cities was also forced out of the cable operating business when it bought the American Broadcasting Co. Cap Cities sold its systems to The Washington Post Co. for $1,000 a subscriber. (Cap Cities CEO Tom Murphy said at the time he was heartbroken to have to sell his cable systems.) But Cap Cities/ABC remained very much in the thick of the cable programming business as the owner of ESPN and an investor in Hearst ABC Video Enterprises.

To the corporate titans who ran American Express, Westinghouse and Dow Jones, 1985 seemed like a good time to get out of the business. But it would prove even better for those who stayed in. TCI, Comcast, Continental, ATC, Century, Cox and others grew enormously as urban systems were built, system extensions and better marketing brought more customers to existing systems, and as other companies, big and small, sold out.

Such growth demanded major changes in the way these companies were operated. At ATC, which served 821,000 customers in 1980 and four million by 1988, the growth prompted chairman Trygve Myhren to institute changes that would almost completely transform the company his predecessor, Monroe Rifkin, had founded and built into the nation's largest MSO.

Myhren's father had immigrated from Norway and gone to work doing technological research for the New Jersey Zinc Co. in Palmerton, Pa. There the young Myhren's best friend was the son of the man who ran the local cable system.

Myhren was something of an Eagle Scout type: president of the high school student body, editor of the newspaper, champion skier. He graduated from Dartmouth College and its Amos Tuck business school in a total of five years. At Dartmouth his major was philosophy and political science. In business school he focused on marketing.

After a stint in the Navy, a job at Procter & Gamble and a spell running his own marketing company, Myhren had been recruited to work at ATC when Rifkin decided his company would need a cadre of well-trained marketing executives to take it into the era of pay television.

As vice president of marketing at ATC, Myhren oversaw much of ATC's premium service strategy, including the implementation of multipay. He developed an in-house pay service, Cinema Plus, to give ATC leverage in dealing with HBO. He launched a pay-per-view service, EvenTV. And he brought to ATC the head of the Literary Guild Book Club, Jerry Maglio, who would be responsible for some of the most innovative marketing techniques in the industry over the next decade for ATC, Daniels & Associates and United Artists.

(When Myhren brought Maglio in to see Rifkin for his first interview, Rifkin opened with a question: "My wife has canceled her subscription to The Literary Guild Book Club a dozen times. Why do you keep sending her books?" Maglio asked if she was paying for the books she received. Told she was, Maglio said "Why should we cut her off as long as she is still paying us money?" He got the job.)

After Time Inc. bought ATC in 1979, Rifkin tapped Myhren to be his successor as chairman/CEO, teaming him with Joe Collins who served as president and COO.

As Myhren took over the CEO duties at ATC he set about to remake the company to deal with its enormous growth, putting more power into the field where individual system managers were closer to their customers and better able to make decisions than corporate headquarters.

Myhren set up a training program for system operators. He brought them to Denver to learn how to write full budgets, plan for capital spending, design and implement marketing campaigns, improve customer service, develop local ad sales, handle public relations, deal with franchise authorities and undertake countless other tasks that had previously been handled largely by the headquarters staff. To run the training program for the ATC managers, Myhren selected one of ATC's first employees, June Travis.

"We had a highly centralized company that had grown like topsy," Myhren said. "We had begun to lose touch with the customer."

He instituted annual surveys to measure the public image of each ATC system. And he tied part of the bonuses paid to each system manager and staff to the results of that survey. By improving customer service and their system's public image, ATC employees could also fatten their own paychecks.

The decentralization strategy also enabled ATC to trim its headquarters staff, largely through attrition. By cutting several hundred jobs, mostly in Denver, the company saved $41 million a year. (The amount that ATC saved through this downsizing plan was more than 10 times the company's annual revenue when it launched in 1968.

Myhren also launched a campaign to develop clusters of ATC systems around major urban areas. The concept was that if cable were to compete effectively with other media for viewers and ad dollars all the cable systems in a particular area needed to be under common control. Television stations, newspapers, radio stations and telephone companies served areas that might have dozens of cable systems, each with a different owner and a different programming lineup.

Clustering systems in a single metropolitan area under a common ownership made it easier to mount effective marketing, program promotion and local ad sales campaigns. It also enabled ATC to cut costs by eliminating staffing and offices.

ATC's strategy, adopted by most other major MSOs, was to acquire systems in areas where ATC already had operations. One of the first such efforts was in Rochester, N.Y., where the purchase of systems owned by Burt Harris gave ATC ownership of both the city and the suburbs.

Maglio, at ATC and later at Daniels, helped introduce more sophisticated marketing techniques to the business. He developed a formula to measure the amount that could be spent to acquire a new subscriber. (For example: a subscriber might last 36 months on a system, generating $10 a month in revenue, half of which would be operating cash flow. Thus, an investment of $20 to acquire a new subscriber would generate a return of 800%. Those kinds of numbers gained the attention of financial officers when it came time to set the annual marketing budgets.)

While ATC was restructuring its systems and gearing up for a new era of cable, just up US Interstate-25 in Denver the folks at TCI were undergoing some transitions of their own. In particular TCI began to make substantial investments in programming in the mid-1980s.

TCI's first programming investment had come in a typically casual way.

In the late 1970s one of the vice presidents of the NCTA was a young, articulate, ambitious quick study named Bob Johnson. The son of Illinois factory workers, he had attended the University of Illinois and the prestigious Woodrow Wilson School of Diplomacy at Princeton University. His dream (which he still held 30 years later) was to become an ambassador.

After the Wilson School, Johnson set out for Washington, D.C., where he landed a job doing public relations for

Helping Hand:
Bob Johnson received money and advice from TCI's John Malone to help launch BET. The money amounted to $500,000 and the advice was "keep your revenues up and your expenses down."

the Corporation for Public Broadcasting. He later moved over to Capitol Hill, working for D.C. delegate Walter Fauntroy. At NCTA he had been in charge of the pay television division, working with HBO and Showtime and major MSOs to help fend off government regulation of pay TV.

There he got to know TCI president John Malone who told Johnson to call if he ever had an idea for a business.

In 1978 Johnson "got this idea," as he later recalled, for a cable channel with programming geared to minorities, particularly African Americans. Although he had no business training or experience, he began to scout out what programming might cost and got UA Columbia chairman Bob Rosencrans to pledge some time on the satellite transponder being used to transmit USA network.

Then Johnson went out to Denver to visit John Malone.

"I told John what I wanted to do, and he like the idea," Johnson later recalled. "He asked me: 'How much do you need?' And I told him it would cost $500,000 to get started."

Malone said he had been thinking about a minority channel as well to help draw new cable subscribers, particularly in the ethnically diverse major cities.

"So he turned to me and said 'I'll buy 20% of your company for $100,000 and I'll loan you the rest. What do you say?'" Johnson recalled. "When I agreed, he called in his lawyer who drew up a one-page agreement and his treasurer who handed me a check for half a million dollars. The whole thing took about 45 minutes. It was more money than I had ever seen. It was more money than I had ever imagined existed. And what John didn't know was that I would have done the deal with the numbers reversed (80% for TCI and 20% for Johnson). But I think he just liked the idea and the idea of helping somebody out."

As Johnson walked out of Malone's office he turned to the TCI chief and said, "You know, John, I really don't have any business training or experience. Do you have any advice for me?"

And Malone, Johnson recalled, replied "Just keep your revenues up and your expenses down. You'll be fine."

Twenty years later Black Entertainment Television and its spin-off networks, BET on Jazz and Action Pay Per View, would be reaching more than 50 million subscribers in the U.S. and millions more abroad. Johnson would be one of the country's most successful programming executives and one of the most successful African American businessmen in any business. The company he had founded with the spur-of-the-moment backing from TCI would be generating around $80 million a year in operating cash flow on revenue of $160 million and be worth, according to Paul Kagan Associate's Cable Program Investor newsletter, in excess of $1 billion dollars.

More than that, BET created opportunities for hundreds of African Americans to gain and demonstrate business skills that could then be transferred to other arenas and other businesses.

But TCI's involvement in programming remained limited until four years later when Malone made a dramatic speech at the NCTA convention. He noted that many of the cable programming networks were foundering, and many were seeking financial support from cable operators. And he reasoned that just as the oil refineries needed to ensure a supply of oil, and got into oil drilling, so must cable operators get in-

Discovery Dreams: John Hendricks channeled his vision of stimulating educational programming into the launch of the Discovery Channel.

volved in supporting cable networks. Without the cable networks, he said, the industry would have nothing to sell.

To spearhead TCI's involvement in cable programming, Malone tapped John Sie, fresh from renegotiating the Pittsburgh franchise. Sie worked on two fronts: to rescue foundering cable networks and to develop what he called "punch-through" programming that would give cable a high profile with viewers.

At the top of the list of foundering programming networks was the creation of a boyish-looking dreamer and romantic named John Hendricks.

Hendricks was born in 1952 in a coal mining town in West Virginia where his father was a builder and contractor. When John was six, the family moved to the booming town of Huntsville, Ala., one of the nation's new centers of activity in space exploration.

"It was a town," Hendricks later recalled, "full of dreamers." He was one of them. He went to school with the children of engineers who built rockets that went to the moon. His sister married the head of the team that built the first Lunar Rover. Hendricks recalled once sitting in on a meeting in his sister's living room when the group decided to use piano wire for the spokes of the lunar vehicle. It was a magic time, and it instilled in Hendricks a love of learning and adventure that never left him. Watching television, he gravitated to the few documentaries that existed, particularly on PBS.

After graduation from the University of Alabama, he took a job at the University of Maryland as director of grants. An acquaintance of his had produced a documentary on the common roots of the world's religions, called "The Children of Abraham." It had been shown on a Washington, D.C., TV station, and Hendricks was asked to help it get broader distribution.

He never did. But the effort prompted him to wonder why there wasn't a home for such documentaries, a channel that would be devoted entirely to films about exploration and science and adventure: "I became obsessed with it."

He visited with Winfield Kelly, head of the cable system in Maryland's Prince George's County, to ask how to start a cable television channel. Kelly hooked him up with Ken Bagwell of Storer Cable. Storer had just come out of the Spotlight pay service and because of that experience and bank covenants preventing entry into additional programming services, took a pass. But Bagwell told Hendricks the idea was a good one.

On the plane back from Miami, Hendricks toyed with some ideas for a name: He wrote down Horizon, Explorer, Vista, Journey and Discovery.

"I just thought Discovery was a name people would like better," he said.

His stockbroker introduced him to some of the people at investment company Allen & Co. They liked the idea and urged Hendricks to develop a full business plan.

"So I went to the library and found a book on how to do a business plan," Hendricks said. The Allen folks had asked him to focus on three issues: access to programming, distribution and advertising.

Hendricks put together a series of letters of intent. The British Broadcasting Corp. and TV Ontario in Canada offered to provide low-cost programming from their libraries. Group W executive Harlan Rosenzweig offered to put up a sneak preview week of programming on its transponder on the new Galaxy cable satellite.

At the Western Cable Show in 1984 Hendricks set up a rudimentary booth to promote the sneak preview. He sent letters to cable system operators asking them to watch and, if they liked what they saw, to send in a letter saying so. He received 900 responses. He took those with him on a trip to Madison Ave. to seek advertisers.

Pretty soon he had "letters of intent" covering all of the elements Allen had asked for: advertising, distribution and programming. None was binding, but it was enough for Allen & Co. to write him a check for $250,000 and raise another $2 million.

That got him started, but wouldn't keep him going. As Hendricks worked to get his service on the bird, Allen scrambled to raise the money to cover the $800,000 a month Discovery would lose in its first few years of operation. (It was free to cable systems and had essentially no advertising revenues. Major costs were $250,000 a month for a transponder and $250,000 a month for programming.)

Financial Backing:
Four MSOs provided the cash to keep Discovery Channel afloat in 1986. On hand to close the deal were (seated): Ajit Dalvi and Nimrod Kovacs. Standing: Harlan Rosenzweig, Thalia Crooks, Jim Robbins, John Hendricks and John Sie.

Then a financial angel appeared in the form of the Chronicle Publishing Co., owner of the *San Francisco Chronicle* newspaper and Western Communications, a mid-sized cable system operator. The man in charge of looking for new investments for Chronicle was very positive about Discovery and would recommend to the Chronicle board that it buy 40% of the new network for $6 million. His name was Leo Hindery.

Hendricks was ecstatic. He had been fending off creditors, promising them almost daily that a check would soon be in the mail. When Hindery made the commitment, Hendricks called the BBC, to whom he owed $500,000 for programming, and promised they would have a check by Tuesday.

Then the Chronicle board changed its mind, overruling Hindery and deciding not to invest in Discovery. (Had they done so, their $6 million investment in 1985 would have been worth roughly $2.5 billion a decade later).

Hendricks was crushed. "I was very down," he recalled. "For the first time in my life I thought 'this thing is falling apart.' My wife still calls it 'Black Tuesday.'"

But financier Herb Allen had one more card to play. He called TCI president John Malone and explained what was up. "John said, 'We can't let anything happen to Discovery. I'll send John Sie down to meet with Hendricks.'"

"We thought Discovery was a great idea," Sie recalled. "It had legs." But he didn't want TCI to be the only investor. He wanted three or four other MSOs to get involved, to provide distribution and to share in the risk.

Sie developed the idea to put together a consortium of cable operators to fund the network. Each would put in an equal amount, and Hendricks would use the money to finance long-term deals with programming suppliers, capping the cost of product. The network also would have to charge cable systems a fee to ensure the revenue it would need to cover expenses. "The strength of cable networks had always been the dual revenue stream," Sie recalled. "We wanted to apply that to Discovery."

But Sie added a twist. If per-sub revenues fees would cover the network's expenses, he reasoned, any advertising revenue would be profit. Part of that, he decided, should go to the cable operators who had agreed to back the network. He set up a deal in which Discovery agreed to rebate part of its profits from the sale of advertising to those operators who signed on as charter affiliates.

It was a complex and innovative scheme, one that reflected Sie's experience on both sides of the operator-programmer relationship.

He quickly drafted a list of MSOs to be contacted. The first call was on Newhouse Communications president Bob Miron.

As Hendricks later recalled, "We met with Miron during the NCTA convention in Dallas at some Chinese restaurant John had picked in one of those suburban shopping centers. We explained the concept to Miron, and he jotted some numbers down on the back of a napkin: 20 million homes times 5 cents per home per month. It would be enough to cover Discovery's operating expenses until advertising kicked in. And the advertising revenue would then all go to the bottom line and to the charter affiliates. Miron put down his pen and picked up his chopsticks. 'This will work,' he said. 'I'll pay a nickel for it.'"

Discovery, Miron reasoned, provided "good, wholesome" programming and could help attract families to cable as well as counterbalance the image cable had in some areas as a service that offered R-rated movies and other programming that was objectionable to some segments of the population.

Game Plan:
ESPN's Roger Williams and TCI's John Sie hammered out a plan for an MSO consortium to bid on the NFL and distribute games through various cable networks. But ESPN's parent, ABC, shifted gears, demanding ESPN bid alone for the NFL rights, which it did successfully in 1987.

Sie and Hendricks charged on. Comcast and Continental said no. United Cable signed on. The final meeting was with Cox Cable at the international terminal at JFK airport in New York. At the end of the presentation, Hendricks recalled, Cox president Jim Robbins "kind of nodded. He would recommend it to the board."

The deal gave each of the four MSO partners 10% of the company for a total investment of $6 million. Each would ensure Discovery carriage on most of their systems, guaranteeing a huge subscriber base and per-sub revenue. That, in turn, could be used to entice advertisers.

The concept, as Sie later explained, was to "flip" the normal cable system relationship with programmers in which the system paid a fee and the network kept all the ad revenue. The Discovery arrangement would allow the operator as well as the network to share in the dual revenue stream of subscriber fees and advertising.

To help Discovery and other cable networks gain the viewers they needed to attract advertisers, Sie persuaded TCI to abandon the practice of charging extra money for additional sets hooked up to cable in a single home. This cost systems revenue since operators would typically charge as much as $3 per set for an additional outlet even though it cost nothing to provide the service once the installation was completed. Sie reasoned that with more people watching in more rooms of the house, more sets would be tuned to cable networks, enabling them to charge more for advertising and enabling cable systems to gain more for local ads. Rate hikes instituted in the wake of the Cable Act allowed TCI to regain much of the lost subscription revenue.

Sie's quest for "punch-through programming" that would give cable something really big to sell on its basic service led to the door of the National Football League. NFL football had been for more than a decade the most-watched programming on television. It was also the most expensive.

Sie persuaded a group of other MSOs to join TCI in a consortium to bid for a package of NFL games that would be shown exclusively on cable TV on Sunday night. Sie convinced both Turner Broadcasting and ESPN, the most likely bidders for the package of NFL games, not to get into a price war with the consortium. The MSOs would buy the games and then divide them up among the networks or carry them themselves.

When Warren Buffett, Cap Cities/ABC's largest shareholder, found out about the arrangement he told ESPN president Bill Grimes that ESPN had to have the rights. When Buffett spoke at Cap Cities, the executives listened. Grimes called Sie and told him that ESPN had to make its own bid. The cable operator then decided to back off so as not to ignite a bidding war. ESPN also had extra leverage because ABC already had the rights to Monday night NFL games. ESPN offered $135 million for a package of Thursday night games for which it imposed a 25 cents per subscriber surcharge on top of its normal affiliate fee.

The price was high, and some cable operators balked. But by the following decade the concept of the NFL on cable had become well accepted.

The concept of an MSO consortium to rescue a programmer came to its apex in 1987. After fending off the challenge of Satellite News Channels, Ted Turner entered a period of enormous prosperity, with both WTBS and CNN gener-

Global Expansion: Ted Turner's empire grew in the 1980s, with the help of 14 MSOs who bought into TBS after Turner overextended the company with the purchase of MGM.

ating profits. But for Turner success was dangerous. It only whetted his appetite for the next adventure, and the next one had to be bigger and more dangerous than the last.

Turner was fully aware of all the huge corporate mergers and leveraged buyouts, many financed by junk bonds, going on around him in the mid-1980s. And he was eager to play. His first effort was a bid to buy CBS, an offer that was so thinly financed it actually sent CBS stock down on the day it was unveiled.

Turner lost his bid for CBS after the company bought back a huge amount of its own stock. The effort left CBS so weakened it was soon bought by Loews Corp. chairman Laurence Tisch. All three broadcast networks underwent ownership changes in the mid-1980s. ABC was acquired by Capital Cities, NBC by General Electric and CBS by Tisch. GE and Cap Cities had both been in the business of owning cable systems, and each had a strategy that involved more investment in cable programming. Tisch understood nothing about cable, and little about television. He would cut off CBS completely from cable.

Rebuffed by CBS, Turner turned his attention to the next best thing to a broadcast network; a Hollywood studio, in particular the ailing queen of the studios, MGM.

MGM was owned by financier Kirk Kerkorian, and its banker was the king of the junk bond: Drexel Burnham Lambert's Michael Milken. Milken convinced Turner that by using junk bonds it might be possible for TBS to purchase MGM. Milken, representing both the seller (Kerkorian) and the buyer (Turner) in the deal, devised a complex scheme under which Kerkorian would loan Turner some $2 billion to buy the studio and then Turner would repay the sum by issuing highly leveraged bonds. If Turner could not make the payments, the agreement stated, he would have to surrender stock in his company.

Turner, eager to marry his cable networks, particularly WTBS, to a producer of television programming and movies such as MGM (with a huge library of classic films and television programs), agreed. But the deal took over a year to get done, during which time MGM issued a series of box office flops. The junk bonds looked less and less attractive. After a desperate attempt to raise cash (including selling some of the MGM assets back to Kerkorian for less than he had paid for them) Turner realized he was, as he later recalled, "overleveraged like crazy."

The major cable operators had been sour from the outset on the idea of having Turner buy MGM. Just prior to the Western Show in 1986, Turner invited a group of cable heavyweights out to the MGM studio lot to explain the deal. Among them were Time Inc.'s Nick Nicholas and TCI's John Malone.

The studio lot, Malone recalled, "was threadbare and dusty and so empty you could have shot off a cannon and not hit anybody."

The group met in the Louis B. Mayer suite. As Malone recalled, "Ted's in there telling us what a fabulous deal this is for the industry and how we all ought to invest in it. He has a set of numbers scribbled down on the back of an envelope. There was a huge negative cash flow." The operators left scratching their heads. There seemed no way to make this deal work. Turner, desperate to raise the cash needed to pay back Kerkorian, began to consider selling assets.

A few weeks later Turner called Malone at home late one night and announced that Time Inc. had offered to buy 51% of CNN. That got Malone's attention. The TCI chief at the time was very protective of the balance of power in the industry and feared giving too much clout, particularly in programming, to HBO parent Time. "Ted," he said, "You can't do that. That's the crown jewels."

"Well," Turner replied. "Either it's that or it will be the KNN (Kerkorian News Network)."

It was, Malone recalled, "crisis time." The TCI chief called a meeting of all the major MSOs in New York and laid down the law. (One can only imagine the scene from "The Godfather" where the heads of the five families meet to iron out their differences.)

"To bail out Ted we need to pony up $650 million," Malone told the group. "We don't have a lot of time to do a lot of negotiating. Anybody who's not serious about this get the fuck out."

"We came out of that room with the money signed up," Malone said. TCI, Continental, Warner Cable, Viacom and Times Mirror had each agreed to kick in. Viacom and Times Mirror, Malone later said, had agreed to contribute only on the condition that Time Inc. not be included. Like TCI, they were wary of giving too much power to Time.

But just weeks later Viacom went into play when Sumner Redstone, owner of a major string of movie theaters, made a bid for the company, topping a proposal for a leveraged buyout made by Viacom's management. "Redstone didn't want to make the commitment," Malone recalled. "We lost a $100 million player. I had no choice but to call Nick up and say 'Okay, you're in.'"

The consortium was expanded to include what finally became 31 different cable companies taking a total of 37% of Turner Broadcasting System Inc. The operators were given seven of the 15 seats on the board of directors. But any major activities of the company, including approval of the budget, had to be approved by 12 of the 15 board members. Effectively the MSOs could veto anything Turner wanted to do. Moreover, to ensure the balance of power between TCI and Time Inc., the two agreed that if either one voted "no" on any proposition before the board, the other would vote the same way. In effect, Time and TCI each had to agree before Turner could go forward with any major new plans.

Turner's days of freewheeling, unrestrained swashbuckling were over. Under the new structure he would need approval from the group of cable operators before he did practically anything. Ted, at last, was chained. Or so it seemed.

But like so many times before when he had been tossed into a briar patch, Ted Turner emerged not only unscathed, but stronger than before. In point of fact he had been a partner of the cable operators from the day he put WTCG on the microwave. His entire business was dependent on the support of the cable

Big Leagues:
In 1983, HBO premiered the first made-for-cable movie, The Terry Fox Story. Four years later, the network won an Emmy for Down and Out in America.

operators who paid hundreds of millions a year in affiliate fees and ensured that the TBS networks would have enough viewers to entice advertisers.

What the MSO buyout of Turner did was simply to formalize and cement a relationship that had been in de facto existence for years. For Turner it provided a sounding board where he could try out ideas on his best customers. It gave the operators a direct stake in the success of TBS and an understanding of its workings they had not had before.

Within five years after the buyout Turner would launch a dozen new cable networks, both in the U.S. and around the world. Many of them – TNT, Cartoon Network, Turner Classic Movies — would make use of the library of films that Turner had acquired from MGM. And the company that had revenues of about $250 million a year in 1986 would grow to a conglomerate with $3.5 billion in revenues by 1994.

Turner wasn't the only programming company to experience rapid growth in the latter part of the 1980s. Premium television, which had suffered badly in the early 1980s, began to find its feet in mid-decade. Stung by the competition from home video, the premium services began to shift the emphasis away from reliance on theatrical motion pictures and more toward original fare. This was particularly true at HBO where Michael Fuchs, who had been HBO's first director of original programming, became chairman in 1984. The first made-for-pay movie, "The Terry Fox Story," premiered on HBO in 1983. And HBO's made-for-cable productions were gaining critical acclaim and a widespread reputation for production excellence. In 1987 an HBO documentary about the homeless, "Down and Out in America," became the first cable program to win an Academy Award.

More and more Hollywood stars lent their talents to HBO, attracted by the creative freedom it offered and by the chance to work on subject matter that the broadcasters were afraid to touch. By the end of the decade HBO had clearly taken on the mantle once worn by CBS as the network most willing to take on controversial, politically-sensitive topics.

While Fuchs was changing the programming fare, Joe Collins, who had come over from ATC in 1984 to stem the HBO bleeding, reorganized the operations. He cut 125 jobs and overhauled affiliate marketing. He introduced "time-locked marketing" to entice the cable systems to spend their marketing dollars at the same time as HBO.

"HBO had been spending lots of dollars on advertising when operators weren't ready to sell it," Collins later recalled. "We organized the market, telling operators that if they ran a sale on HBO and were ready with the phones and telemarketing, we would help with the marketing costs." All the efforts were geared to particular times of the year so HBO could maximize its nationwide publicity just when the operators were ready with local campaigns. The only catch was that the systems had to direct their campaigns only to selling HBO and Cinemax, not other premium services. Showtime howled, but the scheme worked.

Within a year HBO was able to persuade 75% of its affiliates to sign up for the time-locked marketing campaigns.

Premium services were also helped by the rate changes that took place after the passage of the 1984 Act. As operators began to raise rates, they needed to find a way to prevent the total cable bill from growing so large that subscribers would cancel their pay services to pay the higher basic rate.

Many systems dropped the rates they charged for pay services, and more commonly, offered premiums to attract new customers. Operators began to bundle services, offering additional pay channels for a few dollars more than a single channel might cost.

Maglio, by then at Daniels, pioneered in this effort, creating a package he called Showcase and training his customer service representatives to pitch the package first and only sell premium services a la carte if the customer insisted. The strategy created a demand for low-cost premium services dubbed mini-pay.

Bill Daniels and Charles Dolan, two veterans of the cable wars of the 1960s and 1970s, joined together to start a programming company they called Rainbow. They launched two mini pay services: Bravo, a cultural channel, and Escapade, an adult service. Eventually Escapade was taken over by Playboy Enterprises and became the Playboy Channel while Bravo became the foundation for Dolan's programming empire.

Dolan had always loved movies, as he had demonstrated when he developed the idea for Home Box Office. But HBO concentrated on presenting recently released films. Older films, Dolan reasoned, also had an audience and would be cheap to license.

In 1984 he launched American Movie Classics. To run it he hired a young University of Colorado track star and advertising major named Kate McEnroe. After graduation McEnroe worked for a fledgling enterprise that was attempting to start a women's professional basketball league. One of the people she pitched was Bill Daniels. "Honey," he told her, "you couldn't give me another sports franchise."

But Daniels took a liking to the spunky young woman and recommended her to Maglio who brought her on board to market Bravo and Escapade. By 1984 Dolan had put her in charge of AMC.

AMC was innovative in several respects. It was neither a pay service nor a basic. Like a pay service it had no commercials. But like a basic network it carried a very low per subscriber fee (15 cents per sub per month at launch). AMC left it to operators to decide how to use the service. Some used it as a bonus to basic, offering it to subscribers for "free" as a way to cushion a rate increase. Others used it as an incentive for subscribers to take more pay services, offering HBO, Cinemax and AMC in a package that was less costly than simply getting HBO and Cinemax a la carte.

And AMC developed a style of guerrilla marketing, getting as much for its marketing dollars as any other network.

Under McEnroe and marketing vice president Noreen O'Laughlin, AMC hauled out of retirement some of the biggest movie stars of the 1940s and 1950s, taking them on tours of the country. In a given city they would do a dinner for cable operators, appearances on local talk shows and visits to college cinema classes. It was great for the operators, the communities, the actors and most of all for AMC which reaped huge free publicity.

The first to make the trek was Douglas Fairbanks Jr., who visited 50 cities on behalf of AMC. Others who participated included Omar Sharif, Debbie Reynolds and Jennifer Jones. AMC renovated old movie palaces and led the effort to preserve deteriorating films made in the early part of the century. It was all cheap compared to what the broadcast networks, or even the big cable services, spent on marketing. But it did a brilliant job of positioning AMC as the leader in the classic movie arena.

Other programming services also developed clear market niches and became household names in the 1980s. MTV coined a nationally recognized phrase with its "I want my MTV" slogan. It gave away not just money, like radio stations, but such publicity-generating prizes as an island in the Caribbean and singer Jon Bon Jovi's boyhood home. Lifetime's grandmotherly sex advisor, Dr. Ruth Westheimer, became a cultural icon, reminding people in her squeaky voice and European accent: "Always use a condom."

To recognize the growing array of cable-exclusive programming the NCTA board voted in 1978 to create a National Academy of Cable Programming to present Awards for Cable Excellence (ACE) to the best programming. The broadcaster-dominated Academy of Television Arts and Sciences had decided that only shows that reached at least 51% of all U.S. TV homes could be considered for Emmys. This effectively blocked all cable shows from consideration since cable penetration did not reach 50% of U.S. homes until 1987.

The first ACE awards were presented at the NCTA convention in 1978 in the basement of the Chicago Hilton. (The first award went to HBO for its Bette Midler special. Presenter Burt Harris had neither seen the special nor heard of Bette Midler.)

By the mid-1980s the ceremony, under the direction of former NCTA director of research Char Beales, had become a major event. In 1983 Turner Broadcasting System senior vice presi-

Long Tenure:
Chuck Dolan hired Katie McEnroe to run AMC in 1984. McEnroe (center) is pictured here with AMC host Bob Dorian and Lauren Bacall celebrating AMC's 10th anniversary.

dent Robert Wussler arranged for the show to be carried live on WTBS and agreed to foot the bill for the event.

By 1988, Beales recalled, the ACE program had achieved its purpose: to force the Television Academy to allow cable shows to be considered for Emmys. By 1998 cable programming was winning so many Emmy and Academy Awards that the Cable Academy voted to discontinue the national ACEs.

No new programming concept in the history of television created as much frenzy in the financial community as the home shopping craze of 1986. Home Shopping Network had started by accident. Two owners of a central Florida AM radio station – lawyer and real estate investor Roy M. Speer and record store owner and former disc jockey Lowell Paxson – had found themselves stuck with thousands of can openers when an advertiser on their station went broke and was forced to pay in-kind. The two decided to sell the can openers over the air. It was a huge success, so much so that they began to buy up other products to sell on their radio station.

In 1982 they launched a television version of their shopping show (with Paxson doing some of the on-air work). In 1985 the service went on satellite free to cable systems that would carry its 24-hour-a-day pitches for cubic zirconium jewelry, porcelain figurines and other goods.

When HSN stock went public in March of 1986 its price soared from $18 a share to $42 a share on the first day. Within a few months it hit $120 a share and analysts were comparing it with Genentech, up to then the most successful new issue ever launched on the New York Stock Exchange.

It wasn't long before others were in the business. Irwin Jacobs, head of a Minneapolis-based discount merchandiser COMB, called John Malone at TCI to suggest the two team up to form their own shopping service. In April Malone sent Peter Barton to Minnesota to get Cable Value Network up and running.

Barton devised a scheme which would give every charter affiliate of CVN an equity stake in the company according to how many subscribers the cable operators committed. Moreover, he offered each cable system a 5% share of the revenue from any sales CVN made to that system's subscribers.

To allay fears that TCI had some devious plan to outwit the other operators, Barton held a single big meeting of all the operators involved to sign the contracts, all of which were identical and all of which were available for anybody to see. "It was the first time the industry had been asked to trust somebody. It made them very uneasy," Barton later recalled. But it worked. When CVN launched, it had 41 million subscribers.

The network was a victim of its own success. It couldn't handle the volume of orders that flooded in from such a huge audience. Within hours of going on-air all its phone lines, its order-fulfillment system, its billing system and all the other back-office functions were totally jammed. "It was like a giant fire drill," Barton recalled. "This wasn't direct response. This was instant response."

Barton redesigned the order fulfillment system, setting the standard that 30 minutes after an order was placed, the merchandise would be out the door.

Within months CVN affiliates were earning an average of 15 cents per subscriber per month in incremental revenue from the sales made through their systems on CVN. And that revenue came without the need for rate hikes, without any marketing or back-office expenses, with no need for new equipment, without any overhead at all. It was free money, like finding the pot of gold at the end of the rainbow.

HSN soon followed suit with a profit-sharing scheme of its own, and other shopping services jumped into the fray (Sky Merchant, owned by Jones Intercable; QVC, backed by Sears; and Shop Television Network, owned by several other cable operators including Times Mirror and Cox.)

As all the new channels come on line in the 1980s consumers began to thirst for ever more choice. In some areas they were frustrated either because they lived in areas not served by cable or because their local system did not offer as many channels as they wanted. Some were also angered at what they perceived as poor customer service from their cable system. More and more consumers began to purchase satellite dishes to receive their services directly off the satellite without having to buy cable. Initially a phenomenon in rural areas where cable would never be built, the concept of home dish ownership spread into the suburbs and cities as the price of dishes declined and consumers sought a way to get even more programming than the local cable system could provide.

Trygve Myhren, who served as NCTA chairman, told the board in 1985 that there were some 1.5 million dish owners in the U.S. and 60,000 new units being sold every month. Those who purchased the dishes were able to pick up the satellite signals of the cable networks for free while cable customers were being charged $10 a month or more for the same service. The price of a satellite dish, which had been around $10,000 in 1980, had dropped to about $2,000 by mid-decade and was likely to fall even further.

Most unsettling for the cable operator was the prospect that many of the new dishes being sold were purchased by customers not in rural, uncabled areas, but in areas where cable was available. They bought because the picture quality from a dish was better, because they could get more channels off the satellite than from the cable system and, above all, because they didn't have to pay a monthly fee.

Apartment building owners were also discovering that by installing a dish on the roof they could provide cable service to their tenants and keep all the revenue themselves. In Dallas alone some 150,000 apartment units were served by these master antenna systems.

Among the programmers, premium networks were the most concerned about the rising tide of home dishes. Ad-supported services lost subscriber fees,

Growing Competition:
The growing C-band dish market provided competition to cable throughout the mid- and late-1980s.

but gained audience, which could be sold to advertisers. Home Box Office, the biggest premium service, was concerned not just about lost revenue but about the issue of copyright payments they were making to studios for the rights to films, according to Jim Heyworth, who served as HBO president at the time.

As the issue of home dishes became more and more critical, the NCTA at first attempted to form a consortium to develop a technological solution to the problems. The answer was clearly to scramble the satellite signals in much the same way that signals were scrambled by cable systems, so that unauthorized dishes could not pick them up. But the logistics of scrambling satellite signals were nightmarish. There were some 5,000 cable headends that would have to be supplied with equipment that would descramble the signals, and that equipment had to be installed and tested before encryption began. There were also antitrust issues that would almost certainly be raised if a group of cable operators got together to deny programming to a competing technology.

While the NCTA members debated how to proceed, HBO stepped up to the plate. In the summer of 1986 it announced it would scramble its satellite signals starting the following January. It elected to use a scrambling technology developed by defense contractor M/A Com of Burlington, Mass., soon acquired by the General Instrument Corp. The system reversed the color coding and mixed up the lines of the video signals. The audio signal was broken into digital bits and then reassembled by the decoder according to an algorithm that M/A Com assured the industry was as close to unbreakable as could be developed.

Dish Platform:
Rep. Al Gore (D-Tenn.) took on the plight of the small dish owner after signal scrambling took effect in the 1980s and successfully rode the issue into the U.S. Senate.

HBO agreed to pay for the $400 decoder that needed to be installed in each headend for each HBO signal. Other networks quickly followed suit, spurred on by the cable operators, particularly TCI's John Sie, who urged them not to allow their programming to be given away.

When HBO scrambled its signal it set off a firestorm of protest. Home dish owners were livid that they had spent thousands of dollars for equipment they had been assured would allow them to watch cable programming for free. Now they would be shut out or, at best, forced to pay a monthly fee.

Dish salesmen, a feisty lot of mostly independent, largely rural and small-town entrepreneurs, saw their business in severe danger. They held "dealer rallies" in Washington to protest what they saw as their inherent, self-evident right as Americans to receive programming off the satellite for free.

Sixty members of Congress co-sponsored legislation to force a moratorium on scrambling. At the head of the congressional attack was Rep. Albert Gore, then running for the U.S. Senate in rural, mountainous Tennessee where thousands of dishes had been purchased.

Cable operators offered to sell programming packages to dish owners (TCI's package offered 15 channels for $28.95 a month) but the offer only further enraged the dish owners and dealers who saw cable as a monopolistic cabal.

Despite the outcry, scrambling went forward as scheduled. And it worked, at least in the months before computer hackers were able to clone the chips inside the decoders and sell them to other dish owners. HBO even signed up 50 small cable systems as new affiliates in the weeks surrounding the scrambling. These systems had evidently been picking up the satellite feed without telling HBO and then selling the service to their customers without having to send HBO a cut.

Congress rejected a bill sponsored by Gore to force cable networks to sell to dish owners and alternative distribution systems. But the issue would come back in the next decade to haunt the industry.

The political pressure induced HBO and other programmers to develop systems to sell their programming directly to dish owners, at first only those outside the cabled areas. The networks made use of a new division of General Instrument Corp., the VideoCipher Division, based in San Diego, to authorize via satellite signal the boxes of those customers willing to pay for programming. It was the beginning of an industry that would allow the programmers for the first time to deal directly with their customers without having to go through the cable operator.

As more and more networks launched in the 1980s operators began to run short of channels. Even on the huge urban systems of 54 channels and more, there were fewer and fewer open channels as the 1980s drew to a close.

Some programmers developed innovative ways to get around the channel crunch. One was United Video. Started in Tulsa in the 1970s as the microwave subsidiary of United Cable predecessor, GenCoE, United Video had been spun off as a separate company to distribute WGN-TV, the independent television station in Chicago. It had signed on to lease the last available transponder on Satcom I to distribute WGN.

United Video's executive vice president at the time was Roy Bliss Jr., son of the cable operator who had built one of the first systems in the Rocky Mountains. Roy Jr. had grown up climbing poles, changing tubes and selling cable door-to-door. After graduating from the University of Arizona and working for cable equipment manufacturer Ameco, Bliss, at the urging of his father, went to work for Gene Schneider at United Video.

By the mid-1980s United Video was reaching nearly 30 million subscribers with WGN and three other broadcast stations it was uplinking. But Bliss understood that transmission of the broadcast station on satellite used only a portion of the signal available. The remaining part of the signal, known as the sideband, could be use to carry additional audio or text services. These, in turn, could be transmitted to homes over the cable system without using the entire six MHz of bandwidth needed for an additional video channel.

State Of Bliss:
Roy Bliss Jr. helped pioneer the superstation business with WGN as well as the guide programs on cable.

United Video's first use of the sideband was to launch WFMT, a Chicago classical music radio station, as a satellite-delivered audio service. The station was picked up by both cable systems and radio stations. The latter made use of WFMT and additional radio services launched by United Video to eliminate the need for locally programmed service. By using satellite-delivered programming and automatic ad-insertion equipment, it became possible to operate a radio station with no personnel.

In 1981 Bliss developed an even more revolutionary idea for use of the WGN sideband. Guides for cable programming had been a major issue for operators ever since the launch of HBO. The premium services all had their own guides and ATC's Maglio had developed multipay guides that carried the listings for all the pay services offered by a specific system. Publishing companies such as TVSM produced monthly cable programming guides.

But newspapers were very slow in adding cable to their programming logs. And even when they did they often were not complete, partly because a newspaper might have a dozen or more cable systems, each with a different programming lineup, within its circulation area.

TV Guide was also slow to adapt to the cable era. Its first cover featuring cable programming did not appear until 1983 (for HBO's "Passage to India" special). And it had the same problem as the newspapers. *TV Guide* had traditionally printed a separate guide for each TV broadcast market. To serve cable customers it would have to print dozens of different versions of the guide within each city. And distribution was a nightmare. A single supermarket where *TV Guide* was sold might attract subscribers from a dozen different cable systems.

Ultimately the answer was for cable itself to transmit programming information about cable.

There had even been an effort in the late 1970s to start such an on-screen guide, developed by Scripps Howard and transmitted by phone lines to local cable systems. But that guide operated on a page format, with one page coming up on screen at a time and then being replaced by another. And the phone transmission was unreliable at best.

In 1981, Bliss met a fellow from Milwaukee who was using an Apple computer to generate a scrolling on-screen guide to TV programming. The programming information for each network was provided by TV Data, the same company that provided the information for many newspapers and other print guides to TV programming.

Each week United Video would download via satellite all the programming data for the cable networks for that week. Each cable system would then pick up the information for the channels it carried and use an alphanumeric generator to transmit the information to subscribers. To ensure accuracy, UV transmitted each week three separate times. If a given cable system did not receive three sets of identical programming information, it would send off alarm bells in Tulsa.

By the end of the 1980s the EPG had evolved into the Prevue Guide, still featuring the scrolling guide but with a window for video previews of programming as well, and serving some 44 million cable subscribers.

Another network which made use of the sideband to transmit information that could be customized by local systems was the Weather Channel, the brainchild of former broadcast network weatherman John Coleman and launched in 1982 with the financial backing of Landmark Communications. The Weather Channel used the sideband of its satellite signal to transmit thousands of individualized weather reports for individual cable systems. Each cable system could then take the national video feed and intersperse it with alphanumeric local weather forecasts.

As more urban systems came on line, the number of addressable converters expanded. Even TCI, which had long resisted addressability and continued to deploy what president John Malone called "plain vanilla" cable systems, became a convert to addressability when TCI launched the Disney Channel in 1983.

The expanded universe of addressable converters revived an old dream of the cable industry: to deliver programming on a per-show basis and charge viewers only for those programs they actually ordered. By the end of 1985, Paul Kagan told the *Wall Street Journal*, some nine million homes would have addressable converters. Many of those homes were already being offered pay-per-view services through stand-alone operations in which the movie studios would ship tapes to the cable systems which would then deliver them over the cable system to those homes that called to order the film.

It was an unwieldy operation. Yet in 1984, Kagan said, these makeshift systems had generated some $26 million in revenue. The time was ripe for somebody to come in and provide some order to this service.

Two did at almost the same time. One was cable entrepreneur Jeffrey Reiss, founder of Showtime and Cable Health Network and working at his new company, Reiss Media, by the mid-1980s. "A light bulb went off," Reiss later recalled, as he realized that the number of addressable homes had reached the point where a national service made sense.

With funds from his father-in-law, TV producer Norman Lear, and from Paul Kagan, Bill Daniels, John Saeman and Bob Rosencrans, Reiss launched a satellite-delivered pay-per-view service called The Exchange, later renamed Request TV.

Reiss rented out time on his satellite transponder to the various studios and took on the task of packaging and marketing all the films under a single brand name. He worked with operators to upgrade their back-office operations, encouraging them to install automatic number identification (ANI) systems. These would fully automate the order process so that when a customer called in, the system would automatically identify where the call was from, enter a request for a movie, have the film authorized within minutes and then add the cost to the customer's bill.

United Cable, which had pioneered the deployment of store-and-forward amplifiers, moved to the forefront of pay-per-view. Store-and-forward amplifiers allowed the same kind of service as ANI, but worked through the cable system. A cable subscriber would order a PPV program by punching a button on the converter box or remote. A signal would then go to the amplifier, which would forward the information to the headend.

Playboy TV, which had been languishing as a monthly premium service, found new life in a PPV format. Viacom launched its own PPV service, Viewer's

Choice, headed by QUBE veteran Scott Kurnit. Kurnit beat Reiss to the satellite by a single day. In 1988 Viacom brought in as equity partners five other cable operating companies.

By the end of the decade there were some 17 million addressable homes in the U.S. or 35% of all cable homes. Of those, some six million were PPV customers, generating about $210 million a year in revenue. Paul Kagan predicted that the total revenue for PPV would rise to $5 billion a year by 1998.

But all this hype generated yet another set of enemies for cable: the home video store and theater owners who feared that if PPV ever became widespread it would cut into their businesses. They began to work behind the scenes in Hollywood to make sure that films would have a long time in the theaters and video stores before they became available on PPV.

The launch of so many new programming services in the 1980s began to put a huge strain on the old tree-and-branch coaxial cable systems that had been the heart of the business ever since the first one was constructed in 1948.

Upgrading a system to add more channels required adding more and more amplifiers. A 35-channel system required an amplifier every 2,000 feet with a maximum of 40 in a row (or cascade). With 60 channels it was necessary to place the amplifiers every 1,500 feet apart and only 20 could be in a cascade. The farther the signal had to travel, the more amplifiers it had to pass through. Each amplifier degraded the signal, so that by the time the signal reached the home farthest from the headend the picture and sound were often badly distorted.

But new programming services kept launching, almost weekly at times, and cable systems were anxious to add to their lineups to justify the rate increases they were imposing under deregulation. City councils were continuing to demand as many services as possible in return for franchises and franchise renewals.

The solution came from several different sources. But the two key players in the effort to find a way out of the channel logjam were a former center for the Miami Dolphins professional football team and a former crew member on a charter sailboat that plied the waters of Hawaii.

It takes about 20 hours to sail a 50-foot trimaran sailboat from the island of Oahu to Maui. In 1975 Bob Khlopin, a recent graduate of Cornell University, was living out the dream of many a child of the '60s: making a living by taking passengers by sailboat between the Hawaiian islands. But he needed help, particularly for the long trip from Oahu to Maui. So he called one of his Cornell fraternity brothers, now a junior, and asked him to come join him. It was cold in Ithaca, so Jim Chiddix signed on, abandoning the chance to obtain an undergraduate college degree.

Chiddix spent a couple of years serving as crew, until the charter business went broke and he had to look for a job. He had studied electrical engineering at Cornell and signed on to help fix converters and other equipment for a 3,000-subscriber cable system on Oahu's leeward shore.

A few years later he was working for Oceanic Cable, the system in Honolulu headed by legendary system manager Don Carroll. Like many other businesses, cable in Hawaii was different from what it was on the mainland. Far away from any corporate directives or corporate

Sailing Buddy:
Jim Chiddix joined a fraternity brother in Hawaii to crew on his boat before taking a job at Oceanic Cable, where he pioneered ad insertion and AM fiber developments.

help, Hawaiians had to innovate. Chiddix developed his own ad-insertion equipment to create a local ad sales business, later selling the technology to Texscan. He worked on home-grown pay-per-view and pay TV services and joined other systems on Oahu to install the state's first satellite dish in 1978. In 1980 Oceanic was purchased by ATC.

Oceanic was a fertile ground for innovation, and many of its employees — Anne Burr, Carl Rosetti, Tim Evard — went on to much bigger jobs on the mainland. Chiddix attributed this to Carroll, a man Chiddix called "the Bing Crosby of the cable business," because of his laid-back style. "He liked to try new things and he gave us a lot of freedom," Chiddix recalled. "He attracted a group of very creative people."

Chiddix installed a fiber-optic link to hook up the Oceanic headend to the satellite dish on the other side of Oahu's formidable mountains. He installed an FM supertrunk, demodulating each signal and remodulating it as AM when it reached the headend. It worked fine, Chiddix recalled, for sending a dozen video signals from one point to another. But AM transmissions over cable were much more tricky, requiring a very "linear" signal that was not possible using the lasers then available.

In 1986 Chiddix moved to the ATC corporate offices in Denver and formed a research group with fellow ATC engineers Dave Pangrac and Louis Williamson. "We put a lot of our energy into fiber," Chiddix recalled.

The group came across a new kind of laser, called a distributed feedback laser, that produced a much purer optical signal. These had been developed by the telephone companies for use in the transmission of high-speed digital signals used to carry data and voice.

But Williamson used them to transmit 40 AM channels of television over 10 kilometers of fiber with no amplification. It worked. "The implications," Chiddix recalled, "were profound."

Chiddix and his group showed the new device to a meeting of the NCTA engineering committee in Denver in the fall of 1987. The group recognized immediately that the development would allow construction of cable systems capable of carrying many more channels than could be transmitted by the old coaxial systems. It also would make it possible for cable to compete with the phone companies in the delivery of data and even voice.

What Chiddix now needed was a supply of the new lasers. It was just about that time that the head of cable's largest distribution company, Anixter Pruzan, had decided that the time had come to make a radical transformation of his operation.

Anixter had its origins in a telephone equipment distribution company founded in Seattle after World War II by Jack Pruzan and his son Herb. In the 1950s the company moved into distribution of cable hardware, supplying much of the equipment used to build cable systems in the northwestern U.S. In 1969 it was acquired by Anixter Brothers, distributors of wire and cable primarily to telephone companies. The new cable division, Anixter Pruzan, was headed by Pruzan executive Gordon Halverson.

In 1974 Anixter hired as a regional salesman a Boston College graduate named John Egan. Egan, a physics and economics major, had spent two years playing center for the Miami Dolphins professional football team before joining RCA.

By 1980 he had become president of Anixter Pruzan and began to buy up competing distributors. "This is a relationship industry," Egan later said. "Our strategy was to buy up the companies with relationships. Then our only competition would be companies that had no relationships."

As the 1980s drew to a close, Anixter had become by far the largest distributor of cable equipment in the country, with a market share of about 75%. Its parent company also continued to supply equipment to the telephone business, where the Anixters and the Pruzans had their roots.

But the distribution business was becoming more risky as the cable industry consolidated. Cable companies typically needed to buy equipment from 40 or more different vendors. For a small company that might need only half a dozen converters or pedestals at a time, it was difficult to deal directly with big manufacturers who preferred to sell in large quantities. It was much easier for the typical small-system operator to buy all the equipment through a distributor who would carry products made by all the vendors and who would sell in any quantity.

But for a huge MSO it was possible to deal directly with the manufacturers and in effect to establish in-house distribution businesses. And the cable business in the 1980s was consolidating rapidly. So were the equipment manufacturers.

By 1987 "the role of the distributor was getting squeezed," from both ends, Egan later recalled. When he heard about what Chiddix was doing, he had an idea. He had seen the new lasers being manufactured for AT&T, which was still a major customer for Anixter's phone equipment distribution arm.

The AT&T lasers were designed to handle digital transmissions, turning the light on and off to transmit the binary code of ones and zeros that make up digital transmissions. But Egan suggested the system could be modified so that the lasers could raise or lower the intensity of the light, delivering a range of signals that could emulate the analog transmissions used by television signals. (The difference is like the difference between a standard light switch, used to turn lights on or off, and a rheostat that can raise or lower the light level gradually).

Egan called the new device the LaserLink. It was exactly what Chiddix

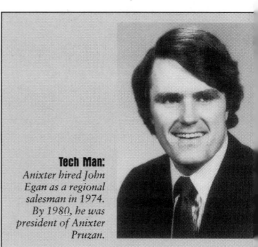

Tech Man:
Anixter hired John Egan as a regional salesman in 1974. By 1980, he was president of Anixter Pruzan.

had been looking for. The first commercial installation of the new LaserLink fiberoptic transmission system took place at ATC's system in Orlando, Fla., in August, 1988.

The fiber was designed to back up an AML microwave system that transmitted signals from the central headend to mini-headends around Orlando. Microwave signals in Orlando were prone to outages or interference because of the violent thunderstorms that frequently hit central Florida.

The first time the microwave went out and the signal switched over automatically to the fiber, the alarm on the microwave had failed, so the system personnel weren't aware of the changeover until they began to get calls from customers. "Don't know what you did, but don't change it," was the message from the consumer. The signals over the fiber were far better than what the microwave had delivered. And the consumers could notice it.

Egan realized that this was the beginning of a revolution, one that would, as he later recalled it, have three elements: "the transformation from copper to glass, from analog to digital and from simple to complex networks."

He proceeded to change his company from a distributor to one that would also become a leading manufacturer in this new environment. To signal the shift he changed the name to Anixter Technologies or ANTEC. And he separated it from the distribution division that would still handle equipment from all the manufacturers, even those with whom the new ANTEC manufacturing arm would compete.

Egan set off on another buying spree, gobbling up Powerguard, Texscan, Re-

Cable Ratings

From the moment Ted Turner launched Atlanta independent broadcast station WTCG on satellite in 1976, advertiser-supported cable networks battled to prove they were being watched and that they deserved a share of the dollars being spent on TV by major advertisers.

When Turner first went on satellite he gauged the size of his audience by counting the mail that came in with orders for proeucts advertised on the channel such as Ginzu knives. He looked at the canceled stamps to determine where his viewers lived. If half the mail was from outside Atlanta, he would figure his audience outside the city was at least as big as it was in the area where it was measured as a broadcast station.

But from the start he also laid siege to A.C. Nielsen Co., the giant audience measurement service that determined how many viewers each network had at any given time. In January of 1981 his battle paid off when Nielsen issued its first report on the ratings for a basic cable network, by now renamed WTBS.

The report was a departure for Nielsen because it measured WTBS only within the network's own universe of potential viewers, not among homes which received only broadcast television or were served by a cable system that did not carry WTBS.

Three years later, Nielsen issued the first quarterly Nielsen Cable Activity Report, covering all the Nielsen-rated basic cable networks, each measured within its own universe of homes where it could be viewed.

But the Nielsen sample size (1,200

gal and other manufacturing companies to enable ANTEC to manufacture a wide range of equipment.

In the meantime Egan continued to work with Bell Labs, the research arm of the phone companies, to develop more reliable lasers. Of the first batch of new lasers produced, only one in a thousand worked properly, and each cost in the neighborhood of $100,000, far more than Chiddix and his cable system colleagues were willing to spend.

But soon ANTEC pared the cost of a system down to $20,000, close to what operators could handle financially. By the mid-1990s the cost of a laser was only about $3,000.

The lasers also were highly durable. The one that Egan installed in Orlando was still working 10 years later when Egan took it out, had it encased and presented it to Chiddix. The serial number was 000.

At that price, it was more cost-effective to install fiber than coax, even without the new services that fiber made possible. Fiber required no electronic gear which needed to be maintained. It could carry far more signals than coax and, because it needed no amplifiers that would degrade the signals, the quality of the picture at the far end of the network was just as good as what could be seen at the headend.

Fiber was used typically to distribute the signal from the headend to nodes that served a group of homes. In the early configurations a node might serve several thousand homes. Later operators began to build systems with nodes serving only a few hundred homes. From the node to the home, a distance of a mile or so, the system would continue to rely on coaxial cable. This so-called hyprid fiber coax, or HFC, ar-

homes when the first cable ratings were issued in 1981) was so small that it was difficult to measure the audience for any cable network that did not reach at least 15% of all U.S. homes. Networks that reached less than the 15% of TV households were not rated. And even for rated networks, with such a small sample, a shift in viewing by even one or two homes could double or triple the ratings.

It was difficult to measure with any accuracy who was watching what programs, a critical piece of information for networks such as Nickelodeon or Lifetime which cared less about total viewing than about whether they were reaching the target audience (children for Nick and women for Lifetime).

The big breakthrough for the cable networks came in 1987 when Nielsen began to use the so-called people meter, a device that not only measured what program a given television set was tuned to, but also who was sitting in front of it.

At the same time Nielsen increased the size of its national sample, first to 4,000 homes and then to 5,000, from the 1,200 it had been using since 1981. Under this structure a cable network that reached only 3% of the total U.S. TV households could be rated by Nielsen. This gave the cable networks the ammunition they needed to lay siege to the big accounts on Madison Ave.

The results were plain to see in the ad revenue generated by basic cable networks. From 1981, when WTBS posted its first Nielsen ratings, to 1989, the total advertising dollars generated by basic cable networks grew from $105 million to $1.5 billion.

chitecture allowed the use of fiber for the big hauls but did not require that the light signal be converted to electronic at every home, just at every node.

When Egan demonstrated the LaserLink system to John Malone, the TCI president looked at him and, Egan recalled, said, "We must be on the side of the angels." Malone invited Egan to make a presentation at the next TCI shareholders' meeting and noted, when he introduced Egan that not a single one of the LaserLinks TCI had installed to date had failed.

Jones Intercable became the first to build an entire system using fiber for the backbone. The system was called the Cable Area Network and was first installed in Augusta, Ga. It cost about $250 per subscriber to install, or about $12,000 per mile. It included six FM fiber supertrunks linking mini-headends and 17 AM links to the coax, cutting amplifier cascades to 10 from the previous 42.

Jones, like many other operators, simply installed the fiber on top of the existing coaxial trunk cable, using the old coax to send interactive signals back to the headend from the home or as institutional networks.

Fiber not only revolutionized the existing cable business, it also raised the prospect that cable systems might be able to deliver data and voice services, previously the exclusive province of the telephone companies. In the late 1980s cable was a business with about $15 billion a year in total revenue. The phone companies were generating about $100 billion. If cable could snatch away even 5% of the phone company business, it would increase cable's revenues by 33%. The numbers were enticing.

But while telephony, data service, pay-per-view, advertising and other new services clearly were the growth areas of the future, basic cable rate deregulation was the locomotive that drove industry growth in the late 1980s.

The Cable Act of 1984 effectively ended rate regulation as of the fall of 1985 (in the meantime operators were allowed to take annual rate increases of no more than 5%).

Operators pounced on the opportunity to make up for years of what they viewed as artificially low cable rates. The average price for basic cable grew from $9.73 in 1986 (the last year of regulation) to $16.78 in 1990, according to Paul Kagan Associates. This represented a hike of 72% in five years.

At the same time the number of basic subscribers continued to increase, largely due to the completion of the new urban systems, line extensions of existing systems and better marketing. The total number of basic subscribers grew from 42 million in 1986 to 55 million in 1990, an increase of 31%.

With more subscribers and higher rates, total revenue from basic cable more than doubled in the last six years of the decade, to more than $10 billion in 1990. And a huge chunk of this new

Quincy:
Veteran Washington cable attorney Jack Cole successfully argued cable's case in the Quincy must-carry decision.

revenue made its way to the cable operators' cash flow line.

Pay revenues also resumed their upward path, driven by a 33% increase in the number of pay units sold. (The average monthly retail price for a pay network actually declined in the period, dropping from $10.25 in 1985 to $10.20 in 1989). Total revenue from pay services increased to $4.9 billion in 1989 from $3.8 billion in 1986.

The number of new cable networks also increased dramatically after passage of the 1984 Cable Act. Some 28 new ad-supported networks launched in the last five years of the decade. And more and more money was being spent on programming. Cable systems, according the NCTA, spent over $3 billion on programming in 1990, an increase of 50% over the 1984 level.

More and more people were watching cable as well. The total day share of audience for cable networks in cable homes doubled to 35% between 1984 and 1990. The broadcast networks' total day share of viewing in those cable homes dropped from 58% to 46% in the same period.

Cable advertising revenue leaped to $2.5 billion in 1990 from $800 million five years earlier.

The 1980s also saw the cable industry win an astonishing victory in the courts over the FCC and the broadcasters. Ever since the late 1960s the FCC had been forcing cable operators to carry all local broadcast signals on their systems. The theory was that such carriage was necessary to ensure the survival of broadcasting and that broadcasting was in the national interest.

But in 1985 the U.S. Court of Appeals in the District of Columbia struck down the must-carry rules. The case had been brought by a small cable company in Quincy, Wash., which had been forced to carry three broadcast network stations each from Seattle and Spokane, taking up half of the 12-channel system's capacity.

The system was represented in court by veteran cable attorney Jack Cole. Cole, a graduate of the George Washington University Law School, had worked for the FCC in the 1950s and then joined the law firm headed by Stratford Smith, first counsel to the NCTA. In 1966 Cole formed his own firm, representing many cable companies in proceedings before the courts and federal agencies.

In the Quincy case, Cole reasoned that a case brought by a small, independent cable company forced by the government to devote half its system to the huge broadcast networks would have a David vs. Goliath tinge to it that sometimes catches the eye of judges.

By the time the Quincy case was decided it had been joined to another filed by Turner Broadcasting which charged that must-carry violated its constitutional rights by giving broadcasters preference over cable networks in the electronic forum of a cable system. This, Turner stated, violated the First Amendment which states that Congress "shall make no law abridging the freedom of speech."

In a ruling that Cole later recalled "astonished everybody," the Appeals Court decided 3-0 in favor of Quincy and Turner. The three-judge panel included liberal Skelley Wright, conservative Robert Bork and moderate Ruth Bader Ginsburg. They all agreed that the FCC had failed to show that must-carry rules were needed to protect broadcasters from serious economic harm. In the absence of such a showing, the court said, the rules were not justified because they did infringe on the First Amendment rights of cable operators and programmers.

The decision presented a dilemma for the cable operators. It was a clear victory, but it also left the door open for congressional action to restore must-carry. And it propelled the broadcasters, petrified at the thought that they might lose carriage on cable, up to Capitol Hill to seek such legislation.

Moreover, most cable systems had no intention of dropping local broadcast signals. They simply objected to a blanket rule that forced duplication of the kind found in Quincy and the carriage of marginally watched stations in some cities.

So the NCTA board voted to seek a compromise with the broadcasters. The solution adopted by the two industries and the FCC hinged on a device known as an A/B switch, which allowed subscribers to switch between cable and off-air reception. The new rules adopted by the FCC and endorsed by the NCTA and NAB required cable systems to carry major local broadcast signals and to install A/B switches so that subscribers could easily receive other broadcast signals off-air.

But even this compromise was challenged by Century Communications, and it was again rejected by the Appeals Court on a vote of 3-0. The court stated, "We conclude that the FCC has not demonstrated that the new must-carry rules further a substantial governmental interest as they must to outweigh the incidental burden of First Amendment interests."

Cole, who also represented Century in its appeal of the second set of rules, said later that the rulings in Quincy and Century, "went a long way to establishing full First Amendment rights for cable."

But the rulings didn't go all the way, and the issue had not yet reached the Supreme Court. What they had done was to stir the broadcasters into a frenzy of activity.

Although cable operators did not engage in wholesale dumping of broadcast signals in the wake of the must-carry decisions, many systems did drop marginal or duplicated signals. And others moved broadcast signals around on the dial, giving the more-watched lower-channel numbers to cable networks.

Broadcasters howled in protest. But cable operators, free at last from years of being forced to carry the broadcasters, were not always very sympathetic. TCI regional vice president Barry Marshall made headlines with a statement that cable operators had paid to build these systems without any help from the broadcasters and could therefore do what they wanted in deciding what signals to carry and where to place them.

The Quincy-Turner and Century decisions offered cable operators hope that they would one day enjoy freedom from government regulation similar to that enjoyed by newspapers. Satellite signal scrambling had ended the ability of home dish owners to pick up cable television network signals for free. Exclusive deals with programmers enabled cable systems to avoid competition from alternative delivery systems such as MMDS. Rate deregulation boosted cash flows.

All this good news was reflected in cable system prices which had hovered around $1,000 a sub in the pre-Cable Act era and soared to $2,500 or more by the end of the decade. Stock prices climbed, and banks were eager to loan.

There were a few storm clouds on the horizon. All of cable's many victories had earned the industry the opposition of a host of powerful interests.

Broadcasters feared that they would continue to lose viewers to the cable networks and that they might lose their carriage or channel positions on cable systems. Hollywood studios continued to grouse about not getting enough copyright payments for their product. Consumer groups and city officials gathered more and more stories about huge price increases and poor customer service. Telephone companies feared competition in the data and voice arenas and itched to get into the business of delivering video services themselves. Dish owners were angered by scrambling. Competitive delivery services such as DBS and MMDS complained about cable systems' exclusive deals with cable networks.

Some predicted that cable's party wouldn't last. Newly appointed FCC chairman Dennis Patrick told a meeting of the NCTA board of directors in May, 1988, that "There is a growing view that cable is an unregulated monopoly. One ought not to underestimate the prevalence of that view in Washington."

Sen. Howard Metzenbaum (D-Ohio), a liberal Democrat, held hearings on rate increases and the monopoly power of cable. Sen. Albert Gore (D-Tenn.) con-

Top 20 Cable Deals of the 1980s

Seller	Buyer	Date	Subs (000)	Price (mil.)	Value/Sub
Warner	Time	3/89	1,583	$3,633	$2,295
KKR/Storer	Comcast/TKR	4/88	1,453	2,880	1,982
United CAble TV	UA Comm.	3/88	1,200	2,446	2,038
Group W	ATC/TCI/Comcast	12/85	1,900	1,730	911
Storer	KKR	7/85	1,550	1,610	1,039
Cooke Media	Consortium	7/89	674	1,548	2,297
Centel	Consortium	3/89	588	1,431	2,434
Viacom	Nat. Amusements	3/87	972	1,361	1,400
Rogers	Houston Ind.	8/88	558	1,312	2,351
Heritage	TCI	1/87	693	1,126	1,625
Cox shareholders	Cox Ent.	12/84	904	795 (60%)	880
Teleprompter	Group W	12/80	1,100	764	695
McCaw Cable	Cooke	1/87	433	755	1,744
American Cablesys.	Continental	10/87	346	750	2,168
BT/Wometco	Cablevision Ind.	5/88	311	720	2,315
Taft Broad.	R. Bass Group	8/87	277	688	2,484
KKR/Wometco	BT Cable	10/86	354	662	1,870
Am. Express	Warner	7/85	614	643 (50%)	1,047
Dow Jones	Continental	10/85	568	562 (51%)	989
Viacom (NY, Ohio)	Cablevision Sys.	8/88	200	550	2,750

Source: Paul Kagan Associates

tinued to push for legislation to require cable networks to sell to home dish owners and to alternative delivery systems.

And NCTA president Jim Mooney found himself the continual harbinger of bad news from Capitol Hill. "We must realize we have a real political problem that largely is premised on rate increases and a perception of poor customer service," Mooney told the NCTA board in September of 1988.

But the cable operators, while concerned about their growing list of enemies, felt that if they simply could tell their story better to the public their political problems would be greatly reduced. It was a public relations issue, most of them reasoned. And besides, cable had a friend in the White House, George Bush, who would almost certainly block any effort by the Congress to undo the 1984 Cable Act and reregulate the business. And even if the very worst happened and some kind of reregulation measure did become law to limit future price increases, it would certainly not roll back the rate increases that had already taken place.

It was a miscalculation of monumental proportions.

Regulation

Daniel K. Inouye belongs to the world's most exclusive Club, that group of 15 or 20 members of the United States Senate who really run the place while their colleagues tilt at ideological windmills, fret about reelection or dream of redecorating the White House.

Membership in the Club has nothing to do with party affiliation or what state is represented. Members of the Club are generally willing to take a pragmatic, rather than a strictly ideological view of issues. They have secure political bases at home, rarely facing more than token opposition when they go before the electorate every six years. They have a sense of how power works. And they understand that a Senator serving 30 or more years wields more power than any President. They are willing to work in the shadows, foregoing the spotlight so many of their colleagues crave. They believe the U.S. Senate is the functional heir to the Senate of Rome, and is in fact as they describe it: the world's greatest deliberative body.

Inouye is a war hero. During World War II he and other Japanese-Americans who volunteered for the Army were assigned some of the bloodiest tasks in the war. In Italy Inouye lost his right arm while his unit was coming to the rescue of a group of soldiers from Texas. In gratitude, the state of Texas made him an honorary citizen.

When he came to the House of Representatives in 1959 as the first Congressman from the newly admitted state of Hawaii (and the first Japanese-American ever to serve in Congress) he became a protégé of the powerful Speaker of the House, Texan Sam Rayburn. It was Rayburn, and his Texas colleague Lyndon Johnson who started Inouye on the ladder to leadership.

Inouye is a man of enormous dignity. With a round face that has aged little after more than three decades in Washington, a deep, James Earl Jones tone and a deliberate cadence to his voice, he commands a respect more often reserved for taller and more physically imposing men.

In Hawaii he wields power greater than any political figure since the death of the Islands' last king, Kamehameha, in 1863.

Hawaii's largest cable system, in Honolulu, had long had an informal agreement with the state, which administered franchises. In return for a low 1% franchise fee, Oceanic Cable, owned by American Television & Communications, agreed to keep rate increases to a minimum. The deal had worked well in avoiding the kind of battles over rate increases that had taken place in other U.S. cities.

When the 1984 Cable Act's rate de-

Club Member: *Rate increases by ATC's Honolulu system in the late 1980s incensed Sen. Daniel Inouye (D-Hawaii), and led to growing congressional pressure to impose rate regulation on cable operators.*

regulation provisions took effect in the fall of 1986, ATC continued to honor the unwritten agreement with the state of Hawaii. It kept rate increases modest and infrequent. But not everybody at ATC's parent company agreed with the strategy. Repeatedly Nick Nicholas, Time Inc. executive vice president and heir apparent to CEO Dick Munro, pressed ATC chairman Trygve Myhren to raise the rates in Honolulu.

Myhren resisted, further aggravating what was already a contentious relationship between the two men. In the spring of 1988 Nicholas ordered that the headquarters of ATC be moved from Denver to Stamford, Conn., effectively forcing Myhren to resign. He was replaced by Joe Collins, who came over from the presidency of HBO.

Almost immediately Nicholas ordered the Honolulu system to raise its rates, by a hefty 10% in June 1988 and another 10% in January 1989. Nobody took the time to inform Inouye or any state officials.

Within hours of the announcement Inouye's long time administrative assistant was on the phone to Myhren, who remained at ATC during the transition to Stamford. "Your people are being absolutely insensitive to the true relationship here," Myhren recalled Inouye's aide saying. "I can't believe this is happening."

"I told Nick he was putting the future of the industry at risk for an additional $250,000 a month in revenue," Myhren recalled. "But he felt we were leaving money on the table."

Nicholas later disputed Myhren's account, denying he had any role in the rate hikes in Honolulu.

Inouye was the wrong person to alienate. He had been a friend of the cable industry, supporting the 1984 Act. He had been in the Senate since 1962 and was chairman of the Senate Communications Subcommittee, which oversaw cable legislation. And he was a member of the Club.

Other Senators had been pounding cable since the ink had dried on the 1984 Act. Sen. Howard Metzenbaum (D-Ohio), backed by the Consumer Federation of America, Ralph Nader and other consumer groups, had held hearings to blast cable operators for raising rates and delivering poor customer service. Senator Al Gore (D-Tenn.) was incensed when MultiVision bought a group of systems, one of which served his family home, and immediately doubled the rates.

But Gore and Metzenbaum were well outside the circle of power. Their tirades were designed more to garner publicity and win points with constituents and potential campaign contributors than they were to pass legislation. (Gore spent most of 1987 and 1988 running an unsuccessful campaign for the Democratic nomination for President).

Double Team:
Sen. Al Gore (D-Tenn.) and Howard Metzenbaum (D-Ohio) led the charge from Democrats in the Senate to rein in cable in the early 1990s.

But when the industry began to alienate Inouye and other members of the Club, it ran into some real trouble. Rate increases weren't the only problem. Well-respected Republican Sen. John Danforth of Missouri was outraged by the heavy-handed tactics employed by some TCI officials seeking a renewal of the franchises in Missouri. Sen. Robert Byrd (D-W. Va.), majority leader of the Senate, found that his flowery speeches on the Senate floor could not be seen by many of his constituents because the systems in several of West Virginia's largest cities did not carry C-SPAN II.

Such incidents served to create a climate of hostility toward cable on Capitol Hill. As Ed Allen, who was chairman of the NCTA when the Cable Act passed, described it in an interview in 1988: "There's an old saying that the cuckoo is the only bird that fouls its own nest. I've got to say that I think we have some cuckoos in our industry which, over the last four years, have fouled our nest, and the Congress is not going to tolerate it."

The political climate was even further poisoned in the late 1980s by the news coming from Wall Street, where cable stocks were the darling of the Street. Almost daily there were reports in the financial press that rate increases had left the cable operators "swimming in oceans of cash," as NCTA president Jim Mooney recalled it. The publicly held companies were happy to tell their upbeat story to the analysts. But every time they did, Metzenbaum and others seized on the news to bolster their argument that the rate hikes were simply lining the coffers of the greedy cable operators.

And the news cable companies delivered to Wall Street was good. In the first nine months of 1989 the price of ATC stock soared 60%, Cablevision's was up 37%, Century's 84%, Comcast 65% and WestMarc 70%.

The cable industry was aware of its growing political problems. An NCTA survey found that rates had increased by 11% in the first six months of deregulation and that in the congressional districts of members of the House and Senate telecommunications subcommittees rates had increased between 30% and 60%. A report by the federal General Accounting Office found that rates increased nationwide by an average of 29% from January of 1987 to October of 1988 and that 40% of subscribers had received rate increases of more than 40%. Mooney kept warning that the combination of double-digit rate hikes and continued inattention to customer service would be a potent combination leading to political disaster.

Cable operators felt that the increases were justified after years of regulation by the cities had kept prices for their service artificially low. They also argued that the increased revenues were needed to pay for better programming and service. They complained that the Congress simply did not understand the economics of the cable business.

But as Mooney told them at one NCTA board meeting, "Congress does not care about economic theory, because economists don't elect them."

Cable's political problems in the immediate aftermath of rate deregulation were exacerbated by a whirlwind of system sales. A survey by the GAO found that some 53% of cable subscribers were served by systems that had changed hands in the three-year period from 1986-1989. And often the new owners would increase rates, restructure the channel lineups, bring in new managers, rename the system and make other changes that would confuse and sometimes anger their customers. In 1988 alone cable systems serving a record 7.2 million subscribers changed hands, some for the second or even third time.

In tiny Alturas, Calif., for example, the system was sold three times from December of 1986 to August of 1988.

And the big companies were merging as well, led by the blockbuster merger of Time Inc. and Warner Communications, a deal that left the combined company with debt of almost $11 billion. A bitter battle left Viacom in the hands of National Amusements chairman Sumner Redstone.

Tele-Communications Inc. gobbled up United Artists Communications and United Cable Television Corp., WestMarc and Heritage Communications. Continental bought American Cable. Cablevision Industries bought BT Cable/Wometco. Cablevision Systems bought the Adams Russell and Viacom systems in Ohio and New York.

TCI president John Malone told the NCTA board in the fall of 1988, "Systems are being bought and sold at high per-sub multiples because of the potential to raise rates continuously. It may be inevitable that there will be some kind of government control or ceiling on rates."

As the perception spread that some kind of reregulation bill would pass, cable operators figured they had better raise rates soon, before Congress shut the door.

As Myhren told the board in 1988, "We may force regulation upon ourselves through dramatic rate increases that may come as a result of the message getting out that the industry is considering agreeing to some form of rate regulation." It was a Catch 22. The more discussion there was about the possibility of a bill passing and the possibility that the cable industry might support a compromise bill, the more operators rushed to get in one last rate increase further angering Congress and creating even more pressure to pass a bill.

Exacerbating all of the problems were the almost continual changes by cable operators in their channel lineups. They were prompted by a variety of factors: retiering in anticipation of reregulation, changes in system ownership, mergers of programming services and new networks coming on line.

One big cause of channel realignment was the federal government. In 1988, the FCC reimposed syndicated exclusivity and network non-duplication rules that had been jettisoned in 1980. These rules gave each local station the right to demand that the cable system black out duplicated shows on imported distant signals.

When the rules took effect in January 1990, some systems dropped the distant signals rather than deal with the complexity of blacking out programs in question. Others installed equipment that would automatically black out specific shows. But both options required extensive explanations to customers.

"By noon on Jan. 2, Bob Rightsell of Comcast Cablevision in Lompoc, Calif., was already hoarse from taking calls from viewers looking for everything from Teenage Mutant Ninja Turtles to Sally Jessy Raphael," *Cable World* reported in its coverage of the second day of syndex.

"These people are irate," Rightsell told the magazine, noting that about 20 hours a day of programming were affected by the new rules.

In Miami one independent station agreed not to seek syndex blackouts if area systems would give it a more favorable channel position. In York, Pa., the system dropped WPHL-TV Philadelphia rather than black out 20 to 30 shows that were duplicated by local stations. The change was part of a general channel realignment that brought American Movie Classics to basic from a tier and added The Comedy Channel. Despite these additions, the system's customers did not welcome the changes. "We had a lot of negative feedback," director of marketing Brad Schofield told *Cable World*.

Eastern Microwave — carrier of New York Independent station WWOR — attempted to solve the problem by offering a separate satellite feed with programming that could replace the blacked-out shows on the station. But while this helped cable operators, it didn't calm viewers who couldn't see their favorite shows.

Another cause of the channel realignments was a move by many major cable companies to tier their services, offering packages of programming at different prices rather than a single package with all the channels.

With so many new channels coming on line it was considered unwise to keep adding to the basic package and having to increase its prices. And operators also figured any reregulation would probably apply only to the lowest, least expensive, tier of service. Typically that contained local broadcast stations, local origination and government channels, and a few relatively inexpensive services such as C-SPAN.

But the biggest motivation for tiering was the increasing prices being charged by the major cable networks to cover

Go To Guys:
ESPN SportsCenter, with anchors Chris Berman and Bob Ley, became a fixture in televised sports among athletes and viewers.

original programming and the cost of outbidding broadcasters for high profile programming such as the NFL. Leading the charge to higher-quality, higher-cost programming were ESPN and USA.

When ESPN launched in 1979 it was free to operators. ESPN even offered financial incentives such as marketing and launch support that made the affiliate relationship a cash-flow negative one for the network.

The bleeding at ESPN was so intense that owner Getty Oil called in McKinsey & Co. to determine whether it even made sense to remain in the cable network business. McKinsey assigned one of its young analysts, Roger Werner, to the task.

Werner determined that only with an affiliate fee could ESPN become a viable financial proposition. In 1982 and 1983 he and company president Bill Grimes and vice president Roger Williams began to approach cable operators about instituting a monthly fee. Without it, they bluntly said, the service would not likely survive.

The first to agree had been Bill Daniels, always a staunch supporter for cable networks and a sports entrepreneur to boot. By the following year ESPN had signed long-term contracts covering 70% of the industry with affiliate fees that ranged from about 6 cents per subscriber per month for the first year up to 12 cents per sub per month.

This base allowed the network to chase after some of the high profile sporting events that had previously been the domain of the broadcast networks. The affiliate fees, Werner recalled, "allowed us to make dramatic improvements in the quality of the product we were offering." That improvement, he noted, enabled cable operators to deliver high quality programming to their subscribers and develop local ad sales.

By the late 1980s ESPN had deals with the major college sports leagues, with Major League Baseball and the NFL and had developed some high profile sports news programs such as SportsCenter.

And ESPN owner ABC had been acquired by Capital Cities Broadcasting, which had been a major cable operator itself and whose chairman and president, Tom Murphy and Dan Burke, were supportive of ESPN's efforts to go after even more sports.

The end of ESPN's three-year contract with the NFL came in 1990, and the bidding war that broke out over the package resulted in a further escalation of prices.

The NFL worked a deal to sell packages of games to both ESPN and Turner Broadcasting. It boosted the cost of the cable rights to NFL games to a total of $900 million from the $135 million ESPN had paid for the original package. The cost per season went from $51 million to $224 million and the cost per game from $3.92 million to $9.2 million.

The operators were faced with affiliate fee increases of around 12 to 14 cents per sub per month from each network to carry the games.

"It's a mess," was the way CableVision Industries vice president of programming Michael Egan put it, reflecting a widespread view that cable had played into the hands of the NFL by allowing Turner and ESPN to get into a bidding war.

By the end of the decade ESPN, which had begun as a free service and had been charging 6 cents a month in 1983, was charging a fee in the area of 30 to 45 cents per month, including the surcharge for the NFL.

USA Network also spent much of the decade improving the quality of its programming, also at a price. In the mid-1980s it broke the lock broadcasters had on the market for first rights to rerun popular programming that had previously aired on the broadcast networks. USA signed deals for the exclusive rights to retelecast episodes of "Miami Vice" and "Murder She Wrote". And in 1988 USA budgeted $250 million to develop original programming including a series of 24 full-length films. Affiliate fees for USA quadrupled during the decade.

In all, the annual expenditure for cable programming, which had been vir-

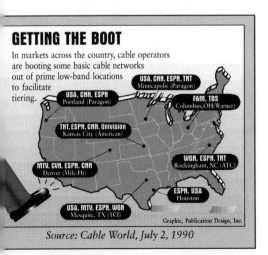

GETTING THE BOOT

In markets across the country, cable operators are booting some basic cable networks out of prime low-band locations to facilitate tiering.

USA, CNN, ESPN, TNT
Minneapolis (Paragon)

USA, CNN, ESPN
Portland (Paragon)

FAM, TBS
Columbus, OH (Warner)

TNT, ESPN, CNN, Univision
Kansas City (American)

MTV, CVN, ESPN, CNN
Denver (Mile-Hi)

WGN, ESPN, TNT
Rockingham, NC (ATC)

ESPN, USA
Houston

USA, MTV, ESPN, WGN
Mesquite, TX (TCI)

Graphic, Publication Design, Inc.

Source: *Cable World*, July 2, 1990

tually nil at the start of the decade, hit the $3 billion mark by the end of 1989, according to figures compiled by Paul Kagan Associates. Affiliate fees charged to cable operators climbed accordingly.

In response, operators began to move ESPN, USA and other networks with high affiliate fees to more expensive upper tiers which were, operators figured, likely to be exempt from regulation.

TCI established a two-tier system in 1990, creating a low-cost basic and a more expensive upper tier with ESPN, USA, American Movie Classics and TNT. Cablevision Systems adopted a tiering system around programming types. It clustered all the sports channels together in one package, all the news and public affairs channels in another and all the entertainment channels in still another.

ATC introduced tiers in the spring of 1990 in many of its systems with the lowest cost cable services and the broadcast stations bundled in the less expensive tier and more expensive services in the upper, discretionary package.

Retiering also caused major upheavals in channel alignments, again igniting customer confusion and complaints.

The battles for carriage also prompted some systems to change their lineups, dropping some networks to make space for others. The result wasn't always consumer-friendly.

The most publicized of the switchouts revolved around the launch of CNBC.

To help NBC devise a strategy to cope with cable and the other new delivery systems, NBC president Bob Wright, who had been head of Cox Cable, hired Tom Rogers, who was the chief counsel on the House Telecommunications Subcommittee and a key architect of the Cable Act of 1984 and other communications laws.

Developing a cable plan for NBC wasn't easy. Within the company there was constant opposition from the broadcast stations and the network whose executives viewed cable as a competitor.

Nor were there any easy entry points into cable itself. Regulations prevented broadcast networks from owning cable systems. "The only major opening was in cable programming," Rogers recalled.

He held discussions with Turner Broadcasting's Terry McGuirk about taking a stake in TBS and had a handshake agreement to buy 25% of that company. But GE chairman Jack Welch blocked the deal, Rogers recalled, fearing that GE as a minority owner would "end up as passive players in an MSO-run programming game."

(That investment of $400 million to buy 25% of Turner in 1988 would have been worth $2 billion to $3 billion in 1998. "We all have regretted not doing it," Rogers later recalled.)

Then Rogers looked at Financial News Network, an upstart operation that was delivering financial news to cable systems nationwide. But FNN wanted too much so he looked elsewhere.

The chance came to buy a network that had been started by Southern Satellite Systems, the distributor for Turner Broadcasting System. SSS had been purchased in 1988 by Tele-Communications Inc., which was willing to sell Tempo TV to NBC. Rogers bit. (TCI president John Malone reasoned it would be better to have the broadcast networks become part of the cable industry than to have them as everlasting opponents politically and economically. Rogers later acknowledged that without Malone and TCI senior vice president John Sie, "There would have been no NBC entry into the cable business.

The Winner Is:
MTV *added another awards show to the nation's consciousness in the 1980s, the MTV Video Music Awards.*

Malone was a guiding force.")

NBC transformed Tempo into CNBC, a financial news service that would compete head to head with FNN. The fight between the two networks for channel space became bloody.

CNBC's first big contract was with Cox Cable, where Bob Wright had maintained good relations. Cox agreed to switch out FNN for CNBC in all its systems. It was a public relations nightmare. Cox simply had not anticipated that dropping a network with such a small audience would create such havoc. At Cox systems across the country, picket lines went up and phone lines were swamped as angry FNN fans protested the move. The biggest problem was that FNN had included a crawl at the bottom of its picture giving the latest stock prices, a feature CNBC did not have at launch.

But CNBC persevered, added the stock quotes and helped other operators with public relations campaigns to prepare viewers for switchouts.

"It taught us a lot," Rogers later said.

FNN, led by former Showtime executive Mike Wheeler, put up a good fight. But eventually CNBC, with its superior resources (and an offer to operators of $3 per sub in launch support money) prevailed, buying out its rival in 1990. That brought NBC firmly and permanently into the cable industry. Because ABC owned ESPN, CBS became the only one of the big three broadcasters with no stake in cable.

Another programming squabble that made waves in the late 1980s involved a spat between Jones Intercable and USA Network. A dispute over a rate increase by USA grew into a personal battle between Jones CEO Glenn Jones and USA chief Kay Koplovitz. Jones abruptly dropped USA from all its systems, and the network retaliated by filing lawsuits (including one charging Jones with violating anti-racketeering laws) and by testifying against Jones at franchise hearings. The dispute was eventually settled, but a lot of blood had been spilled in a very public way.

MTV Networks stirred operator ire when it launched HA! TV Comedy Network by running a sneak preview of the network on Nickelodeon, MTV and VH-1 on April Fools Day, 1990. Nobody had told the cable operators. The stunt backfired in some areas as angry consumers called to complain about programming such as "Saturday Night Live" showing up on the channel normally devoted to Nickelodeon.

The MTV stunt was an example of a growing trend in which cable networks would promote a new service by running ads on existing networks. HA! was promoted in spots on MTV, Nick, and VH-1. WTBS and CNN were full of ads for new Turner networks such as TNT. On USA, viewers were exposed to a barrage of ads for the new Sci-Fi Channel. The ads frequently urged viewers to call their cable operators to demand that the new network be carried. Operators, short of channel capacity, were not pleased.

But whatever the cause of programming changes, the result almost always produced consumer outrage. Americans had a personal, almost intimate, relationship with their television sets that went far beyond what they experienced with such other services as electricity, water, mail, or telephone. Messing with somebody's favorite channel was akin in many consumers' minds to breaking into their home and rearranging their furniture. When it happened, consumers felt violated.

And cable operators were ill-equipped to deal with customer complaints or even queries. Most cable system offices in the late 1980s operated on a 9-5, five-day-a-week basis. Customers calling with complaints or questions outside these hours were greeted with a recording at best or no answer at worst. And with the increased demand for more channels and service, customer service reps and installers were swamped. Service calls sometimes went unanswered and installers were often late for appointments.

Amos Hostetter, whose Continental Cablevision systems were considered to be the best-run in the nation, urged his fellow operators to improve. "Operators need to make sure their phones are being answered and take a look at the pattern of rate increases," he told an NCTA board meeting in 1988.

Another consequence of the mergers and acquisitions that took place almost weekly during the 1980s was an increasing gap between the big operators and the smaller ones in terms of what they paid for programming and equipment.

Nearly every vendor in the business — networks and equipment suppliers — offered volume discounts, giving as much as 25% to 50% off the "list price" to customers willing to purchase in large quantities or deliver large numbers of subscribers. In the case of programming, the contracts also sometimes included both maximum and minimum payments. The practice enlarged the gap between what large and small cable companies might be paying per subscriber for a programming service.

For example, a programming service might have a list rate of 20 cents per subscriber per month for its service. It might additionally offer big cable companies a discount of 20%, bringing the rate down to 16 cents per sub per month. It might also have a cap of $25,000 per month as the maximum any one company would pay and a minimum of $75 per month (an amount the network would need to justify sending out a bill every month and providing other services to a new client.)

A company with 500,000 subscribers who had this type of deal would be paying 5 cents per sub per month, an effective discount of 75% off the list price. But a system with only 100 subscribers would have to pay the minimum, making its effective rate 75 cents

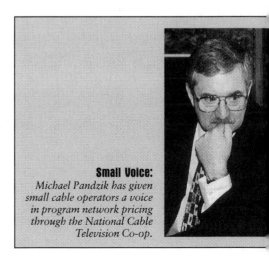

Small Voice:
Michael Pandzik has given small cable operators a voice in program network pricing through the National Cable Television Co-op.

per sub per month, 15 times as much per sub as the bigger company.

That kind of disparity played a key role in spurring even more mergers and acquisitions as bigger companies found they could pay the interest on a loan to buy a smaller company simply by applying the corporate discount to programmers and equipment suppliers.

It rankled the smaller operators who did not want to sell out. Nowhere was the discontent more palpable than among the members of the Mid America Cable Association, a trade group of cable systems in Kansas, Oklahoma, Nebraska and Missouri. Particularly concerned was John Thompson, an Oklahoma cable operator and the grandson of a movie theater operator. Thompson remembered what had happened to his grandfather's theater business in the early part of the century when the movie studios began to gobble up theaters and show their films only in the theaters they owned. He feared the same fate for the small, independent cable operator.

At Thompson's urging, the Mid America Association in 1984 proposed forming an association to act as the collective buying agent for all the small cable systems in the country willing to join. The operators kicked in enough money to conduct a study of the legal aspects of the proposal and formally incorporated in 1985. To run the operation they hired Michael Pandzik, a University of Nebraska graduate who had run the Kansas City office for Home Box Office and later worked in HBO's new business development unit.

The idea was that Pandzik would negotiate a common contract for all participating systems with each vendor, taking advantage of the same volume discount that would be available if the member systems were all under common ownership.

The buying group, called the National Cable Television Cooperative Inc., would then handle all the billing for its members and send each vendor a single check each month. This would save the vendors the task of mailing out hundreds of bills each month and keeping track of each account, some of which were systems serving only a few dozen subscribers.

It seemed like a good idea to the cable operators. Hardware vendors readily accepted the idea. They preferred to sell in volume in any case, avoiding the need to warehouse equipment to respond to a request for a half dozen amplifiers or 300 yards of trunk cable.

But it didn't sound so good to the programmers, particularly those who would realize a huge drop in the per sub fee they would receive from the member systems.

Pandzik set out to land his old employer, HBO, as the first to make a deal with NCTC. "They led us to believe they would be willing to negotiate with us," Pandzik recalled. But after many trips to New York to discuss the matter he sensed he was running into a stone wall. "At the end of the day," he remembered, "they said 'We don't want to do a deal with you.'"

It would be another 15 years before HBO was finally willing to come to the table and then only after NCTC had completed deals with HBO rivals Encore and Showtime.

But Pandzik kept up his quest, and by August had signed his first contract with a programmer, the Weather Channel. The Weather Channel was seeking to expand its carriage, particularly in smaller systems, which found it tough to justify spending the $3,500 it cost to purchase one of the Weather Star units that would download the local forecast as part of the channel's service.

The Weather Channel offered to provide a Weather Star for free for any system with more than 3,000 subscribers, but systems under that level were required to pay.

Pandzik worked out a deal under which the NCTC members could meld their systems to take advantage of the offer. He would match a system with, say, 1,000 subscribers with another system serving 5,000 subs and qualify for the Weather Star as if it were two systems with 3,000 subs each. NCTC members would enjoy a volume discount on TWC rates just as if they were big MSOs, and the Weather Channel, in turn, was able to land dozens of new affiliates.

Creative thinking of this type helped NCTC grow and reach deals with many of the major ad-supported services during the late 1980s. Members of the Co-op were required to pay an up-front fee to join, but no dues after that. The NCTC made its money by taking a commission, typically 2%, from all the funds it collected from its members and paid out to the vendors.

After 15 years of operation the NCTC had grown from an organization serving 12 cable companies with 120,000 subscribers to one which had 900 members serving more than 10 million subs. The median size of the system remained the same, though, at about 300 to 350 subscribers.

Although some major programmers such as Disney and ESPN continued to refuse to deal with the cooperative, Pandzik estimated that over its first 15 years of operation, the NCTC was able to reduce by an average of 25% the

CableLabs:
Richard Leghorn used lessons learned in the Defense Department to push for the establishment of CableLabs.

payments its members had to make for programming and equipment (more for the smaller operators that had been subject to a minimum monthly fees imposed by many programmers). These discounts almost certainly were the critical factor in enabling many smaller cable systems to maintain cash flow margins and credit at the bank and avoid having to sell out to bigger MSOs in the 1980s and '90s.

The mergers and acquisitions of the 1980s also had another impact, this time on the big companies that were so rapidly growing bigger. As they acquired more and more systems, the MSOs found that their equipment was more and more diverse and that many of the devices that worked in one system would not work in another.

Companies such as TCI and ATC would have to deal with dozens of vendors making incompatible equipment. Adjacent systems acquired by an MSO as part of a clustering strategy often did not have compatible equipment and required separate crews of installers and engineers, negating some of the benefits of clustering.

This dilemma added momentum to a

drive that had begun in 1984 when a cable system operator in Cape Cod, Mass., decided to apply to the cable business the techniques he had learned first as a student at the Massachusetts Institute of Technology and later while doing work for the Defense Department that included the top-secret Corona Project which developed high-altitude surveillance techniques to keep watch on the Soviet Union in the 1960s and 1970s.

Richard Leghorn had entered the cable business by accident. Frustrated by his inability to pick up Boston TV signals at his home on Cape Cod, he built his own cable system in 1966, later expanding to run systems in five states. He sold out in 1985.

Cable TV was always a side business for Leghorn, albeit a lucrative one. But he realized some of the lessons he had learned in developing equipment for the Defense Department could be applied to the cable industry.

In particular he advocated the use of systems engineering, first developed at Bell Labs in the 1920s and taught in a course at Leghorn's alma mater, MIT. Under systems engineering, a particular industry or company develops its equipment so that as many elements as possible will be compatible with other elements of the system, even if they had been made many years before or by different manufacturers. Systems engineering also attempts to predict future industry needs so that new equipment will not prove to a hurdle to progress in years to come.

To implement systems engineering many industries had formed cooperative ventures to study new technologies and set standards for development of new equipment. BellLabs, the Electric Power Research Institute and the Gas Research Institute were among the best known examples in the U.S. Such groups were even more widespread in Europe, where the antitrust laws were less stringent.

Leghorn first made his pitch for some kind of consortium to practice systems engineering in cable television at an NCTA board meeting in 1984. But it wasn't until three years later that the industry began to respond. Leghorn had three important allies.

The first was the federal government. In 1984 the Congress enacted a law that reduced many of the antitrust barriers that prevented companies within an industry from getting together to conduct common research and set standards for equipment suppliers.

Leghorn's crusade got a further boost when he offered to provide funding for a study by the Rand Corp. to explore the future of cable and to determine the need for a common research and development entity.

Finally Leghorn won the support of TCI president John Malone who had started his career at BellLabs, the fore-

Show And Tell:
One of the offshoots of CableLabs was its exhibit of new technology each year at the Western Cable Show.

most practitioner of systems engineering. Malone agreed to back the Rand study and urge his fellow operators to do the same.

The task of persuading different MSOs and equipment companies to join a common group was a daunting one. The aim was to foster interoperability of equipment in the cable industry without reducing the innovations and inventions that had propelled the industry forward in the past and without diminishing the fierce manufacturing competition that kept the price of products low.

Leghorn decided the new entity should focus primarily on one phase of the manufacturing process. In manufacturing, the theory went, there were three phases: invention, innovation and dissemination.

Leghorn's view was that invention and dissemination would best be left to the manufacturers while CableLabs could play a role in the middle, the innovation, stage.

The Rand study, which concluded that the cable industry would face stiff challenges from the telephone companies and from direct broadcast satellite companies in coming years, helped galvanize industry thinking on the issue.

In Decembe,r 1988, the NCTA announced the formation of a Research and Development committee, chaired by Malone, to lay the groundwork for what would become CableLabs.

On May 11, 1988, CableLabs was incorporated following a meeting of cable executives that included Malone, Leghorn, Comcast Corp.'s Brian Roberts, ATC's Gary Bryson, independent cable operator Joseph Gans, HBOs' Ed Horowitz, Cox Cable's Jim Robbins, and TeleCable's Richard Roberts. The group issued invitations for other MSOs

Diplomat:
CableLabs CEO Dick Green successfully walked the political minefields between operators and vendors during his tenure at CableLabs.

to join. Dues were set at two cents per subscriber per month. By the end of September cable companies serving some 75% of all U.S. subscribers had signed on as members.

The next task was to find somebody to head the organization. It would have to be someone with extensive engineering credentials but also with the ability to manage the delicate political job of dealing with intensively competitive equipment vendors, fiercely independent MSOs and such related and sometimes hostile industries as the TV set manufacturers and broadcasters. The new CableLabs president would also have to be focused on the market and not interested solely in academic or pure research.

Finally, the new CEO would have to keep in mind what ATC chief engineer Jim Chiddix described in a memo to the founding committee. The structure of CableLabs, Chiddix wrote, would have to "build in disincentives to guard against the empire-building and thirst for self-perpetuation which lie in wait for all organizations not regulated by the harsh realities of profit and loss."

On Aug. 1, 1988, Leghorn announced that CableLabs had found its man: Richard Green, a self-effacing but clearly brilliant engineering Ph.D. who had worked at Boeing and Hughes Aircraft, ABC Television Network and who had run the CBS Advanced Television Technology Laboratory and served as senior vice president in charge of all technical and operational activities for the Public Broadcasting Service.

Green's credentials were outstanding, but so was his personality, a kind of "aw shucks" style which lent itself to consensus-building and cooperation rather than confrontation.

After deciding to locate its offices in Boulder, Colo., (home state of many of the largest MSOs and of cable's strongest political ally, Senator Tim Wirth) CableLabs set down to work.

It adopted a process for standard setting built around Requests for Information and Requests for Proposals. These were at the heart of the equipment procurement process for most major corporations and government agencies.

The process essentially invited vendors to submit proposals for equipment that met the CableLabs specifications, allowed CableLabs to determine from this process what the standards should be for that piece of equipment and then allowed cable operators to continue to buy a wide range of equipment from a variety of vendors but with the assurance that all the equipment fell within the parameters set by CableLabs.

The result was to ensure that the diverse products of the various vendors would, to the degree possible, be compatible with other devices that cable operators were using or were likely to use in the future.

Among the first projects undertaken by CableLabs was the development of a standard configuration for the new fiberoptic cable systems which operators began to deploy in the late 1980s. CableLabs recommended that within a metropolitan area all cable systems be connected together in a "ring" configuration that would require only a single headend. Fiber backbone would then distribute the signals throughout the region. This kind of architecture was also considered optimal for such future services as telephony, regional advertising and regional programming.

In 1989 CableLabs also became involved in the debate over high-definition television systems, a battle that had at one time pitted the entire Japanese electronics industry against a single cable television executive, TCI senior vice president John Sie.

In the 1980s the giant Japanese broadcast network, NHK, together with a group of Japanese electronics manufacturers, had begun to demonstrate a new kind of television system, dubbed Muse, that was designed to replace the existing NTSC and PAL standards em-

Diplomacy:
Turner's Terry McGuirk and TCI's John Sie and John Malone view a Sputnik exhibit in Moscow at Turner's 1986 Goodwill Games, which were designed to promote better relations between the U.S. and Soviet Union.

ployed in the United States and Europe, respectively.

The NTSC standard had been adopted by the U. S. in 1940s in order to ensure that all TV sets would be compatible with all TV signals. Each TV signal would be designed to broadcast to a set that scanned 525 lines 59.94 times a second on a screen that had an aspect ratio of 4:3 (meaning that if it were, for example, 20 inches wide it would be 15 inches high).

The NTSC systems worked very well as long as TV sets were relatively small. But as the demand began to grow for much larger TV sets, the old standard seemed inadequate. The 525 lines simply not did deliver a clear enough picture to be blown up to a size much larger than a 35-inch set.

Muse was pitched as a way to deliver a signal capable of creating a picture on a wide screen TV set that was as clear as what one might see in a movie theater.

When the Muse system was demonstrated at the convention of the National Association of Broadcasters in 1987, it created huge excitement, and many broadcast television engineers came to believe that it would indeed soon replace the NTSC standard.

Muse quickly won the endorsement of the NAB's Advanced TV Systems Committee, the Society of Motion Picture & Television Engineers, and the American National Standards Institute. The State Department, which represented the United States on various international committees working on common standards for electronic equipment, also went along with Muse.

But it had several drawbacks. To produce such a clear signal, it required 1125 lines of scanning, versus the NTSC's 525 lines. To deliver the extra information needed for such intensive scanning, the Muse signal required well over the 6 MHz of spectrum that the NTSC signal occupied. And Muse produced a picture with an aspect ratio of 16:9 versus the NTSC's 4:3. These were differences with enormous implications.

Adoption of a Muse standard by the U. S. would have required the replacement of all production equipment, transmission equipment, television sets and videocassette recorders in the county, at a cost of hundreds of billions of dollars.

And that, Sie argued, was exactly what the Japanese had in mind.

In an article in *Broadcasting* magazine in 1988, Sie wrote "NHK is aggressively pushing the Muse system to the rest of the world, and it's not just because it is a vastly improved technology. Japan, as they say, needs a hit. 1987 marked the first time that the Japanese consumer electronics industry experienced a decline in revenue. To stimulate the marketplace, Japan must come up with new or improved technologies that consumers will both want and buy."

The Muse system, Sie charged, was simply a way for Japan to persuade the world to adopt a new standard that would require the replacement of every piece of television equipment in the world. All of it would have to be replaced with new, Muse-compatible equipment, most of it made in Japan. The total cost would be in the hundreds of billions of dollars.

Sie took his case to the broadcast networks, to the Congress and to the FCC. And he began to make some headway, particularly after he discovered a new kind of NTSC system, developed by a California-based company called Faroudja Laboratories, that could double the number of lines delivered by the NTSC signal without requiring additional bandwidth or a change in aspect ratios.

The United States, Sie told anyone who would listen, should adopt something like the Faroudja system as an interim solution to the desire for enhanced television services while it waited for development of a much more promising technology, one that would clearly dominate the communications industry in the 21st century: digital television. TCI's chief engineer, Tom Elliot, had persuaded Sie that digital television was likely to develop much sooner than most people thought, perhaps by the middle of the 1990s.

But to most of the world, including most of Elliot's colleagues, the concept of digital television seemed like pretty blue-sky, high-tech stuff. As he visited the politicians, cable operators, network brass and others in his crusade against Muse, Sie took the position that even if digital were still decades away it would make more sense to focus in the near term on improved NTSC standards such as those developed by Faroudja, while the long term aim should be to "leapfrog" the Japanese by developing digital transmission systems that would dominate television in the 21st century.

By 1990 ABC and NBC had both dropped their support for Muse and within a few months ANSI reversed its decision. (It was the first reversal of an ANSI standard in the history of the organization.) The tide had turned. The FCC agreed to delay any decision on a standard for high definition until the completion of a series of tests of various different formats. When the agency called for entrants into the test, it had a surprise last-minute contender from the General Instrument Corp. It was called DigiCipher. When DigiCipher was un-veiled it ensured that the window of opportunity for Muse had closed forever.

It is difficult to determine how many billions of dollars the Japanese electronics industry might have earned had the U.S. adopted the Muse standard for high definition television. But whatever it cost the Japanese, it was an irony of history that the loss came at the hands of a refugee who, as a boy 50 years before, had fled the Japanese attack on Nanking, just escaping the massacre that followed.

But dreams of high definition television, digital TV and other modern wonders were set aside for a period in the early 1990s, as were plans for more mundane projects such as those to upgrade and rebuild existing cable systems. The cause was a series of regulations governing the banking industry collectively known as the highly leveraged transaction or HLT rules.

Like so many government rules, these were not designed to have an impact on cable television in particular. In fact those who drew up the rules knew little about cable. The rules were aimed at quite another area of the economy: the banks.

In the late 1980s the American savings and loan industry was plagued with a series of failures and bankruptcies that required a huge bailout by the federal government. To try to prevent such disasters in the future, the government adopted a set of regulations to limit the amount of money that banks could lend to so-called highly-leveraged companies.

The question remained as to how to define a highly-leveraged transaction. In October of 1989 the Comptroller of the Currency redefined highly-leveraged transactions. Prior to the ruling, any company with debt equal to 75% or more of its assets was considered highly leveraged. The ruling lowered that figure to 50%. It also defined as debt subordinated debt and redeemable preferred shares, instruments that had been considered equity prior to the ruling. The ruling also required that all subsidiaries of a company be lumped together for purposes to determining debt-to-equity ratios.

At the time of the ruling, 10 of the cable industry's largest companies had debt-to-equity ratios greater than 75%, ranging from the 76% ratio of UA Entertainment to the 136% of Adelphia. In fact, the ruling snared almost all the big cable companies, and effectively prevented them from obtaining additional financing from banks. For the big, publicly held companies with access to the stock market, bonds and foreign fund-raising instruments, this was a problem. For smaller companies that relied almost entirely on bank loans, it was a disaster.

The first impact of the HLT rules came in the stock market. Adelphia, among the most highly leveraged and therefore most vulnerable to the new rules, saw its stock price plunge from $28.50 a share in late 1989 to $10.50 a share in May, 1990. The MSO stock index maintained by Paul Kagan Associates recorded an average drop of 40% for a group of cable stocks during that time.

As the stock prices declined and the value of systems also plummeted, Salomon Brothers warned that several major cable companies were about to reach a point where their debt exceeded the breakup value of the company. Some smaller firms were forced to file for bankruptcy protection, the first time that this had ever happened in the cable industry, according to records kept by Daniels & Associates.

Cable operators cut back on costs to generate more cash flow and reduce

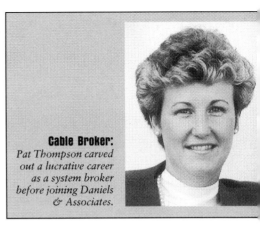

Cable Broker:
Pat Thompson carved out a lucrative career as a system broker before joining Daniels & Associates.

debt. The impact was felt in the construction business. In February, 1990, TCI announced it would reduce its capital expenditure budget to about $160 million from a planned $500 million. The stock of hardware supplier General Instrument plunged $3 a share, or almost 10%, on the news.

The cutbacks quickly hit the construction companies. Cable Constructors Inc. was forced to lay off 170 of its 350 employees, and executive vice president James Bandkt said that the period from mid-1990 to the end of 1991 "was the worst in our history, and we've been around for 30 years." Cable Masters Corp. dropped from 130 employees to around 50.

The effect also was devastating for brokers who depended on a robust market for cable systems and on the hardware manufacturers who suffered when the funding for rebuilds and upgrades dried up.

In the first five months of 1990, system sales totaled 40 versus 161 deals that had been completed in the same period a year earlier and 187 in the first five months of 1988. The dollar value of the deals fell to $751 million in the first five months of 1990 versus the $8.1 billion in the same period the year before and the $9.0 billion in the first five months of 1988.

The boom days after deregulation had produced an even bigger boom in the brokerage business. The biggest brokers — Daniels & Associates and Communications Equity Associates — added staff, beefed up their investment banking capabilities and began to expand into the foreign arena and adjacent industries such as cellular telephone and wireless cable systems.

A host of smaller firms had also emerged. John Waller, a baby-faced former affiliate sales rep for HBO, found that he could parlay the contacts with operators he had made while selling HBO into a list of possible acquisition targets for large MSOs. His firm, Waller Capital Corp., soon shot up into the top ranks of brokers.

Another such "boutique" broker was Pat Thompson, the first woman to start a brokerage and one of the first to focus on sales of small systems.

Thompson was working as a bookkeeper for a commercial ice machine company in Denver in the late 1970s when one of her colleagues, Charles Martz, left to go to work for Jones Intercable in its in-house brokerage.

In 1978 Martz brought Thompson over to meet Glenn Jones who created a new position for her: systems analyst. Her job was to do the financial projections for systems Jones was seeking to buy. In a couple of years Thompson moved up to become a broker.

"Sometimes we would talk to a small system about doing an acquisition and for one reason or another Jones would not do the deal," Thompson recalled. While working for Jones, she couldn't seek another buyer for the system and would sometimes leave prospective sellers hanging. "So I saw this niche and got brave and decided to start my own brokerage." This she did in 1982.

By the late 1980s Pat Thompson Co. had 13 people working full time and was doing 10 to 15 deals a year. Brokers were paid on commission, typically 5% of the first million dollars and then smaller percentages above that. The sale of a system with only 1,000 subscribers at $1,000 per sub could bring in a commission of $50,000. It was a pretty sweet business.

But the highly leveraged transaction rules hit Thompson and other small brokers like a two by four.

"We had six or seven sales under contract," she recalled. "Overnight the banks started to back out of the deals. They were even willing to pay the fees and penalties that were required for them to back out at such a late date."

At first, Thompson recalled, "I thought 'this can't go on forever.'" But it didn't have to. A year after the rules came out the situation had deteriorated to the point where Thompson figured she would have to shut her doors and find another job.

Bill Daniels then called and asked her to join his company. But Pat Thompson's days as a independent broker were over, and many of her fellow boutique brokers faced the same fate.

The impact on the hardware side of the business was equally devastating. C-COR Electronics, the State College, Pa.-based equipment manufacturer that had been founded by Jim Palmer, laid off 270 of its 790 employees in the first nine months of 1990. Another 50 positions were eliminated by retirement or attrition. The company's net sales dropped from $13 million in the third quarter of 1989 to $10 million in the same period of 1990, and net income plunged from $1.4 million to a loss of $244,000.

General Instrument Corp., newly acquired by investment firm Kohlberg, Kravis Roberts, laid off some 65 out of 165 workers at its headquarters. Times Fiber announced layoffs of 30% to 40% of its workforce.

It was in this gloomy atmosphere that Tele-Communications Inc. sought a way to increase the value Wall Street placed on TCI's non-cable system assets, particularly its investments in programming companies. TCI had in the prior decade invested in dozens of programming ventures, some of which were now worth billions. It also had minority interests in dozens of cable companies.

But these assets were not reflected in the TCI stock price. The price of the stock, Malone later recalled, was calculated on Wall Street simply by taking the value of a subscriber and multiplying it by the number of TCI subs. Programming and off-balance-sheet assets were ignored.

The answer was a spin-off company called Liberty Media. Its impact would go far beyond just increasing the value of TCI stock.

"Liberty was born in the humid, fetid atmosphere of the pool house at John Malone's home," outside Denver, recalled Peter Barton, the TCI senior vice president Malone would tap to run the new venture. "We sat around with the smell of chlorine and bromine and made a list of TCI assets that might be spun off into the new company." (The name Liberty was one TCI had acquired when it purchased a cable company by that name.)

The idea was simple: take all the non-cable-system assets of TCI, plus a few minority interests it had in cable systems, and put them all into a new company. But the structure was enormously complex, largely because Malone wanted it to be a tax-free deal in the eyes of the Internal Revenue Service. At the same time he wanted each shareholder to be able to change as many or as few TCI shares for Liberty shares as they chose. In this respect the deal was different from a traditional spin-off where each share of the parent company receives a certain number of shares in the spin-off.

TCI shareholders were to be given a right that entitled them to convert some of their TCI shares into Liberty stock. Because it was uncertain how many shares would be tendered, it was also uncertain how Liberty would be structured. If a relatively small number of shareholders elected to tender, the amount of debt Liberty would have to undertake to TCI in return for the assets it was receiving would be large. If a large number of shareholders tendered, the debt load would be much smaller.

The analysts' meeting at which Malone and Barton were scheduled to

Different View:
Encore Chairman John Sie (far right) took a different view of the pay TV business, launching Encore and then Starz! in the early 1990s under the Liberty Media Group umbrella. Sie is shown here with actors James Woods, James Garner and Jack Lemmon.

explain the new venture was held in Denver one weekday morning in October. Most of the analysts had flown in from New York the night before and took a few minutes to focus their bleary eyes on the phone-book-thick prospectus they were handed as they entered the meeting.

Malone opened with a discussion of the "Big Bang" theory of the beginning of the universe, likening the classes of Liberty stock to some of the subatomic particles that existed for only a microsecond after the first explosion that created the universe. He described the deal as a defensive one, positioning TCI so that there would be a minimum of disruption if Congress forced the company to split up its programming and operating businesses. "We are drawing a dotted line and saying to the regulators 'Cut here'," Malone said. *Cable World* described his presentation as "somewhat downbeat."

The analysts fogged over.

There was no way any of them would be able to comprehend this enormously complex deal let alone recommend to their clients that they tender their stock in the "safe" TCI for the much more risky Liberty, which appeared to them to be just a collection of flotsam and jetsam that the company had collected along the way to becoming the biggest cable operating company in the business. (The initial Liberty holdings included part ownership of American Movie Classics, QVC Network, Black Entertainment Television, Christian Broadcasting Network, a series of regional sports networks and a group of cable system partnerships. TCI continued to hold its shares of Turner Broadcasting System Inc., and The Discovery Channel.)

Moreover, Malone noted, the new company would have to do some heavy trading in the first year to ensure that it was not classified as a holding company rather than an operating company for tax purposes. There was, therefore, no real way to know what Liberty would look like a year out. And the value of the assets TCI was contributing to Liberty was uncertain, given the minority status of most of the investments. Malone pegged the total at $605 million.

As the analysts' meeting ended, Barton stood by the exit door to get a sense of the reaction. "I talked to 21 analysts, and 19 of them said 'no way.' I was stunned," he later recalled.

But Malone, TCI chairman Bob Magness and a handful of analysts (including Gordon Crawford of major TCI shareholder Capital Research Co.) saw the issue differently.

What they saw was an opportunity to take a group of assets that had had little value in terms of TCI stock and transform them into some real jewels.

When the tenders were counted, only a few major shareholders had opted to trade their TCI shares for Liberty. Among them were Malone and Magness, who ended up with a majority ownership in the new venture. There were those in later years, after Malone's shares of Liberty increased in value by 500% and after he transferred them back into TCI, who suspected that he had been planning all along to downplay the deal so that he would end up with a larger stake in the venture.

But all the plans of Liberty were clearly outlined in the prospectus, there for anyone to read. And among the tiny type in the back of the prospectus was a plan for a new pay television service. It would be called Encore.

Malone had always had a touch of

Otto von Bismark, the great German chancellor who believed that the best way to preserve peace in Europe was for Germany to maintain the largest army and to use it to maintain a balance of power between the major nations on the continent.

Malone believed in a balance of power in the cable industry. When Time Inc. and Warner Communications merged, he was concerned that the balance had been upset. His concerns increased when Japanese electronics firms Toshiba and C. Itoh invested heavily in Time Warner, helping to reduce its staggering debt.

Malone in particular worried that Home Box Office, the largest pay service, had been joined with Warner Studios, a major producer of theatrical films and television shows. And the two were now allied with a huge cable operating company.

Also concerned about the Time Warner merger was Sumner Redstone whose Viacom Inc. operated Showtime and The Movie Channel. In 1989 Redstone agreed to sell half of the Showtime/TMC venture to TCI for $250 million. But a year after Redstone and Malone shook hands, the deal still had not been approved by the Justice Department's antitrust division. And the partners, in the meantime, had discovered some differences of opinion as to how the company should be run.

Pay television had had another tough year in 1990. Many of the top cable operators saw an increase in the number of basic subscribers during the year, but a decline in pay subscriptions. Comcast, for example, increased basic subs by 4.2% in 1990, but saw pay subs decline by 3%. Times Mirror posted an increase of 5.4% in basic subs but a drop of 4.2% in pay. According to Paul Kagan Associates, pay subscriptions in 1990 grew only 850,000 versus 1,970,000 in 1989 and 3,560,000 in 1988.

Marketing executives at the MSOs were convinced that the drop was caused by poor promotion. A Jones Intercable survey found that 80% of the time its customer service reps failed to mention pay television services to a prospective new customer.

But Sie and Malone had a different opinion. In their view the four major pay services — HBO, Cinemax, Showtime and The Movie Channel — were too much alike, too expensive and too full of films that might offend some viewers, particularly families with small children.

And, Sie later recalled, cable operator marketing efforts were also at fault. Whenever an operator fell short of budget he would be able to run a "special" on pay services, selling them at a steep discount often using marketing money supplied by the premium network. The tactic worked to pump up numbers in the short run. But after the discounts ended, the number of subs would decline again and the system would wind up even deeper in the hole.

Well aware of the success of American Movie Classics — which kept its prices low and programmed 100% old films — the TCI brass reasoned that there might be an opportunity for a service in the middle offering a menu of films from the '60s, '70s and '80s for more than what AMC was charging but below the price for the maxi pays. Malone, Barton and Sie urged Viacom to transform The Movie Channel, the weakest of the four networks, into such a channel.

Showtime management disagreed. Frank Biondi, newly hired as Viacom president, and Tony Cox, installed by Biondi to run Showtime, were both veterans of Home Box Office. They had grown up believing that the prime mission was to protect the mother ship. Their research told them that a low-priced pay service would only cut into the subscriber base of the existing services, cannibalizing the current market.

Sie began playing around with the notion of starting his own network when he got a letter from an old Hollywood friend, George Stein. Stein was running a small premium network that served primarily home dish owners, SMATV systems and wireless cable operators. Called Starion, it had about 350,000 subscribers, and had been funded by Amway, the big household products company that hoped to use its network of sales people to sell pay TV along with its toilet brushes and furniture polishes.

But Starion was struggling; it couldn't compete with HBO and Showtime for first-run feature films. Stein was in a cash crunch.

Sie offered to buy out Starion if it would first secure contracts for the libraries of major Hollywood studios. All of a sudden George Stein, who had been broke just a few weeks before, was showing up at the studios with checks in hand to buy old films, many of them little in demand. The only catch was that the contracts for the product had to be long-term, assignable and enforceable. The studios happily took his money for films they weren't selling anywhere else. Only later did they realize they had actually been selling their product to a pay service that was owned by TCI.

With product in hand, Sie got the green light from Malone to launch. The strategy was to keep the price low and to get the service into as many homes as possible. For those who already took pay services, Encore was offered for only $1 a month. For those with no services it was priced at $4, about a third of the retail price for HBO.

Then Encore hit a speed bump. Eager to gain a wide audience for the service, TCI decided to offer it to every TCI sub for free for a month, after which those who said they did not want it would be free to disconnect at no charge. Those who did nothing would be billed. It was called a negative option.

The plan also helped out on the operational level. Assuming that a majority of subs would take the service, it was far easier to send the signal in the clear and then trap out those who did not want it than it was to scramble the signal and roll a truck to every home that wanted the service.

The negative option was an ideal solution. The only trouble was that it was illegal—at least in the opinion of a host of state attorneys general who filed suits all across the country to stop the TCI plan.

"We had been meticulous in checking it out in every state," Sie later recalled. "The direct mail letter was written entirely by the lawyers."

Yet when the project was announced, the attorney general in Florida declared it was unfair to make consumers in his state reject something they had not ordered in the first place. Faced with lawsuits in 35 states and an avalanche of negative publicity, TCI abandoned the plan. It was the rockiest launch of a pay service in history. It wasn't until years later that TCI would actually win the court challenges to the negative options.

But Sie persevered, and within months was touting research that showed that Encore, rather than cannibalizing existing services, actually helped retain existing pay subs and attract basic subscribers who had never before signed up for a pay service.

HBO and Showtime reacted swiftly. Each launched a "multiplex" of its existing service, offering a similar network with the times of the movie showings shifted. The multiplexes would be sold at a slight incremental cost to the maxi core service. HBO brought in a new president, chief financial officer Jeffrey Bewkes, to manage the increasingly complex financials involved in a business which had grown from selling one network to selling half a dozen.

But while the strategies did appear to pay off where they were tested, many cable systems did not have the channel capacity to add new premium networks — maxi, mini, multiplex or whatever.

What saved the premium channels in the early 1990s was not cable. The real savior of pay in the 1990s would be direct broadcast satellite services, hungry for product and eager to offer DBS subscribers as many premium services as possible.

Basic cable, in the meantime, continued to grow, fueled by the new expenditures by such networks as USA and ESPN. Also helping to boost ratings were such events as the 1991 Gulf War when huge audiences tuned in to watch CNN and its three brave correspondents — Bernard Shaw, Peter Arnett and John Holliman — who were trapped in Baghdad as the war opened and managed to report live during the first days of the conflict direct from the enemy capital.

New ad-supported networks did not fare as well. Sports News Network, a venture of TV syndicator Vic Piano Jr., ran out of cash after spending millions to build a state-of-the-art studio where it planned to deliver 24 hours a day of sports news.

Big Night:
The world was riveted to CNN's coverage of the Gulf War in 1991 as Bernard Shaw, John Holliman and Peter Arnett broadcast from inside a Baghdad hotel as U.S. missiles hit the city.

Operators could see little point in taking on a new network that was doing essentially the same thing as ESPN's well-established SportsCenter. (Years later, however, ESPN and CNN would both launch sports news channels.)

And the Christian Science Church flopped in spectacular fashion when it attempted to create a video version of its well-respected newspaper, the *Christian Science Monitor*. Church members were appalled when the *Boston Globe* reported that the channel had lost $500 million in just one year of operation. Church officials angrily denied the report, saying the loss was only about half that much.

As in $250 million, a quarter of a billion dollars.

The sum was staggering. Even CBS Cable, the highly publicized gilt-edged entry into cable by CBS, lost only $50 million before it closed down. Church officials said the bulk of the money had been spent on studio and production equipment and on salaries. At its peak, the network had employed 300 people who occupied almost an entire five-story building in Boston, another 30 in its Washington, D.C., office and dozens of

Triple Play:
NBC-Cablevision's Olympic Triplecast was an artistic success and a financial failure. Pricing and channel capacity issues doomed the project. Shown here are executive producer Terry Ewert and Triplecast VP Marty Lafferty at Barcelona Stadium.

others in bureaus around the world. It would remain, through the rest of the century, the largest loss by a programming network in history.

But the most spectacular failure of the early 1990s came in the pay-per-view arena. For years the trade publications had run headlines such as "Pay Per View: Has Its Time Come?" and "Pay Per View: Is This the Year?" The answer had always been "no."

Cable operators were slow to invest in back-office changes, marketing and other expenses needed to make pay-per-view a reality. TCI remained on the pay-per-view sidelines throughout the 1980s, cutting off about 10% of the nation's subscribers from access to PPV products. Studios, frustrated at the inability of cable to market pay-per-view, gave an earlier window on product to the video rental stores. Warner Bros.' Ed

Bleier became a regular on the cable convention panel circuit, scolding the industry for failing to make the necessary investments in marketing, channel capacity and customer service. But operators found it more lucrative and much easier to devote any open channels to home shopping services or mini pay networks.

All that was supposed to change when a partnership between NBC Cable and Cablevision Systems Corp. announced in 1988 that it had secured the rights to telecast portions of the Olympics on a PPV basis in 1992.

The deal seemed a natural. NBC had won the television rights to the Olympics after an intense bidding war. It would have to cover all the events taking place in Barcelona but would be able to air only a small portion of them on the broadcast network and only some of those in prime time. Less widely watched events such as fencing, equestrian, and the pentathlon were unlikely to make it on to the network.

These events did have small but loyal followings in the U.S. And many who followed such sports were in the upper income brackets. The theory was that the incremental cost of telecasting these additional events would be relatively small because NBC was covering the Olympics in any case. And the hope was that if the less-popular games were offered on a PPV basis they might attract a small but enthusiastic following.

The plans were grandiose. The partners wanted each cable system to devote three full channels, 24-hours-a-day to the 15-day "Olympics Triplecast." Subscribers would be offered a choice of viewing packages priced from $95 (for weekends only or the first seven days) to $170 for the full Triplecast plus a variety of premium items. Later, consumers were allowed to purchase a single day of the telecast for $29.95. Systems were allowed to vary the price by as much as 20% in either direction.

The Triplecast had an incredibly complex system for determining how much revenue would go to the system and how much to the programmer. The split was anywhere from 30% to 55% for the operator depending on how many channels were devoted to the project, how much marketing was done and how the price was set.

Triplecast executives predicted the event would be purchased by about 5% of the nation's 40 million addressable homes for a total audience of around two million. It planned to spend around $50 million to promote the venture.

The logistical problems were horrific. Many cable systems that wanted to participate in the event did not have the addressable equipment necessary to make it happen. They were forced to rely on traps that had to be installed outside the home of each Triplecast subscriber. Because there was no way of knowing how many orders would come in at the last minute, there was no way of knowing how many traps would be needed.

Other systems did not have the billing and back-office setups necessary to handle the complex ordering process. And few systems were able to train their customer service reps to answer the myriad of phone queries about the event.

But the biggest problem was how to clear three channels for 15 days. Few systems had so many blank channels. The Triplecast executives attempted to work out an arrangement with other basic networks — Discovery and C-SPAN in particular — to allow their channels to be used for the Triplecast, but the other networks weren't anxious to go dark for two weeks in the summer. And cable systems could not simply drop a channel for that long without violating carriage contracts.

NBC allowed operators to drop CNBC for the Triplecast and Cablevision volunteered American Movie Classics. Still, many systems did not carry those services or had reservations about offending viewers by taking them off for the Olympics.

Just weeks before the event TCI wrangled a deal which allowed it to offer a single channel of Olympics pay-per-view, complicating the marketing and operations issues even more.

In the end the venture was a huge disaster, at least from a financial point of view. Rather than the two million subs that had been projected, the Triplecast attracted less than 500,000 subscribers, less than 1% of the potential audience. NBC and Cablevision came nowhere near taking in enough revenue to cover the cost of the event. The partners split a loss of around $150 million.

But the Triplecast was not a complete failure. The event did prove to be a catalyst for many cable operators, including TCI, to get into the PPV business. And it taught cable operators a lot about how to market and stage PPV events.

The Triplecast also received generally favorable reviews from the consumer press. *The Los Angeles Times* said the "Triplecast is triple terrific," and the *Wall Street Journal* called it "a delight for the avid sports fan."

The losses didn't faze Charles Dolan, Cablevision's indomitable chairman. Dolan, after all, was the guy who had continued to believe in Manhattan Cable in the 1960s when it was bleeding millions of dollars a year and his partner was desperate to bail out. And he was the one who continued to believe in a service he launched called HBO even as his partner, Time Inc., was considering killing it.

The day after the Triplecast Dolan took out two-page ads in the trade press praising his fellow operators for putting up with all the logistical headaches and "for your characteristic willingness to take a chance with new product to give your subscribers yet another viewing option."

Dolan immediately set to thinking about how to do it better next time. "There has to be a way that cable can do this and hit a home run," he said.

The Triplecast highlighted a problem that had grown considerably worse for cable operators during the 1980s: customer service. With inadequate phone systems, limited office hours, poorly trained and compensated customer service representatives and lack of installers and phone people, the industry simply was not able to deliver the kind of service that consumers had come to expect from other business giants that burst on the world in the 1980s such as Federal Express.

As the decade progressed cable operators became increasingly serious about improving their dealings with customers. Tele-Communications Inc., which as the largest MSO had borne the brunt of criticism about customer service, began to attack the problem in the late 1980s. It established a system of regional customer service centers open beyond normal business hours. If a customer in a small TCI town called after business hours, the call would be switched to the regional center where at least it would be answered instead of forcing the customer to wait until the next morning to call.

Many MSOs began to change the way they compensated their system managers to place greater emphasis on customer service, abandoning compensation systems that tied bonuses solely to annual cash flow objectives, which had encouraged managers at times to cut corners.

Falcon Cable, for example, increased the amount of bonus paid to system managers able to demonstrate increased customer satisfaction. Warner Cable changed its plan to place equal emphasis on customer service and cash flow growth.

But the most dramatic effort was the attempt by the NCTA to establish a nationwide set of customer service standards.

The standards included these pledges:
- Phones would be answered within 30 seconds;
- Systems would respond to service interruptions within 24 hours;
- Customers would be notified of rate or channel changes 30 days before they took effect;
- Refund checks would be issued within 44 days.

The standards were not paper tigers. United Artists Cablesystems president Marvin Jones estimated his company would have to hire 450 new employees to meet them in UA's systems.

Still, the efforts did not have much immediate impact on customer perception. One reason was the very nature of the relationship between a cable company and its customers. Subscribers did not generally have occasion to call their systems regularly. If they received a poor response, it might be months or even years before they had a reason to call again, meaning it might be months or years before that customer had a chance to see service at the system had improved. And it would take even longer for that changed perception to trickle up to the elected representatives, who continued to receive complaints from disgruntled customers.

But cable's biggest political problem continued to be rates.

Operators continued to raise fees. In March 1990 they went up 1.3%, versus an increase of only 0.5% in the overall consumer price index. The increase followed a 1.8% increase in February. A GAO study found that cable rates jumped 10% in 1989. It also found that just a handful of operators were responsible for the bulk of the rate hikes: 12% of the cable operators increased rates by 20% or more. The study also found that 53% of cable systems changed ownership between 1985 and 1989.

And the pace didn't seem to slow as time went on. Paul Kagan Associates found that rates for basic cable increased by 16% in 1990 just when many operators were desperate to increase cash flow to meet the new highly leveraged transaction rules.

Locally the rate hikes made big news and an easy target for politicians. A plan by Century Communications to raise rates in its huge Los Angeles system by 13.9% — to $24.20 a month for basic — in November, 1991, prompted city mayor Tom Bradley to call for federal legislation to control rates.

"The citizens of Los Angeles should no longer be held hostage by an industry which operates virtually unchecked by competition or by government," Bradley said.

As the drumbeat of publicity about rate hikes went on, dramatic shifts took place in the public perception of the relationship between the price of cable and its value.

In an annual survey of cable subscribers conducted by polling firm Talmey Drake Research for *Cable World*, the per-

Bill Of Rights:
Viacom Cable president John Goddard, along with Robert Miron and other executives, led NCTA's development of customer service standards.

centage of consumers saying cable rates were "way too high" climbed continually from 17% in 1989 to 37% in 1992.

Cable's customer service image lagged behind the electric company and the phone company. It was worse, customers said in the 1992 survey, than the U.S. Postal Service.

All of this was not lost on the lawmakers considering legislation to reestablish regulation of cable rates.

The effort began as a series of hearings held by Sen. Howard Metzenbaum (D-Ohio). But it picked up steam in the late 1980s as senators such as Danforth and Inouye joined the call for reregulation of cable rates.

Cable's marketplace rivals were quick to pile on. Telephone companies called for legislation to allow them into the cable business. DBS, private cable and multichannel MDS companies demanded equal access to satellite-delivered programming. City officials and consumer groups called for reregulation of rates.

And, most important, broadcast television stations sought legislation to reestablish the must-carry rules that had been struck down by the Supreme Court in a startling decision in 1984.

Broadcasters wielded enormous clout on Capitol Hill. Each congressman and Senator was well aware that the local broadcast news operation could ensure that his or her picture would be seen by millions of constituents.

Against this array of opposition cable had two high cards. The first was Senator Tim Wirth (D-Colo.), cable's longtime ally whose campaigns had been funded by Denver-based cable operators for years and who had been the author of the 1984 Cable Act which deregulated rates.

Cable's other big ally lived at the other end of Pennsylvania Avenue.

When Ronald Reagan's second term as President came to an end, a host of Republicans vied to succeed him. Considered the heavy favorite was Vice President George Bush. But Bush received a nasty surprise in the first presidential contest of 1988, the Iowa precinct caucuses. The winner was Sen. Bob Dole of Kansas. The outcome was labeled an all-but-fatal loss to the Vice President by many of the journalists covering the race.

The next day, Bush attended the largest political fundraising event ever held in the state of Colorado. It gave his campaign a badly needed boost, both in terms of money and morale. It enabled him to go on to New Hampshire where he beat Dole and went to on to win both the GOP nomination and the general election.

That key fundraiser was held at the home of Bill Daniels. It was an event Bush was not likely to forget.

As the drumbeat for legislation grew louder, the NCTA board of directors debated what to do. In 1989 they came to the conclusion, at the prompting of Mooney, that some kind of legislation was likely to pass the Congress and that the industry would be better off attempting to shape the new law more to its liking than to try to kill something that was seen as inevitable.

By 1989, Mooney recalled, "the political situation was starting to get out of control. People couldn't understand why cable rates had to rise at multiples of the CPI year after year at the same time they were receiving bad customer service. Congressional staffers, journalists, FCC employees — all had some experience with bad service."

As Mooney trudged around Capitol Hill seeking to gain some leverage on the bill, he found that it was an almost impossible task. Unlike 1984 when the issues had been fairly simple, the situation in 1989-90 was far more complex.

Each time Mooney would find a compromise acceptable to his board and to the Hill on one issue, another would arise. Particularly difficult were the issues of programming access and must-carry.

Mooney had a tough time keeping his own troops in line. TCI president John Malone shocked Sen. Metzenbaum when he testified in 1989 that he could live with a bill that would reregulate rates. But as it became clear that the legislation would include far more than limits on rates, TCI backed off and opposed the bill.

Time Warner, on the other hand, was concerned that the legislative uncertainty was cutting into its stock price. It was far more willing to accept a compromise just to settle the financial markets.

And while NCTA was expressing a willingness to work with the key members of Congress on a bill, the Community Antenna Television Association was opposed to almost anything that was offered.

Even President Bush complicated matters. Bush had chaired President

Reagan's deregulation task force and had staked out the issue as a key one in his campaign for president. While the NCTA had the threat of a Bush veto as its trump card in negotiations on the Hill, it was far less certain that it could persuade Bush to sign a bill even if it was acceptable to cable. Such a change would have required Bush to reverse positions on a major campaign pledge.

Nor was it clear that even if the NCTA agreed to a deal most operators would go along. Many of them, including Bush's friend Bill Daniels, were opposed to any kind of reregulation bill.

As the process moved forward, the industry found it had another problem. Cable's strongest ally in the Congress, Rep. Tim Wirth, had been elected to the Senate in 1986. His replacement as chairman of the House telecommunications subcommittee was Rep. Edward Markey (D-Mass.). Wirth was a new kind of Democrat, elected from a marginal district in a Western state. Markey was an old-fashioned liberal from Massachusetts where winning the Democratic primary for Congress was tantamount to being elected for life.

While Wirth believed that marketplace forces were the best way to ensure diversity and competition, Markey believed that government should play a major role in regulating the industry.

To replace Tom Rogers, Markey hired as the subcommittee's chief counsel David Moulton, a former aide to Ralph Nader with a deep distrust of big business of any kind.

By the time the full House and Senate had voted on the bills in late 1990, they included provisions that would:

- Allow the FCC to regulate the rates charged by cable systems.
- Force satellite-delivered cable networks to sell their programming to alternative delivery systems such as DBS and wireless cable.
- Reinstate the must-carry rules.
- Impose limits on the resale of cable systems to curb what the lawmakers viewed as "trafficking" in systems.
- Instruct the FCC to determine if there should be limits on the size of cable companies.

As the bill reached the Senate floor in the fall of 1990, the programming access provisions were the ones that caused the most trouble for cable operators. They reasoned that because they had supported cable networks as startups, they should be allowed the exclusive rights to distribute them.

The issue also divided operators from programmers. The networks, after all, were happy to find audiences wherever they could, and although most of them continued to say that they supported exclusive deals with cable systems, most preferred having alternative delivery systems.

A filibuster by Sen. Wirth at the end of the session forced Sen. Gore and other backers of the access provisions to back down and allow exclusive deals between operators and programmers so long as they did not "impede competition." To secure the deal Wirth incurred the ire of many of his colleagues who were impatient to get back to their districts for reelection campaigns and resented this Senator tying up the works for a single constituent.

(Two days after Wirth ended his filibuster, the cable systems in the suburbs of Denver announced a rate increase. The local papers savaged Wirth as a mouthpiece for a greedy industry.)

Despite the Wirth-Gore compromise the bill died as a host of Senators offered last-minute amendments and Bush continued to threaten a veto. Many backers of the bill felt that cable had changed its mind at the last minute and killed a bill it had pledged to support.

When the new Congress convened in January, 1991, the backers of cable legislation set to work with a vengeance. The 1990 fight had "left us in pretty bad shape in the Senate," Mooney recalled. "There was a bad taste left. We were seen as giving the (legislative) system the finger. And what happens then is that the system breaks your finger."

While the NCTA argued that the 1990 measures it opposed went far beyond what had been contemplated when it expressed willingness to compromise, many lawmakers were not inclined to agree. And many were still angered over rate hikes in their own districts.

Among them was Sen. Inouye, still smarting over the rate increases imposed in Honolulu in the wake of the 1984 Act.

Inouye introduced legislation that would not just reimpose must-carry, but would actually allow each local station the option to choose either must-carry or retransmission consent. With must-carry, the station would have the right to carriage on the lowest tier of service on every cable system in the station's area. But the station could not charge the cable system. Under retransmission consent the station could elect to charge cable systems for carriage. The system would then have the option to pay the fee or drop the station.

The plan gave broadcasters the best of both worlds. Weaker stations would opt for must-carry, ensuring them widespread distribution. Stronger stations, particularly network affiliates, could charge for retransmission consent, confident that no cable system could afford to drop them.

Inouye even added a provision that would have prevented cable systems from passing on to consumers the cost of retransmission consent payments.

A frustrated Mooney declared that he would be willing to work with the broadcasters on retransmission consent or to work with them on must-carry. "It's when they want both that it gets me going," he said.

The industry continued to fight the bill, but the legislative locomotive was moving faster and faster. The one hope was that President Bush would veto the bill. This looked like a pretty good strategy in 1991 and the beginning of 1992 as Bush appeared headed for a reelection victory on the heels of his soaring approval ratings after the Gulf War.

But Bush began to slip in the spring after nearly losing the New Hampshire primary to Pat Buchanan. By summer he was in deep trouble. Democrats on the Hill began to search for ways to embarrass him politically. One way to do that was to override a veto, something Congress had not been able to do in the four years since Bush took office.

With the cable bill they had their chance. It was a great opportunity for the Congress to go on record as pro-consumer and to embarrass Bush at the same time. And the bill was easy to vote for in part because it pleased so many constituents at once (broadcasters, consumer groups, local politicians, DBS and MMDS companies) and because it was vaguely worded, leaving much of the final rulemakings up to the FCC.

Bush did indeed veto the bill, but the political pressure on the Congress was too great. In the House of Representatives, just hours before it was set to adjourn, the motion to override Bush's veto passed by the margin of a single vote.

Markey called the legislation "the most important piece of consumer legislation in this Congress." But the measure left to the FCC the job of drafting rules on rate regulation, programming access, leased access, protection of children against adult programming, limits on MSO size, technical standards equal employment rules and a host of other issues.

"This bill has been crudely hewn by the Congress, but the final sculpting will be done by the FCC," said Joe Collins, Time Warner Cable's chief executive and NCTA chairman. "Only then will its final shape and its real impact on us and our customers be known."

Collins was right. But neither he nor any of his colleagues could predict that fall of 1992 just how bad the impact of the new law would be. The defeat of George Bush in the presidential election of 1992 brought into office Bill Clinton and his vice president — cable's most vehement opponent on the Hill, Al Gore. (Gore referred to John Malone as the "Darth Vader" of the communications industry.)

Clinton would allow Gore to direct much of the administration policy on telecommunications, including choosing the new chairman of the FCC.

Gore's choice would be a little-known Washington lawyer with almost no background in telecommunications law. Reed Hundt's chief qualification seemed to be that he had been Gore's high school classmate. By the time Hundt left office he would have done more damage to the cable industry than any single individual in history.

The Door-To-Door Billionaire

Even as the highly leveraged transaction rules wreaked havoc with the expansion plans of major cable operating companies, tanked dozens of deals and forced widespread layoffs at brokers and equipment companies, a few hardy souls still found it possible to expand their companies. One even found a way to launch a new MSO.

Jeffrey Marcus started his career in cable selling service door-to-door while an undergraduate at the University of California in Berkeley. "I had been driving a garbage truck to make money for tuition and college expenses, and my roommate had been selling cable door-to-door. I was getting up at 5 a.m. to go to work and getting home at 4 p.m. He was going to work at 6 p.m. and getting home at 10. And he was making more money. So I decided to give it a try," Marcus later recalled.

"In my first day selling I worked two hours and made six sales. Each sale earned me $3 commission so I was making $9 an hour. It was like I had struck it rich."

After college Marcus sold cable full time and soon formed his own company which contracted to handle the marketing for major MSOs. The company, Marcus later recalled, was a victim of its own success, going under after expanding too quickly. Marcus received a call from an executive at Sammons Communications asking if he would be interested in going to work for the MSO as vice president of marketing.

"I remember my interview with C.A. Sammons," Marcus recalled. "He quizzed me closely and then asked 'Jeff, for years you've been coming down here telling us we shouldn't do our own marketing, that we are much better off contracting it out to you and your professionals. Now you tell me you want to do it in-house. How do you square those two?'"

"Mr. Sammons," Marcus replied, "when I was selling you the contracted service ..... I lied."

Sammons laughed and hired him. Marcus boosted sales at Sammons and instituted a professional marketing effort that won a major NCTA award. He then got a call from Marc Nathanson, newly hired as vice president of marketing for TelePrompTer, asking Marcus to come to New York to help the nation's largest MSO claw its way back from the edge of bankruptcy.

It was 1973 and TelePrompTer owner Jack Kent Cooke had promised his bankers that by year-end, the company would increase its subscriber numbers by one-third, to one million subs. Nathanson and Marcus were hired to keep that promise.

Every day Cooke called the 27-year-old Marcus into his office or asked him to walk him back to his hotel. And every day Cooke pounded the young man to get more subs. After hiring 600 new sales people, the effort paid off and saved TelePrompTer from bankruptcy.

When new management came into TelePrompter, Marcus left and joined former Times Mirror executive J. Patrick Michaels to form a new cable brokerage company, Communications Equity Associates.

Although Daniels & Associates had for 20 years maintained a huge share of the cable brokerage business, Marcus and Michaels decided there was room for a competitor. "After all," Marcus later recalled, "We figured Daniels couldn't handle everything."

Marcus knew nothing about being a broker and had no financial training. With no financial resources, his wife Nancy took on the job of earning the income by selling magazine ads while Marcus went in search of deals.

His first listing came from an insurance company that was partly owned by Marcus' old employer, Sammons. Marcus first thought about buying the system himself, but couldn't raise the money.

Then he got a call from John Booth, the multimillionaire former publisher of *Parade*. Booth was looking for some systems to buy. But Marcus was reluctant to tell him about his prospect because he didn't yet have a signed listing agreement and would risk losing his commission. Marcus called Booth back and worried the whole time about how much the long distance call, charged to his home phone, was costing.

"Look," Booth said, "You don't know me. But I am a man of my word. And I am a cash buyer. I'll have my man go up there this week with a check for the down payment. If he likes it we will do a deal."

"So there I was," Marcus recalled. "I was calling him long distance and was worried about how much the phone call was costing. And I was trying to raise the money to buy this system myself. And I didn't have a signed listing and had no way to protect my commission. And here this guy was, telling me to trust him."

Marcus said okay.

The next week the deal was done and Marcus collected a commission of $50,000, more than his annual salary had been in any prior position. Marcus said he went back to his hotel that night and wept.

By the mid-1980s Marcus was spending a lot of time with TCI president John Malone, selling him systems throughout the country. And always in the back of his mind Marcus had the notion of becoming an operator himself.

In 1979 Marcus found a system in Wisconsin he wanted to buy and put $100,000 down for the $632,000 purchase price.

A few weeks later Marcus and Malone were in Acton, Mass., trying to buy the Acton MSO for TCI. Late one night, after a well-lubricated dinner with the owners of Acton, the two got to talking and Malone told Marcus, "You know, Jeffrey, you ought to sell me that system of yours in Wisconsin. You don't know anything about running a cable system and I have lots of systems in that area. It would fit in nicely."

Marcus recalled that he replied "Fine. But I won't take a penny less than one million dollars. And Malone said okay."

The next morning Malone got up and told Marcus, "You know I think I made a mistake last night." (Malone was already building a reputation for making a deal and then trying to make it a little sweeter. It was a great negotiating tactic since the other party had already mentally made the deal, and often mentally spent the money, and

Money In The Bank:
Jeff Marcus displays the first check he received for Markit the cable marketing company he founded. He went on to serve in marketing positions at Sammons Communications and TelePrompTer before buying his first cable system.

was likely to try to accommodate Malone's last-minute demands.)

"I don't think I can buy that system unless you can get a rate increase from the city council," Malone continued.

Marcus replied that he would get the rate increase if TCI would promise to upgrade the system to 24 from 12 channels. Malone agreed and the deal was done, leaving Marcus with a profit of $370,000 on an investment of $100,000 in just a few weeks.

That is the way life sometimes worked in those days.

By 1982, although CEA had grown to where it was challenging Daniels for the position of biggest broker in the business, Marcus continued to want to own systems. He approached Malone about buying back the Wisconsin systems. Malone suggested the two become partners and offered to help finance the acquisitions if Marcus managed the venture. It was an arrangement Malone would employ many more times over the years (with such veteran operators as Jerry Lenfest and Bill Bresnan, among others). The deal allowed TCI to list some of its cable holdings off its balance sheet, employ the talents of some experienced operators, keep the TCI operating staff from getting completely overwhelmed, make TCI look smaller to Washington regulators and allow for an easy breakup if new legislation or rules required it. It also allowed Marcus and the others to take advantage of TCI's bulk discounts on programming and hardware.

By 1988 Marcus' company, after merging with TCI subsidiary Western Communications, served 550,000 subscribers. Marcus stepped down to spend more time with his family. (He continued to live in Dallas although WestMarc was based in Denver).

But two years later, the itch hit again. This time Goldman Sachs & Co. offered to finance his efforts to start a new MSO and he began again, once again with 15,000 subscribers from the same area of Wisconsin, which he acquired in exchange for part of his ownership interest in WestMarc.

Marcus also had his eye on another group of Wisconsin systems, owned by Star Cable TV group. He had begun to negotiate with Star in 1989 and reached a handshake agreement that fall. But then the HLT rules hit, and the agreement collapsed when several of the banks that were to provide subordinated financing backed out.

But Marcus persisted. Eventually he was able to hammer out an agreement by putting up 30% of the purchase price from the Goldman Sachs money, financing part of the balance through a group of banks led by First National Bank of Chicago and persuading the sellers to carry the subordinated debt that he was unable to find a bank to handle.

And he managed to persuade Star to drop the price from $2,160 a sub agreed to in fall 1989 to $2,000 a sub when the deal was finally put together in August, 1990.

The business plan presented to the banks was, by necessity, very conservative. Marcus vice president Louis Borelli at the time noted that "These are plain vanilla systems with little addressability. I didn't project hardly anything in ad revenues. I think there's upside here, but hell if I know how to do it yet." Marcus would figure it out. Eight years later, almost to the day, he would sell those systems, and dozens of others, to computer billionaire Paul Allen for $2.8 billion.

A few others were able to swim against the HLT tsunami and complete purchases. Marc Nathanson closed the purchase of a group of systems owned by Cooke Media Co., and Intermedia partners bought systems serving 41,500 subs from US Cable.

But such deals were the exception rather than the rule.

The Dawn Of Digital

Long before Columbus, Native Americans developed the first digital communications system in the Western Hemisphere. A series of smoke puffs transmitted simple messages over great distances. Later, Samuel F.B. Morse invented another digital transmission system, this one using dots and dashes transmitted electrically over wires, for his Morse code telegraph.

Digital systems were also at the heart of computers from the very start. A binary system of ones and zeros enabled users to store and retrieve information and perform a variety of other tasks.

But television developed as an analog system in which the signals are transmitted using a radio wave. Each wave has a different height and different spaces between the peaks. The television receiver measures the heights of the waves and the space between them to decode the information necessary to determine how to scan the TV screen to bring up the appropriate dots of color.

The prospect of using digital signals to send TV transmissions took hold in the cable industry in the mid-1980s when Home Box Office scrambled its satellite signal to thwart pirates who were picking up HBO for free off the satellite.

HBO issued a request for proposals for an encryption system for its signal. The system it picked was provided by a Massachusetts-based company called M/A Com.

M/A Com had developed several systems for sending encoded digital communications for the Defense Department. It modified some of those for HBO, providing a system that scrambled an analog video signal and a digital audio stream.

The system, taken over by General Instrument Corp. when GI bought M/A Com, did not prove to be unbreakable. Within a few months, a group of hackers had found a way to clone the chips in the decoder boxes provided to legitimate HBO subscribers and then sell those chips on the black market.

The problem created tension between GI on the one hand and the programmers and operators on the other. More than a few in the latter group grumbled that GI was in no hurry to fix the problem since every cloned or pirate chip had to be installed in a GI box. The more boxes that were sold – pirate or legit – the more money GI made.

It took GI until the late 1980s to solve the problem and develop a new system it called VideoCipher II Plus that proved largely invulnerable to attack. By the end of the decade GI's Videocipher division had a monopoly on the system used to

Digital Domain:
Marc Tayer, Jerry Heller, Woo Paik, Ed Krause and Paul Moroney led the General Instrument Corp. team that developed the company's DigiCipher digital compression technology.

scramble satellite signals and the system used to authorize descramblers.

When a consumer wanted to subscribe to satellite services directly off the bird, he or she would buy a backyard dish and a set of electronics, including a descrambler. After the equipment had been installed, the consumer would call one of a number of networks and purchase either a single service or a package of networks. The consumer would give the programmer the number of the descrambler, and the network would transmit that information to the VideoCipher Center in San Diego where a signal would be sent to the satellite and down to the box in the consumer's home, authorizing the decoder to receive the signal.

GI's monopoly on the business of authorizing satellite reception didn't make it popular with the programmers who hated to depend on a single vendor for such an important function. But it made GI a bundle of money.

Then GI began to face another threat. The Japanese broadcaster NHK, together with a group of Japanese electronics manufacturers, had introduced a high definition television system called Muse, which produced a picture far more clear and sparkling than the standard NTSC system could deliver. When Muse was unveiled it sent shudders through the VideoCipher division.

"We knew that if this system took off it would be big trouble for our company," recalled Marc Tayer, then the manager of new business planning for GI's Videocipher Division.

Because Muse required more bandwidth and greater power to deliver than the NTSC signal, it would likely require a realignment of satellite transponders (each capable of carrying only 6 MHz of analog signal) and force most home dish owners to buy much bigger dishes (five or six meters versus the three meters then the standard). That would send a jolt through the entire satellite business.

GI began to seek a way to cram a lot more signal into 6 MHz of bandwidth. At the head of the effort was Jerry Heller, who had done some of the pioneering work on digital transmissions for M/A Com subsidiary Linkabit. GI began to scour the halls of the Massachusetts Institute of Technology where a group of electronics engineers had been playing with digital transmissions and to hire them away as quickly as possible.

In 1990 the company formed a core group to develop a new type of television transmission, which it christened DigiCipher. The group leader was Tayer, a 32-year-old Wharton School graduate who had headed up the international sales for VideoCipher. System engineering was under the direction of Paul Moroney. Others on the team were Woo Paik and Ed Krause.

The initial idea was to find a way to transmit a high-definition signal from an earth station to a satellite and back down to earth using only six MHz of bandwidth. The effort was built around the notion that the major portion of a television picture does not change very often. A scene from "The Bill Cosby Show," for example, may focus on actor Bill Cosby talking. His face changes, and he may move around, but the background remains the same.

Under the NTSC analog signal the TV scanner would scan the entire 525 lines on the TV set 59 times every second, directing each dot on the screen what to do. Digital transmissions had to be converted back to analog in order for the TV scanner to recognize them.

The idea behind the new kind of compressed digital signal was to send instructions concerning only those portions of the screen where a change was necessary. When the signal was reconverted to analog it would produce an analog signal that was the same as before except in those areas where the picture had to be changed.

This system required transmissions of only that information needed to make the changes in the scene rather than the information required to create an entirely new scene every 59 seconds. The amount of information sent, therefore, could be far less than what had previously been needed.

The GI team set to work to develop the algorithms, or mathematical equations, needed to translate a picture into digital form, transmit it using as little bandwidth as possible, and then reconstruct it as an analog signal for viewers to see on their TV sets. By the time the mathematicians had finished they were able to transmit digital signals using less than 2% of the information that had been required to produce a similar picture with analog transmissions.

Then, Tayer recalled, it occurred to the team that the new system would enable not only the transmission of a high-definition television signal in 6 MHz, but would also allow the transmission of multiple standard television signals using that amount of bandwidth as well.

The implications were enormous. For satellite-delivered networks, the highest cost of operations, particularly in the months and years immediately after launch, was their satellite transponder. Programming could be inexpensive re-runs of already produced shows. Uplinks and studios could be leased at a reasonable cost tailored to the needs of the programmers. But a transponder was a transponder, and there was no way to get one cheap. Their rent could amount to millions of dollars a year. Moreover, each satellite carried only 24 transponders, and the real estate on the most widely watched cable birds was hard to come by.

But if the new digital transmission system would allow each transponder to handle eight or 12 or more signals, then the cost of launching a new network would plummet. The satellite providers charged by the transponder, not by how many signals were transmitted off each transponder. It was as if a single person had rented a house for years by himself and then discovered he could take in eight or 10 roommates and divide the rent.

The same was true for the direct broadcast satellite services. Here the numbers were even more important. A DBS provider launching two 24-transponder satellites could provide only 48 channels of service, about what a local cable system could provide. But with compressed signals, a DBS service might be able to provide hundreds of signals from those same two satellites. With that kind of programming inventory, DBS, which had been a blue-sky technology unable to compete with cable, might become very real indeed.

For the cable operator the stakes were even higher. If programming services would multiply because of the new economics of compression and if DBS could deliver all those channels to the consumer, cable would have to follow suit or risk losing millions of customers to DBS. If, on the other hand, the digital system could be transmitted not just from the satellite uplink to the headend, but from the headend to the home as well, the cable system would be suddenly capable of increasing its capacity from 54 channels to 200 or more almost overnight.

And the digital signals sent by the cable system were inherently better than analog signals. Analog signals by their very nature are only representations of a picture in electronic form sent to a receiver which decodes the signals and then forms a picture close to what was originally sent. (The word analog comes from the word analogous). Moreover, the farther each analog signal is sent and the more devices it must pass through on the way, the more it is distorted when it finally arrives at its destination.

Digital signals, by contrast, consist of simple ones and zeros which do not degrade over distance and produce a picture at the end that is almost an exact replica of the original.

But cable operators looking at digital faced one huge obstacle: the cost of the unit that would have to be installed in the home to change the signal from digital back to analog so that it could be read by the analog scanners in existing TV sets. A digital signal sent down the cable would have to be reconverted to analog to be used by the television set.

Performing that task was an immensely complex procedure requiring enormous computing power. To put such a device in the homes of millions of consumers would be expensive, far more expensive than the revenues such services could be expected to generate.

For the DBS provider this was not a problem. Every DBS customer was new and would have to make the choice to buy a digital-to-analog converter along with the dish and other electronics that made up the DBS package. And the package could be paid for over time.

When the folks at General Instrument began to make contact with their cable customers about digital services, the biggest issue was the converter. But there was at least one believer, and he, by chance, worked for the world's largest cable operator.

Tom Elliot was a down-to-earth Montana native who grew up on a ranch where he quickly became more interested in how the machinery worked than in how the livestock was getting along. But his hunger to understand how things worked was tempered by the very fundamental need to be, as he later recalled, "just gut-practical." Elliot was a dreamer, but he also knew that he needed to get things done.

After graduating from the Colorado Technical Institute, he got a job with a company that helped test nuclear devices in the Nevada desert. He was like a kid in a candy store with unlimited credit.

"We used to get anything we wanted. We would order all the latest test equipment, and we would take them apart and modify them before we even turned them on," Elliot recalled.

But he had a yen to return to Montana and when, in 1964, he heard about an opening at a Montana-based com-

Practical Engineer: *Long-time TCI engineer Tom Elliot proselytized the digital compression gospel in the late 1980s and early 1990s.*

pany called Western Microwave, he went after the job. The company was owned by Bob and Betsy Magness, and its primary business was to deliver distant TV signals via microwave to cable systems throughout the West.

By 1980, long after he had moved to Denver, Elliot was persuaded by John Malone to move over to the cable side of the business. Malone, he recalled, was an ideal boss in many ways because his background as an electrical engineer enabled him to grasp quickly the information Elliot and his colleagues passed along.

"It was a great advantage that he understood what we were talking about," Elliot recalled. "Of course the downside was you couldn't bullshit him."

As Elliot, the General Instrument group and others began to play with digital television, the great question for cable operators was the cost of the in-home device that would transform the digital signal back into analog.

That cost related to the huge amount of computing power that would be needed to make such a conversion as quickly as required to run a TV set.

Most engineers were convinced that such computing power would not be available for a very long time, if ever. Many suspected that the silicon base of computer chips simply could not handle such huge processing demands and that some new kind of base for the chips would have to be found.

But Elliot had visited the foundries where the chips were made, and the manufacturers had assured him they could build such powerful chips. "When you went out and talked to the guys in the trenches," he recalled, "they would tell you 'we can do that today.'"

By the late 1980s Elliot "felt comfortable predicting that the devices to convert digital to analog could be available at a reasonable cost by the middle of the next decade. "Everybody thought I was crazy. The most common reaction was 'not in my lifetime.'"

All of these factors were floating around the elite of the engineering community when on Jan. 21, 1990, the General Instrument Digital VideoCipher division invited a few of its most valued customers to take a look at the new DigiCipher technology.

First to see the new device was a group from Viacom, led by senior vice president of technology Ed Horowitz. Viacom owned Showtime and MTV Networks and was one of the largest cable operating companies. The company's technical brass quickly grasped the significance of what they saw.

Even more blown away was the delegation from Hughes Aircraft, which was in the process of developing a new direct broadcast television service to be called DirecTV.

Senior vice president of marketing Jim Ramo watched the demonstration and declared it was "black magic."

Back at TCI, Elliot warned a special meeting of TCI's board of directors that digital was on the way and that the company had better get ready. He told them that digital would make direct broadcast satellites and multichannel MDS services real competitors to cable and might even make it possible for the telephone companies to deliver video signals over their lines. He urged the company to go public with its plans for digital transmissions to get ahead of the competition. "If we don't go public, they will," Elliot recalled telling the board. "If we do that we can lead and shape it. If we don't it will happen anyway."

General Instrument wasn't the only

Chief Tech:
Viacom's Ed Horowitz foresaw the value of digital compression technology and what it would mean for Viacom's networks.

company working on the problem, and it was still unclear which of various competing digital transmission systems would be the best. After the GI system was made public, Scientific- Atlanta unveiled a system called VQ (vector quantization). AT&T had two different systems of its own. Malone asked Elliot to broker a deal between S-A and GI to develop compatible systems. "It was like getting the Arabs and Israelis to agree," Elliot later recalled, but in four months the deal was done.

During the NCTA convention in March, 1991, Malone, Jerrold president Hal Krisbergh, S-A CEO Bill Johnson and CableLabs president Dick Green announced they had formed a consortium to jointly develop digital technology for cable systems. "Then the ball started to roll," Elliot said. "John got enamored with it."

By the Western Show in December of 1992 Malone was ready to announce that he had placed orders for one million digital set-top converters from General Instrument and AT&T along with the uplink, headend and other equipment needed to deliver digital signals down the cable plant. As part of the deal GI was required to license the DigiCable technology to two other vendors, S-A and Hewlett Packard.

At the press conference announcing the deal, Malone was asked to quantify how many additional channels of programming would be available. An offhand calculation led to a front-page story in the next day's *New York Times* predicting that cable systems with 500 channels of programming would be available within a few years. The story also noted that Malone had said TCI would be ready to roll out the new technology within a year to 18 months.

Meanwhile, at Time Warner, top engineer Jim Chiddix was keeping chairman Steve Ross up to date about the progress toward high-channel-capacity cable systems.

Ross at the time was looking for partners for the heavily leveraged Time Warner and was holding a series of meetings to pitch prospective investors about what a great business cable was. Chiddix was called on to help.

At a rehearsal for one of those meetings Chiddix rattled off all the new technologies in the pipeline and concluded with an offhand remark that with equipment currently available it was possible to build a system that would deliver 150 channels of programming.

"Ross stopped me and said 'Wait a minute. We could do that right now?'" Chiddix recalled. "We could, I told him, but it would be kind of expensive."

"I don't care what it costs," Ross answered. Chiddix said he could see Time Warner Cable CEO Joe Collins across the table. "He kind of blanched."

In March, 1991, just as GI was unveiling its DigiCable system, Time Warner announced plans to upgrade its 77-channel system in Brooklyn/Queens,

New York, to 150 channels by the end of the year. The new system, using analog signals, would have dozens of channels of pay-per-view and allow experiments with interactive video games via cable and other advanced services.

The best part, noted Time Warner vice chairman Jerry Levin, was that the new system would cost only about 20% more to build than a conventional upgrade, largely because of an innovative scheme to bring fiber to "nodes" that would serve 500 to 1,000 homes and to use a powerful new 1 GigaHertz amplifier that had been developed by C-Cor Electronics. The set-top boxes, supplied by long-time Warner Cable vendor Pioneer, would include on-screen ordering menus to make ordering pay-per-view as easy as possible.

By December the company had its first subscribers hooked up and was hosting dozens of high-powered financial and communications executives who flocked in to see the cable system of the future. "Queens became a Mecca for visiting people," Chiddix recalled.

Other cable operators also were working on their own advanced cable systems. In January 1992, Viacom Cable began to rebuild its 18,000-subscriber system in Castro Valley, Calif., outside San Francisco, to test some of the new services that might be possible with the advanced technology. On the drawing board was video-on-demand, which would allow viewers to choose from a wide menu of movies and then have the movie start, and pause, at times selected by the customer. Conventional pay-per-view required viewers to wait for a movie to start at a pre-set time.

Viacom's aim was to determine which services customers wanted and what they might be willing to pay for them.

Like the Time Warner Queens system, the Castro Valley operation would use C-Cor amplifiers and would deliver fiber to nodes that would serve about 800 homes each.

The biggest announcement of a new kind of cable system came in January, 1993, when Time Warner unveiled plans for what it called the "full service network," to be built in Orlando, Fla.

The idea was to build a system that would combine a fiber backbone, digital transmissions and file servers to deliver such services as video-on-demand, video phones, sophisticated interactive games, teleshopping, telemedicine, advanced on-line news, and other services that the company hadn't yet dreamed up when it announced the project.

Jerry Levin, who had become Time Warner chairman on the death of Steve Ross, declared that the new system "is a really profound aggregation of concepts. It's truly infinite. I would describe it as a perfect medium."

As the project moved along, Time Warner announced ever more exotic services that it would deliver. For example, advertisers were pitched on the notion that a viewer watching an ad for, say, Ford trucks, could push a button and get more information, both video and faxed, about the trucks and then could schedule an appointment for a Ford salesman to come out to the home with the very make and model the viewer selected.

Children could play video games against a friend across town, and each could see the other's moves in what was almost real time. Direct mail could be delivered via fax only to those who had expressed an interest in getting it, and who would receive a price break on cable from Time Warner. Advertisers then could be sure that those people who

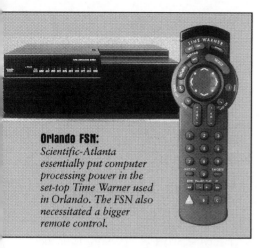

Orlando FSN:
Scientific-Atlanta essentially put computer processing power in the set-top Time Warner used in Orlando. The FSN also necessitated a bigger remote control.

were getting the mail would pay some attention to it and could avoid the cost of paying the post office.

To mollify the U. S. Postal Service, the system also gave consumers the chance to shop for stamps and other postal services electronically. FSN subscribers could order their stamps over the cable system and they would be mailed to them the next day.

When the system was unveiled, in December, 1994, Levin hailed it as "the most powerful tool ever devised" for communications. In a display of its power, he ordered a movie, put it on pause after a few minutes, played a game of cards with a family in another location, and wandered through a video mall.

The cost of the project raised some doubt on Wall Street. Each set-top box reportedly cost thousands of dollars to construct. But Time Warner and its partners – Silicon Graphics Inc., Scientific-Atlanta and AT&T – maintained that the costs of the equipment would drop rapidly.

Some other technology was truly of Star Wars quality. For example, the remote control had to be completely reconfigured. When a viewer pushes a remote control button it sends a signal to the TV set ordering the channel to change or the set to go on or off. The reaction seems to the viewer almost instantaneous because the signal has to travel such a small distance.

But in the Full Service Network, commands such as asking a movie to pause or rewind had to travel from the remote to the converter, hundreds of yards up the cable to a node and then back again. Yet viewers would not likely be willing to wait a couple of seconds for a response. They would want something that responded as quickly as their TV remotes.

Time Warner and its partners had to design an entirely new remote control that that would respond to all viewer commands nearly instantaneously.

The system was the first operational cable system to deliver both analog and digital signals in an asynchronous transfer mode (ATM) over a hybrid fiber-coax network. ATM was the system designed to send packages of bits that would permit delivery of voice, data, and multimedia services over the same wire at what to the consumer appeared to be the same time. Somebody could be watching a movie, therefore, while somebody else was using the same wire to make a phone call and a third person was using it to communicate with the computer.

The use of file servers combined with fiber was another major step in the evolution of the cable system. File servers were devised to store data, such as movies, that could be delivered to customers on demand. They were also the basis of what was becoming known as the Internet, a global network of file servers that allowed computer users, initially via the telephone lines, to connect to a huge range of information sources. But the data from file servers could be sent over fiber/coax cable systems much more rapidly than over the phone lines.

With the enormous carriage capacity and reliability of fiber, and the coming arrival of digital transmissions, it became clear to operators that they would be able to offer their customers a whole new array of services without impinging on the video delivery. One of the most promising areas was voice and data, until then the exclusive province of the telcos. Chipping away at the telephone business was an enticing prospect for cable operators.

Telephony in the U.S. generated more than $100 billion a year in revenues while the cable industry was under $30 billion. By taking just 10% of the telco revenues, the cable industry could increase its size by one-third. With fiber rebuilds already justified by the savings they would make possible in the core cable business, the addition of telephony could come at a relatively low cost.

The local phone companies seemed ripe for the picking. Businesses, particularly in the financial sector, were increasingly dependent on lightning-fast, highly dependable communications between remote sites.

Yet in many major markets, including New York City, the phone network was notoriously unreliable and service was even worse. It often took weeks or even months to get a new phone line installed, and breakdowns in the network were all too frequent.

In the late 1980s an upstart company called Teleport Communications Group Inc. began to construct an alternative fiberoptic communications system in the New York City area to provide more reliable service to such institutions as banks, stock brokers and stock exchanges. When Teleport began to get demand for links to offices outside the area where it had lines, it contracted with the local cable systems to use their fiberoptic lines to hook up the Teleport customers.

The Teleport system operated on highly reliable fiber lines and had several backup systems so that service would not be interrupted even if an individual line were physically cut. In addition to the point-to-point service, Teleport also provided its customers with a cheaper way to hook up with long-distance providers than the local telcos.

In early 1992 the company caught the eye of Cox Communications and Tele-Communications Inc., who joined to buy out Teleport owner Merrill Lynch. Cox took the majority stake.

Each company had already been experimenting with telephone service. Cox had been a pioneer in the development of personal communications services. PCS was a service that relied on phones similar to cellular telephones. But the idea was to provide each consumer with a single phone that could be used at home, at work or in the car or while out walking. It would communicate with a box attached to a cable system which would then link it to the local phone company network or to other PCS users.

Because the cable system ran throughout most major metropolitan areas, the PCS unit could be much lighter than a typical cellular telephone which required a strong signal, produced by a large battery, to communicate with cellular sites often miles away.

In February, 1992, Cox Enterprises chairman James Kennedy placed the first personal communications service phone call using a cable television plant, ringing up FCC chairman Al Sikes in Washington, D.C., from a site in San Diego.

All Smiles:
TCI's John Malone and Bell Atlantic's Ray Smith were all smiles at the October 1993 announcement of the TCI-BA merger. By early the next year, the deal would fall apart.

In July, 1993, Cablevision Systems demonstrated a PCS service that worked on its cable system in Long Island.

The telephone companies were well aware of the threat cable posed and were also eager to get into the business of delivering video services. For years the telcos had argued that they ought to be the ones to provide the main conduit into the home for voice, data and video. Trouble was their plant, built over a 70-year time frame to handle voice traffic only, would not support such a huge task.

Coaxial cable simply could deliver far more information than the twisted pair of copper wires that linked each home to the phone lines outside. While it was possible for a phone company to rebuild its trunk lines with fiber, it would be a gargantuan task to rewire every home with coaxial cable to replace the twisted pair. It was in the last 100 feet to the home where cable had an enormous advantage.

The various regional bell operating companies (RBOCs) took different approaches to the problem, with some of them playing out several different strategies at the same time. Some claimed that new compression techniques would enable them to offer a full array of video, voice and data services over their existing wires. Others announced plans to rebuild their entire plant with hybrid fiber-coax plants to match cable systems' capacity. Some began to buy up wireless cable companies to compete with cable.

Several major U.S. telephone companies, including U S West and Southwestern Bell, had obtained first-hand experience in running cable systems when they invested in systems being built in Great Britain. There the Thatcher government had dismantled many of the regulatory barriers that had prevented progress in communications. Cable companies were allowed to offer both video and voice services, and U.S. companies were welcomed. Moreover, the U.K. also had a thriving DBS (or Direct to Home as they called it) business.

Malone called the U.K. "The Spanish Civil War" of the communications industry, a reference to the 1930s conflict where Germany, Italy, France and other powers tested some of the weapons they would later use during World War II.

In the end the telcos realized they had no real choice. If they wanted to play in the new information age they would simply have to buy cable systems. Federal law allowed them to do this so long as they did so outside their own service areas.

The first shot was fired in February, 1993, when Southwestern Bell announced plans to buy Hauser Communications for an astonishing price of $2,900 per subscriber, or 11 times cash flow. (Hauser had been started by former Warner Cable chairman Gus Hauser and served 225,000 subscribers in areas around Washington, D.C.) To run the system, Southwestern Bell selected the man who had managed its cable/telephony systems in the UK.

The Hauser-SWB deal opened the floodgates for telco purchases of cable systems. Phone companies were prohibited from owning cable systems within the telco's own service area but could buy cable companies outside those regions.

In October, 1993, BellSouth announced the purchase of 22.5% of Prime Cable, operator of systems serving 525,000 homes, with an option to buy the rest of the MSO. In December, Bell Canada bought a 30% stake in Jones Intercable with options to buy more. Nynex announced an investment of $1.2 billion in Viacom, and Southwestern Bell followed its Hauser purchase by announcing plans to buy Cox Cable. U S West bought a 20.5% stake in Time Warner Cable.

The biggest deal by far, in terms of subscribers, money and impact on the telecommunications world came in September, 1993. Bell Atlantic, the world's largest local telephone company, announced it would purchase TCI, the world's largest cable system operator. The value of the deal was set at $11.8 billion, or 11.75 times cash flow.

The plan was to form a huge telecommunications company that would rebuild Bell Atlantic's telephone plant and TCI's cable systems into full-service networks offering a wide range of telecommunications services. The structure would leave BA chairman Ray Smith as the chairman and TCI president John Malone as vice chairman of the new company. The value of the stock Malone himself would receive in the deal would top $1 billion.

The huge sell-out of cable companies to the telcos caused Ted Turner at the 1993 Western Show in December to comment, "It's been a great ride for the last 20 years. This is (the cable industry's) last cable show. We'll manage to adjust to our new telephone company owners."

His eulogy was premature.

A combination of technological delay, looming competition and a huge hammer blow from the federal government combined to scare off most of the telcos before their deals were complete.

When Congress passed the Cable Act in 1992 it left to the Federal Communications Commission the responsibility for implementing many of provisions of the new law. As the FCC waded into the task, it quickly became apparent that the new law would have an impact far different from any envisioned by its framers, and far worse than the cable companies had thought possible.

In fact, the implementation of the Cable Act of 1992 would become a case study in how the federal government, an 18th century institution designed to dilute power and delay decisions, is inherently unable to regulate a modern

Breaking Ground:
Cox Cable President Jim Robbins goes over the Cox-SBC deal points with James Kahan, SBC senior-VP, strategic planning and corporate development. Months later, the merger would fall apart.

industry where new technological developments and business deals can change the landscape overnight.

The first part of the new law to take effect were the rules governing the relations between broadcast stations and cable systems. At first the drafters of the law had simply intended to reinstate the must-carry rules, struck down by the Supreme Court in the 1980s. Must-carry had guaranteed that every local TV station would be carried by every cable system within the reach of the station's signal.

But for the big broadcast stations, must-carry wasn't much of a benefit. Cable systems would carry these stations with or without a law. They had to. Viewers would not want to subscribe to cable if it did not include the major stations. In many areas cable still was sold primarily as a way to get better reception of the broadcast signals.

What the big stations wanted was to be able to charge cable systems for the right to carry the local broadcast signal. But the smaller stations, including religious and foreign-language stations, wanted compulsory carriage.

Congress decided to let the broadcasters have their cake and eat it too. The law set up a system under which a station could choose between must-carry and retransmission consent. Those picking must-carry would be ensured of carriage on the local systems but would not be able to charge for carriage. Those stations picking retransmission consent would be able to charge for carriage, but would not be ensured a spot on the cable system if the operator and the broadcaster could not come to terms.

Cable operators were determined not to allow the big broadcasters to force systems to pay for the right to carry over-the-air signals that were free to anyone else. Cable carriage, the operator reasoned, was simply a way to improve the broadcast signal and allow it to reach more viewers, increasing the broadcasters' audience and boosting their ad rates.

Continental Cablevision, led by Amos Hostetter, took the most militant public stance. In November, 1993, it placed orders for one million A/B switches that would allow its customers to switch between cable and antenna service. In its request for proposals to suppliers seeking to sell the switches, Continental noted that the Cable Act "requirements are likely to necessitate removal of certain broadcast stations from Continental cable systems." It was ordering the A/B switches, it said, to allow its customers to have access to the off-air signals of those stations.

Big broadcast stations insisted they would settle for nothing less than cash payments from cable systems. Gary Chapman, chairman of group station owner LIN Broadcasting, said that his stations should have the same kind of payments received by satellite-delivered cable networks. "I think ESPN talks cash (when seeking affiliate fees). Turner

MTelco Visionary:
U S West Chairman Dick McCormick engineered the telco's investment in Time Warner Entertainment.

talks cash. Viacom talks cash. Believe me, when I want to buy *Oprah*, King World talks cash. We are talking cash."

But in the marketplace, unlike in Congress, cable operators had the upper hand. Broadcast stations at the local level and broadcast networks at the national level were intensely competitive with one another. Cable systems were de facto monopolies within their service areas. It was therefore possible for each cable company to deal separately with a number of different broadcast stations, dividing and conquering.

At the national level the situation was complicated because several of the big station owners had interests in cable systems or networks. Viacom president Frank Biondi, for example, said his stations would not seek cash payments, but would opt for partnerships with local cable systems that could benefit both.

Among the broadcast networks, which also owned groups of stations, two of the three (ABC and NBC) had major cable programming interests.

Finally, all broadcasters were facing a new threat: a fourth network that had been launched by Fox Broadcasting, owned by Australian media mogul Rupert Murdoch and run by veteran broadcast executive Barry Diller.

Many of the Fox affiliate stations were UHF or VHFs with weak signals. And in some areas of the country there were no Fox affiliates. Fox, the weakest of the networks, had the least to lose and the most to gain from cutting a deal with cable. It was the first to break ranks.

In May, 1993, Murdoch struck a deal with Malone. Fox agreed to launch a new cable network and to invest $100 million in original programming for it. It also agreed to drop its retransmission consent requests for all TCI systems. TCI in return, agreed to carry the new Fox cable network on all its systems, paying a total of $30 million to Fox in affiliate fees over five years. It also agreed to carry all Fox-affiliated broadcast stations on its cable systems on the VHF tier, giving the stations effective parity with the stronger stations affiliated with the other broadcast networks. Finally, Fox agreed to pay five cents of the 25 cents per sub per month affiliate fee it was getting for the new cable channel to its broadcast station affiliates and to give them a 25% equity stake in the new network.

It was a clever deal that worked for both sides. TCI was able to strike the first retransmission consent deal with a broadcast network and was able to do so without having to make cash payments to the stations, at least directly. The deal put heavy pressure on the other broadcasters to come to the table. Finally, it opened the door for Fox to become a cable programmer, giving it and its local stations a stake in the industry and giving TCI another supplier of programming.

It also formed the beginning of a long-term relationship between TCI's Malone and the wily Australian maverick Murdoch.

The Mexican standoff between operators and broadcasters continued right up until the last week before the Oct. 6 deadline by which agreements had to be in place under the new law.

Less than a week before the deadline ABC reached agreements with most major operators to forego retransmission payments for its owned and operated stations in return for carriage of a soon-to-be-launched ESPN 2. NBC struck the same deal for its new cable network, America's Talking.

Only CBS continued to hold out for cash payments. Operators refused. But with all three of the other broadcast networks having reached agreements, CBS had no real options. It couldn't afford the ratings hit that would result if cable systems dropped CBS stations from their lineup, even for a few days, while continuing to carry ABC and NBC and Fox.

Just days before the new rules were to take effect CBS caved. It waived its rights for retransmission consent payments for a year, winding up with nothing to show for its years-long crusade to persuade the Congress to enact retransmission consent and then to get the operators to pay.

Talks between cable operators and broadcast stations continued right up until the midnight deadline, and in some cases beyond. While there were scattered reports of stations being dropped from cable systems, in most cases agreements were reached that allowed both sides to claim some sort of victory. But in the overwhelming majority of cases the cable system did not end up paying cash. Even LIN Broadcasting, which had been among the most militant in seeking cash payments, blinked at the last minute. It dropped its demands for cash if the local systems would carry a new local weather channel LIN planned to launch.

Must-carry and retransmission consent forced cable systems to make huge changes in their channel lineups. They had to make room for the weaker, lesser watched stations (many of them fundamentalist religious, or foreign-language) that demanded must-carry. And they had to make room for the three new cable networks (from Fox, NBC and ABC) they had agreed to carry so the networks' stations would not seek retransmission consent payments.

The impact of must-carry was increased when the FCC ruled that even broadcast stations that devoted full time to shopping channels were entitled to carriage on the local cable systems.

In hundreds of cable systems across the country operators were forced to drop satellite-delivered signals to make room for the broadcasters. The Viacom system in San Francisco had to combine 12 cable networks into six channels to make room for the must-carry signals.

Among the biggest losers were C-SPAN and C-SPAN II, which carried the proceedings of the U.S. House of Representatives and the U.S. Senate respectively.

Neither channel generated any revenue for a cable system, and neither one had a huge audience. Together the two public service networks lost more than 500,000 viewers as a result of must-carry and retransmission consent. The loss of C-SPAN was particularly noted by lawmakers who had not anticipated that the new law would mean that cable systems would drop the channel that carried their debates and speeches in order to add home shopping or foreign language stations. It was not the only unanticipated result of the Cable Act of 1992.

The turmoil surrounding must-carry was a minor irritant compared to the upheavals caused by the rate regulation provisions of the act. The law directed the FCC to draft many of the final rules that would be needed to implement the act. The agency's actions in carrying out that mandate provided a textbook study in governmental ineptitude.

One problem was a shifting cast of characters. President Clinton, elected in 1992, was slow to fill the vacancies at the FCC, and during the first few months of regulation only three of seven seats on the Commission were filled. Among the vacancies was the chairman's seat.

When the administration did appoint a new chairman, he was a man who knew next to nothing about many of the issues he was confronting. Reed Hundt, who had been a classmate of Vice President Gore at the elite St. Albans prep school in Washington, had done some work for the Wireless Cable Association and worked on some antitrust cases involving telecommunications companies. That was the extent of his experience in areas related to cable.

Below the top ranks, the Commission staff was woefully shorthanded, even after Congress granted an emergency appropriation of $11.5 million in mid-1993 to help hire extra staff to deal with the issues.

On the cable side, the resignation of James P. Mooney as NCTA president in July of 1993 left the organization without a permanent leader for six months. The search for a new president finally ended up on the NCTA's own doorstep when the board appointed as president Decker Anstrom, who had been executive vice president under Mooney and had served as acting president during the search.

But the biggest problem was the law itself. The 1992 Cable Act had directed the FCC to ensure that rates charged by cable systems were "reasonable." In determining what constituted reasonable rates, the law had also directed the agency to take into account such various factors as the rates charged by cable systems in cities where there was direct competition to cable, the cost of programming and equipment and the taxes imposed by franchising authorities. The law also directed the commission to allow the cable companies to make a "reasonable profit" from their businesses. Reading the law, it sounded like the re-

Cable Guy: NCTA President Decker Anstrom steered the industry through tough times after the FCC regulated rates and during passage of the 1996 Telecommunications Act.

sult might not be so bad. The exact formulation of reasonable rates was left entirely to the FCC.

Framers of the law had anticipated that cities would undertake much of the burden of regulating rates. The act allowed municipalities to file with the FCC for permission to regulate rates. But in many communities the city councils did not want to take on the headache of rate regulation, particularly since the new federal law provided no financial resources to cities which did undertake the task. By November of 1993, a year after the act had passed, only 3,780 of some 33,000 franchising authorities nationwide had filed to regulate rates.

For the rest of the cable companies, the FCC would have to determine what was a reasonable rate for basic service. Other than the general guidelines, the law made no provision for how the FCC was to determine a reasonable rate for cable service.

Left largely on its own, the Commission decided to establish a formula to determine a "benchmark" rate for basic cable service. Each cable system could then figure out what its own benchmark was and whether the rates it charged were above or below that benchmark.

While this process was underway, the FCC froze all cable rates and ordered a rollback of rate increases that had taken place after September, 1992, when Congress passed the law. The order also abolished fees for second sets and remote controls and mandated that rates be uniform for the same kind of service throughout a system's service area. Finally the FCC allowed operators to raise or lower rates for particular customers so long as the average rates remained the same during the freeze or under the benchmark after it was determined.

Complaints by consumers about unjustified rates would trigger an FCC review of the cable system rates and a possible rollback.

The recipe seemed like a good one on paper to the chefs at the FCC and on Capitol Hill. But when the separate ingredients were mixed and placed in the oven of the marketplace, the concoction simply blew up, leaving those in Washington covered with political soot.

By ordering that cable rates be uniform throughout a franchise area, the Commission effectively ended discounted cable service for such groups as senior citizens and low-income families. Many of these discount programs had been instituted at the suggestion of local city councils. As the rules took effect, millions of senior citizens and low-income families saw their cable bills rise.

Because the FCC rules required uniform pricing on a per-channel basis, it also forced those systems which had been offering low-priced or even free basic service, to charge.

In Omaha, for example, Cox Cable offered all its customers a free basic service consisting of broadcast signals, public, access and government channels. Only a few thousand customers chose this service, but when the new law took effect Cox was forced to begin charging them for a service they had been getting for free, some of them for as many as 18 years.

The biggest snafu with the FCC's rate scheme was the fact that it allowed the operator to raise the rates of some customers and lower those of others so long as the average rates remained the same throughout the system.

In its issue of Aug. 2, 1993, *Cable World* reported on the impact these provisions had had in Aspen, Colo.; " Prince bin Sultan Bin Abdul Aziz of Saudi Arabia, who built a $26 million, 55,000 square-foot house overlooking Aspen will see his rates fall at least $400 a month when his cable operator's additional outlet charges are wiped off the books later this year. Conversely, residents in nearby Rifle, Colo., may see their expanded basic rates rise if TCI chooses to lift those rates to match the FCC-set benchmark. 'Essentially, what the law means is that the poorer people here will now subsidize the richer residents,' says Tom Coleman, TCI's area manager."

The scenario was repeated to a greater or lesser degree throughout the country. Homes with multiple sets — often owned by high income families – saw their rates slashed when second-set fees were eliminated. And operators were permitted to raise the rates for basic service to cover the losses, increasing the charges for one-set homes, typically in the lower-income demographics.

The Wall Street Journal editorial page picked up the *Cable World* article and used it to blast Rep. Markey in a lead editorial entitled "Cable Malarkey."

"The cable bill is a case study in what happens when Congress decides to run an industry," the editorial said. "Mr. Markey pledged during Congressional debate that cable prices would fall after the bill passed. But it turned out he didn't know how the industry works. It now appears that at least 40%, and maybe 50% of the nation's 57 million cable consumers are going to pay more."

When the rates took effect it was impossible to determine exactly how many subscribers had been hit with rate increases. *The Washington Post* reported that "most" cable customers in its area would see rate hikes but amended that the next day to say only a third of bills would go up.

The FCC, understaffed and baffled by the reports that its rules — designed to reduce rates – were having the opposite effect, was unable to get a handle on what was really happening. "Some of these things," said acting commission chairman James Quello, "are caused by the press."

It was the worst of all worlds. The impact of the new rules was unquestionably to reduce overall revenue for cable operators. But the public image was that rates were going up, and rates were indeed increasing for many cable subscribers – particularly those least able to pay. And those whose rates increased let their congressmen know. And the congressmen passed along the complaints to the author of the bill, Rep. Markey.

Most devastating from Markey's point of view was an editorial cartoon in his hometown newspaper, *The Boston Globe*.

The *Globe* cartoon depicted Markey, with a goofy grin, walking a dog with the leash wrapped all around the Congressman's body. The dog was labeled "cable".

Markey was furious. Here he was

Hill View:
Rep. Ed Markey (D-Mass.) kept the pressure on the FCC to make sure rate regulation worked in the eyes of Congress.

posing as the great populist, the crusader for consumers. He had hailed the Cable Act as the most important piece of consumer legislation in the Congress. But his new law had slashed costs for a Saudi prince while increasing rates for busboys and carpenters. He blamed the FCC and he blamed the cable operators. And he threatened to take action.

"Any cable operator that raises rates for programming service must be carefully monitored," Markey said. "Rates should not be going up for anyone unless they are receiving a higher level of service."

Then, to rub even more salt in the would, cable operators began to find that even among those subscribers whose rates had been cut, the difference was being used to buy new pay services. News accounts began to appear that the resourceful cable operators had managed to persuade many subscribers to use the money they were saving from basic rate cuts to buy more premium services.

TKR Cable reported selling 1,000 new pay subscriptions in homes where the new law had forced it to drop the additional outlet fee of $6 a month. Continental Cablevision in St. Paul launched a new multiplex pay offering in conjunction with reregulation and sold 1,000 new pay units. And because premium services were not included in calculating the FCC benchmark rates, it was possible for some systems to pick up much of the lost revenue twice – by selling more pay services and by hiking basic rates to cover the losses of fees for remotes and second sets.

It began to seem that the cable operators who had been portrayed as greedy and arrogant by Markey and his allies in the debate over the Cable Act, were going to get away largely unharmed by the new bill.

But the government would have the last word. The rate freeze imposed by the FCC had been scheduled to expire on Nov. 15, 1993. After that cable systems would be free to raise rates as long as they were under the benchmark and to pass along increased costs for equipment and programming and for inflation. It appeared as if another round of rate hikes was at hand.

Ten days before the freeze was to end the FCC extended it, claiming it needed more time to find out the impact of its rate rules. "What they are really doing is stopping an industry dead in its tracks while they figure out what to do," charged CATA president Steve Effros. At the same time the FCC refused to extend the Nov. 15 deadline by which systems choosing to apply for cost-of-service exemptions to the rate rules had to submit financial information, even though the FCC would not issue its guidelines for cost-of-service rulings until December.

"It's really a bizarre system when you have to file and the rules are not finalized," said Times Mirror vice president Dick Waterman.

Within six months many operators would look back fondly on those days when the actions of the FCC were merely bizarre. The FCC was about get a leader whose actions, the cable industry felt, bordered on the sadistic.

Reed Hundt was confirmed as the new FCC chairman on Nov. 20, 1993, and immediately was greeted by a barrage of letters from Capitol Hill demanding action on cable rates. A group of 35 Senators wrote, "The FCC must take additional action to reduce cable rates and step up its enforcement activities. We continue to receive complaints from our constituents indicating their displeasure with the new charges for their cable service."

Hundt wasted little time and gave no false signals about what to expect. He appointed as head of the resurrected Cable Bureau a New York state assistant attorney general who had built a reputation as a consumer advocate but who had almost no experience with cable television. As the commission's chief economist he appointed a professor from the University of California at Berkeley. Hundt's own economic views were clear. Rate rollbacks, he opined, would actually help the cable industry

Gore Man:
Reed Hundt's tenure as FCC chairman was marked by onerous rate regulations and high tension with the cable industry.

because lower prices would bring in more subscribers. It was an argument lost on most cable operators and on the bankers who loaned them money.

As Hundt began to put his stamp on the Commission, Congress continued to toss grenades his way. Markey and Rep. Chris Shays (R-Conn.), the ranking Democrat and Republican on the House Telecommunications Subcommittee, wrote Hundt offering full support for his effort to redraft the regulations "to more accurately mirror competitive rates as promised under the Cable Act." Sen. Inouye wrote the FCC chief expressing concern that the first set of rate regulations "did not go far enough to accomplish the purposes of the act."

The only question was how big the additional cuts would be and how the FCC would justify them.

Hundt, prodded by Markey, argued for a rollback totaling 28% below what the rates had been when the bill passed. That would require an additional 18% cut below what the first benchmarks had achieved. But some commission members balked at issuing new regulations that so clearly contravened what they had just issued and had certified as fair and reasonable. In particular, Commissioner Andrew Barrett held out for no changes.

In the end the commission agreed to compromise on a rollback totaling 17% below what the rates had been prior to the act, or 10% below what the first set of benchmarks had achieved. Then came the task of how to justify that figure.

The answer was to jimmy the equation. In developing the original benchmarks, the FCC had discovered that in many systems with low penetration, the rates were fairly high. It speculated that this was because operators with fewer customers needed more revenue per subscriber to cover costs.

In setting the new formula it simply decided to toss out many of the low penetration systems before it did the averaging that determined the benchmark rates. The result was a significantly lower benchmark. When the math was completed, lo and behold, the new benchmark produced rates that were 17% below what the rates had been prior to the act.

The FCC order mandating the additional rollback was the most severe government action taken against a particular industry's rate structure in the history of the country. In the past the government had acted against particular companies for fixing prices and had moved to enforce antitrust laws that effectively reduced prices for particular products. And it had, at various times, imposed wage-price regulations on the entire economy. But, outside of wartime, it had never taken an action that so dramatically reduced the rates charged by an entire industry so sharply without first finding that the industry had violated antitrust or other laws.

Not only were the new rules arbitrary and punitive. They were also immensely complicated. The formulas for figuring out what the benchmark should be for an individual system required, among others, the following calculation:

"The May, 1994, benchmark formula is the following:

$LAR = .204 + .07(MSO) + 8.14(RSS) -1.45(RTC) + .253(PNB) + .103(PAO) + .172(PRM) + .057(PT2) + .353(PTC) +.069(LIN)$

where

LAR = *natural logarithm of the benchmark average monthly regulated revenue per subscriber*

MSO = *MSO status (1=MSO, 0=non MSO)*

LMS = *natural logarithm of the number of systems in the MSO*

RSS = *1/number of systems in the MSO*

RTS = *1/average total channels (ATC) where ATC is the weighted sum of the number of channels on each regulated tier weighted by the proportion of total subscribers who subscribe to that tier, for the franchise*

PNB = *proportion of the weighted sum of non broadcast channels to the weighted sum of total channels, both weighted by the proportion of subscribers to each tier*

PAO = *proportion of remotes to total subscribers for the franchise*

PT2 = *proportion of tier 2 subscribers to total subscribers for the franchise*

PTC = *proportion of tier changes to the total subscribers to the franchise*

LIN = *natural logarithm of media household income*

In all, the regulations and the forms that each cable operator would have to fill out to figure the benchmark and to comply with the reporting requirements amounted to more than 1,200 pages, weighing more than five pounds.

It made the tax code look simple.

Inevitably there were mistakes which the FCC acknowledged when they were pointed out but which often took weeks or even months to repair.

The agency continued to add staffers who knew little or nothing about the industry they were supposed to regulate. By the time the second set of regulations were issued only two members of the Cable Bureau were left who had worked on the first set.

The entire process was made even more difficult by Hundt's style. With a kind of superficial, at times stupefied, smile pasted on his face, he listened to the complaints and suggestions of cable operators. But he did not hear them. He marched to his own tune and a tune set by Rep. Markey. A lawyer accustomed to the process of litigation, Hundt set up a system in which he was presented with options on either side of an issue and then made a ruling, just like a judge, in favor of one or the other.

He evidenced none of the people or political skills of some of his predecessors who had attempted to find a middle ground in disputes so that all sides could prosper.

Above all Hundt was beholden to Gore who had nothing to gain from giving a single inch to the industry he had spent a career in Congress bashing.

The impact of the Hundt rules was devastating and immediate. The first casualty was the merger of TCI and Bell Atlantic. Just a day before the new rate regulations came out, TCI and Bell Atlantic had agreed on the final terms of their merger. But the FCC ruling forced them back to the table. When they got there they found that the new numbers simply didn't work.

"We both tried diligently to come to terms but when we put it in context with the marketplace and the regulatory uncertainty facing the cable industry we realized this deal just wasn't meant to be," said TCI president Malone. "We agreed on the low end of our values and the high end for their tolerance of pain. But the impact of the FCC rules was so severe rather than continue in agony we terminated the deal."

The two companies had found some differences, particularly in corporate culture, during their attempt to merge, and these certainly played a role in the final decision to tank the deal. But the overwhelming factor was the $300 million reduction in cash flow that the new rules would impose on TCI. At the same time the share prices of Bell Atlantic stock had declined, as investors who had been drawn to the telco because of its stability, realized that the new company would be subject to much more marketplace and regulatory volatility.

In the end Malone was simply unwilling to take a lower price for TCI, and Bell Atlantic was unwilling to stick by an offer it had made for a company that would now produce much less cash flow.

Southwestern Bell and Cox reached a similar impasse, and their deal was canceled as well.

Hundt was so isolated from and ignorant of the business world that he refused to acknowledge that the FCC rules played any role in the collapse of the TCI-BA or Cox-Southwestern deals. The unflappable FCC chairman continued to maintain that the rules would be good for cable.

That wasn't the view of cable operators staring at financial projections with severely depleted cash flows.

Continental CEO Hostetter explained the amplified impact the regulations would have on cash flow, used to determine the value of a cable system and the amount of money it could borrow.

If a cable system were taking in $100 per year per subscriber in gross revenue from regulated services, and the expenses for the system amounted to $58 it would leave a cash flow of $42. When rates were rolled back by $17, the revenue would decline to $83 per year. After subtracting expenses, which would not change, the result would leave cash flow of $25, a reduction of almost 40%.

"We don't want to say the sky is falling yet," Hostetter said. "But it will be hard to operate under these conditions. A lot's been said that indicates the commission doesn't understand our industry."

Particularly hard hit were the smaller operators. Having clawed their way back to solvency after repeal of the Highly Leveraged Transaction rules, many operators now found themselves sliding back to the brink of bankruptcy or beyond.

A 76,000-subscriber system in Prince Georges County, Md., filed for bankruptcy just weeks after the rules were issued, saying the FCC regulations had made it impossible to renew bank loans. The filing was the largest ever in the cable industry.

"Although the FCC's rate regulations did not directly cause the bankruptcy, they made it virtually impossible to negotiate the financial restructuring of the senior and subordinated debt without a bankruptcy filing," the MSO said.

Operators slashed capital expenditures, bringing construction projects to a halt all across the country. TCI cut by half the $1 billion it had planned to spend on system upgrades and extensions in 1994, and Time Warner lopped $100 million off its $700 million capital budget for 1994. Cablevision Systems held to its construction budget but planned for layoffs of 160 people at corporate headquarters.

By May, said Falcon CEO Marc Nathanson, "the financial markets have all but shut down for operators."

"If you were an operator and went out on a road show today," said longtime cable investor Gordon Crawford of Capital Research Co., "investors would hoot you out of their offices. The outlook for cable equity and bonds is dreadful. They've done just about everything they can to discourage investment in the industry."

Wertheim, Schroeder vice president Tom Donatelli said the FCC had unleashed a "regulatory reign of terror on the cable industry."

Cable stock prices, which had held up reasonably well after passage of the bill and even after the first round of rate cuts, went into a free fall. The Kagan Cable MSO Average which had been over 1500 in November of 1993, dropped to 975 by May of 1994. The price of Falcon stock dropped by 41%, TCI 34%, Cablevision Systems 38% and Century 34%.

Nor was there any end in sight to the regulatory nightmare. The FCC continued to monkey with its formulas, issued a set of "going forward" rules that were designed to provide incentives for cable systems to add channels. But these proved just as complex as the original benchmark rules and provided even less relief to the bottom line.

The understaffed FCC was unable to deal with the thousands of complaints filed by individual cable customers unhappy with rates, programming or any other aspect of their cable service. The unresolved complaints added to the uncertainty on Wall Street about how to calculate what future cable system cash flow might be.

Then the Congress, which in the elections of 1996 had been taken over by the Republicans, began to consider a huge comprehensive telecommunications bill designed to overhaul all the rules covering all sectors of the industry. While the bill offered hope that rate regulation would one day be ended, it also would have opened the doors to competition from the telephone companies, something stock analysts, confused about all the technological jargon, feared.

The telcos, almost as soon as their deals to buy cable systems collapsed, resumed their crusade to be allowed into the business of delivering video in their areas. U S West announced plans to rebuild its entire plant, beginning with Omaha, with a hybrid fiber-coaxial network that would allow it to deliver cable services as well as voice and data.

But the biggest challenge to cable from a Bell operating company came in Toms River, N.J.

There Bell Atlantic asked the FCC for permission to build a fiber-coax plant capable of delivering 64 channels of analog video as well as data and telephony. The video services would be provided by a company, Futurevision, that would lease the plant from BA in an arrangement reminiscent of the old leaseback schemes that the telcos had first hatched in the 1970s. Futurevision was run by Marty Lafferty a cable industry veteran who had been in charge of the ill-fated Olympics Triplecast pay-per-view venture.

Futurevision offered consumers a

Defense:
Bell Atlantic put Adelphia Cable on the defensive in Toms River, N.J., but the Rigas family, James, John and Michael, was equal to the task.

"lifetime pledge" that their prices would always be 20% lower than those offered by the local cable operator. It mounted a well-financed campaign of door hangers and direct mail marketing to win customers away from the incumbent cable operator.

Bell Atlantic and Futurevision thought they had picked their opponent wisely. The local cable system was owned by Adelphia Cable Communications, one of the most highly leveraged of all the major MSOs and one which had been most deeply wounded by the rate rollbacks mandated by the FCC.

But Adelphia was run by the cable industry's equivalent of the Cartwright clan. John Rigas and his three sons weren't about to let some big-city dudes run them off the ranch they had been working for 40 years, at least not without a fight.

While the telcos marched forward in their quest to conquer cable, the industry received bad news about progress in its secret weapon factory.

The rollout of digital compression, which had seemed on track for a 1994 time frame when John Malone made his announcement at the 1992 Western Show, had been delayed.

Two factors were to blame. First, said TCI vice president of engineering Tom Elliot, the problems associated with systems engineering were much more difficult than had been anticipated. Part of this was due to the difficulty in making sure the new digital signals would work with the existing cable plant. But much of it related to the desire of cable operators to have a highly flexible system that would accommodate future services as well as the core business of delivering video.

The DigiCable system was geared to delivering video, and any other information that was sent had be made to look like a video frame, Elliot recalled. He asked the company to go back and re-engineer so that the system would be simply a "bit pipe" and would make no distinction between data and video and voice.

"From a competitive perspective," Elliot recalled, "We didn't want to get this out there (with a system built to handle video) and then find all kinds of auxiliary (services) that would have much more value."

The second major reason for the delay was the continuing effort by the Moving Pictures Experts Group (MPEG) to develop international standards for the delivery of digitally encoded material.

The group had decided that the best way to transmit digital signals was in packets which would contain a wealth of information and then be followed by another packet. It might be possible, for example, to send a packet of information to tell the TV set what to do for the next few minutes followed by a packet of information that would contain data signals that could be used by the computer.

The MPEG group got hung up trying to decide how big the packets should be and how much information they could contain. It took two years to figure this out.

In the meantime the credibility of the industry had been stretched. Reporters for newspapers and general interest magazines, who knew little about technology, knew only that the cable had promised 500 channel systems and had failed to deliver on time.

While all these regulatory and technological snafus were under way, the industry managed to get into a particularly nasty internal war. When Para-

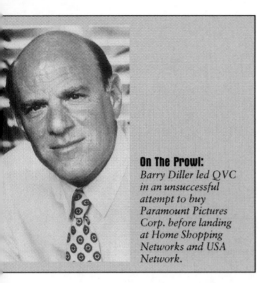

On The Prowl:
Barry Diller led QVC in an unsuccessful attempt to buy Paramount Pictures Corp. before landing at Home Shopping Networks and USA Network.

mount Pictures came into play, bids came in from QVC and its allies TCI and Liberty Media. Attempting to outbid them was Sumner Redstone of Viacom.

The Viacom-TCI relationship had already been strained by the launch of Encore, TCI's low-cost premium service, which Showtime officials felt was aimed directly at them. In the middle of the fight over Paramount, Redstone filed a lawsuit against TCI, charging it with antitrust violations and singling out Malone for "suppressing competition, restraining trade and coercing suppliers, customers and competitors."

"Malone's monopoly power has intimidated the entire cable industry, but we could no longer sit idly by as he and his controlled companies excluded competing cable programmers from his cable systems and undermined our service and those of other programmers," Viacom said. The suit charged TCI had refused to renew its Showtime contract, had extracted unfair terms from programmers and was attempting to monopolize the digital encryption and compression technologies under development.

It was not the first time Redstone had resorted to a huge lawsuit. In 1989 he had sued Time Warner, charging HBO with monopolistic practices. What made the suit particularly juicy was the fact that three former top executives of HBO – Frank Biondi, Tony Cox and Matt Blank – now worked at Viacom. Some of the practices that Viacom was challenging as antitrust violations had been created under the leadership of Biondi, Cox and Blank.

The suit dragged on until 1992 when it was settled out of court. Part of the settlement provided that HBO would drop its time-locked marketing campaign and would promote jointly with Showtime. Another result of the suit was to gain carriage for Showtime on Time Warner's systems in New York City.

The Viacom suits caused huge upheavals in the offices of cable operators and network executives who were forced to spend countless hours answering requests from lawyers on all sides for information and documents.

"It got so that I couldn't go for a walk down the hall without taking my antitrust lawyer along with me," recalled Liberty president Peter Barton.

The lawsuits were only the tip of an iceberg of programmer discontent that grew almost daily during the mid 1990s. The must-carry laws caused major restructuring of cable system channel lineups, at times angering programmers who were bumped to upper tiers, cut to half-time or jettisoned from the lineup altogether. And as rate rollbacks took effect, operators became much tougher in seeking rate breaks from the networks (the going forward rules did allow operators to pass along programming rate hikes and to increase rates to cover the addition of new networks.)

Above all, cable operators became

suspicious that the programmers favored DBS as a distribution system that gave them some leverage with cable and that allowed them to deal directly with their ultimate customers. And there was no doubt that many programmers did welcome the advent of a technology that gave them some way to reach viewers without having to go through cable.

But lawsuits and battles with programmers were the least of the operators' worries as the first half of the 1990s drew to a close.

The full weight of the FCC's rate regulation rules and the delays in rolling out digital transmissions were much more serious. So, too, was the advent of the first real competition to cable television in a half century.

While the cable industry continued to attempt to iron out kinks in the digital transmission of cable signals, DBS operators moved ahead rapidly.

During the 1980s, half a dozen would-be DBS operators had emerged and then retreated in the face of the daunting cost and logistical problems inherent in launching such an enormous project. Among them was Home Box Office, which took a $35 million writeoff of its Crimson DBS project in 1989 after it failed to sign up other programmers to join the effort.

By the early 1990s two main ventures remained. The first was a joint effort by nine cable operators and GE Americom, called PrimeStar. The service used a medium-powered satellite to deliver service to customers who would buy dishes ranging in size from three to six feet in diameter, at a cost of about $700 per dish.

This was a huge improvement over the existing C-Band service which required consumers to shell out as much as $2,000 for a dish that might measure 10 feet in diameter. (Such large dishes were banned as unsightly in many neighborhoods and developments.)

But the PrimeStar project did not go as far as some DBS dreamers hoped. The ultimate goal was a service that would deliver a high-powered signal to dishes as small as one foot.

In February, 1990, NBC, Cablevision Systems, Rupert Murdoch's News Corp., and Hughes Communications joined to announce that they would provide just such a service. The partners pledged to raise $1 billion to launch a service that would deliver 108 channels of programming to a flat dish measuring 12-18 inches across. The venture was called Sky Cable. Malone dubbed it "The Death Star."

Both Primestar and Sky Cable hit snags. Primestar by the middle of 1990 had come under the scrutiny of the Justice Department and a host of state attorneys general who suspected it was simply a way for the cable companies to monopolize the delivery of all video

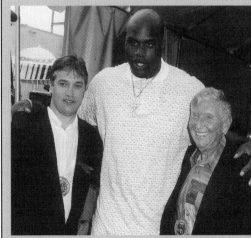

Viacom Crowd:
Nickelodeon president Herb Scannel, L.A Laker Shaquille O'Neal and Viacom Chairman Sumner Redstone.

Death Star:
Rupert Murdoch's interest in Sky Cable and ASkyB sent shivers through the spines of cable operators, but the international media visionary became a programming partner of cable by decade's end.

services to the home. Sky Cable fell apart when the partners failed to hammer out a definitive agreement.

But Hughes kept its interest in DBS services, and by 1992 had found that the time was right. A provision of the 1992 Cable Act had required that all cable networks make their services available to alternative delivery systems at a "reasonable" price. This guaranteed that DBS providers would have access to all the channels that cable could deliver.

And the quantum leaps in digital compression technology had made it possible for a DBS provider to deliver four or more channels per satellite transponder. Hughes signed on as a partner the huge electronics firm Thomson to provide the ground electronics. The two in turn contracted with California-based Compression Labs Inc. to do the encoding and with C-Cube Microsystems to create the integrated circuits that would be needed to handle the compressed satellite signals. Hughes called the new venture DirecTV.

Unlike the cable operators, DirecTV decided not to wait for MPEG to decide on a worldwide standard for digital transmissions of television. By the summer of 1994 it was ready to launch with a service that offered a basic package of 31 channels of programming for $21.95 a month and access to dozens of channels of premium television and pay-per-view movies.

Teamed with Hughes was a visionary, almost messianic broadcaster from Minneapolis named Stanley Hubbard. Hubbard had dreamed for almost a decade of launching his own DBS service to compete directly with cable systems and had spent much of the early 1990s predicting that DBS would one day put cable under. (One year he was the guest speaker at the Rocky Mountain cable convention. He walked up to the podium, held up a small dish and told the assembled cable operators, "This is going to put you out of business.")

While DirecTV had signed deals with programmers such as ESPN, USA, Discovery, Turner, The Weather Channel and The Nashville Network, Hubbard had snared the rights to sell HBO and the Viacom networks.

Although initial marketing was complicated by the need for a consumer to deal both with the DirecTV and Hubbard to gain access to all the programming, and by a shortage of dishes in some markets, the DirecTV launch in June of 1994 promised a bright future for the service.

Programmers loved it. On DBS there was no shortage of channels. DirecTV and Hubbard together offered 150 channels. Premium services loved the fact that there was no requirement that consumers purchase basic cable before being allowed to subscribe to pay networks.

DBS, said Showtime chairman and CEO Tony Cox, "is a tremendous opportunity to give consumers quality entertainment programming at affordable prices."

Encore chairman John Sie later recalled that without DirecTV his vision of a multiplex of premium networks would have been delayed for years.

With the price of a transponder dropping because of compression, more and more new programming services launched, some of them to very few cable homes. Turner Classic Movies, which launched in April, 1994, was the first network to launch to more viewers with satellite dishes than cable homes. But it would not be the last.

Offering more channels and a clearer picture than most cable systems, DBS took off. By the end of 1996, according to industry newsletter *DBS Digest*, DirecTV had amassed a base of 2.34 million subscribers, just short of the three million it initially needed to break even. Primestar, still in the medium-powered satellite, was serving 1.64 million subs and Echostar, a late entry into the business, had signed up another 356,000.

Dishes were selling for as little as $199 and with three national services, the barrage of consumer advertising was relentless. Moreover, Rupert Murdoch, who had been Hughes' partner in the ill-fated SkyCable project, had announced plans to launch his own DBS service, ASkyB, modeled after his successful European DBS service, BSkyB.

Most troubling for cable operators was the news that over half of the DirecTV subs were coming from areas served by cable systems, and that many of those switching to DBS were cable's best customers. Those who made the switch liked what they found.

Research conducted by Horowitz Associates and Liebmann & Associates for a group of cable operators found that 80% of cable subscribers who switched to DBS would recommend

DBS Duo: DirecTV's Eddy Hartenstein (top) and Echostar's Charlie Ergen gave cable operators a run for their money with their respective DBS services.

DBS to a friend who still had cable. And the survey found DBS had taken away many of cable's best customers. Fully 77% of former cable subs who had switched to DBS had been premium service subscribers, with an average monthly bill of $38.

While the DBS services were getting up and running, other competitors were on the prowl as well. Multichannel MDS also benefited from the 1992 Cable Act provision which required programmers to make their services available to all distribution technologies at a fair price. By the middle of 1994 nine wireless companies had filed for public offerings, and the industry's first high yield bond issues had been completed.

The Wireless Cable Association estimated that the number of subscribers for

MMDS would reach 3.2 million by the end of the decade from a level of 500,000 in 1994. And it projected that revenues would expand to $1.5 billion from $120 million in the same time frame.

With real competition literally on the horizon, with the FCC showing no signs of easing up on rate regulations, and with looming costs for installing fiber and digital services, the mid 1990s seemed to more and more cable companies like the right time to sell.

TCI purchased the 18th largest MSO, TeleCable Corp., for $1.4 billion in TCI stock, or $1,736 per subscriber. Cablevision Systems purchased three systems from Sutton Capital Associates for $423.4 million or $2,392 per subscriber. Newhouse agreed to merge its systems serving 1.4 million subs into Time Warner Cable, and Summit Communications traded its systems serving 160,000 subs for $340 million in Time Warner stock.

"Joint ventures have to happen," said Falcon Communications CEO Marc Nathanson. "The FCC's indecisiveness is making for a lot of uncertainty and that means the smaller guys, even those with a million subs, are hurt the most. Only the biggest operators will survive this."

Continental Cable purchased the Colony Communications division of the Providence Journal Co. The deal, which added 750,000 customers to Continental's roster, was valued at about $1.4 billion, or $1,870 per subscriber. It also made Continental, for the first time, a public company.

Comcast purchased the U.S. holdings of Maclean Hunter, serving 550,000 subs for a total of $1.3 billion. And Marcus Cable Partners gobbled up Crown Cable and a host of smaller properties, boosting its sub count to 560,000 from 146,000 in 1994 alone.

And, in one of the biggest deals of the period, Cox and Times Mirror agreed to combine their cable holdings into a new company that would go public in a deal valued at $2.3 billion. The majority of the new entity would be held by Cox.

The deal added 1.3 million Times Mirror customers to the Cox rolls, making the new company the third largest in the country. Times Mirror shareholders had been unable to realize the value of their cable holdings because the parent company was primarily a publishing concern. The new arrangement would give them much greater liquidity for their investments.

But in the frenzy of 1994-1995 the merger, large as it was, was only the deal of the week. In the first nine months of 1995 alone systems serving some 10 million customers, or 17% of U.S. cable subscribers, changed hands, according to Paul Kagan Associates.

Viacom exited the cable operating business to focus on its programming ventures. It agreed to sell its systems serving 1.1 million subscribers to TCI and InterMedia Partners for $2.3 billion. But Congress found a way to kick cable operators even on the way out of the cable business.

The Viacom deal was structured to take advantage of the minority tax certificate program, drafted in the Carter Administration. The tax certificate provided tax breaks for sales that involved minority-owned companies. The Viacom sale was structured so that it would be made to a consortium with TCI and InterMedia Partners as limited partners. The general partner would be Mitgo Corp., owned by Frank Washington, an African-American businessman who had drafted the minority tax certificate plan when he had been an official in the Carter

Administration. The minority tax provisions would have saved Viacom $300 million in taxes.

But the Republican Congress voted overwhelmingly to rescind the tax break, forcing the deal back to the table. It was resurrected without the tax certificates, with most of the Viacom systems going to TCI. The sale also included the end of Viacom's bitter lawsuit against TCI and Malone and provided for, among other things, long-term carriage deals between TCI and the Viacom programming networks.

Century Communications gobbled up the 135,000-subscriber Southern California division of Multivision, giving Century a total of 350,000 subs in the Los Angeles area.

Most telling was the decision of CableVision Industries' chairman, founder and CEO Alan Gerry, to sell out in a deal that swapped his 1.3 million subscribers and $2 billion in debt for about $1 billion in Time Warner stock, or $2,300 per subscriber.

Gerry had founded his company in 1956. He built it on bank debt, never going public, never taking on partners and never giving up an ounce of equity. He rode out every downturn in the business, always managing to keep paying his loans on time and taking advantage of the upsides to expand, mostly through new franchising in the small- and medium-sized communities in the northeast where he had his roots.

During four decades in the business, Gerry had turned down dozens of buyout offers. He simply loved the cable business and, as he later recalled, "There just wasn't enough money in the world to take that away from me."

Gerry also had a loyalty to his employees, some 3,000 of them by the mid-

Liberty Man: *Cablevision Industries Alan Gerry spent nearly 40 years in the cable business, before selling his MSO to Time Warner in 1996.*

1990s, many of whom had worked for CVI for 20 years or more.

But after passage of the 1992 Cable Act and the rate rollbacks ordered by the FCC, he began to change his mind. "I had lots of conversations with Hundt, Markey and (Rep.) Billy Tauzin (D-La.)," Gerry recalled. The atmosphere, he said, was "very anti-cable, very politicized against cable."

Gerry had hoped to leave his business to his children, all of whom had worked for CVI since they were old enough to reach the doorbells of potential customers. But the poisonous atmosphere in Washington, Gerry said, led him to the conclusion that "this was not a future I wanted my kids to face in running the company."

Time Warner, he knew, was much better equipped to fight political battles. And, Gerry said, he was "very comfortable" with the company, which ran systems in many areas where CVI also had operations.

Finally, Gerry said, "I was very fond of Jerry (Levin) personally. He is a deep thinker and a futurist and he is always willing to listen. He is receptive to new ideas and doesn't take the attitude that it has to be his idea to be right."

All the purchases left Time Warner, TCI and the other remaining MSOs even more deeply in debt. Many of them found their stock prices under assault.

TCI, in an attempt to goose its stock price, divided itself into four "tracking stocks" to focus Wall Street's attention on its non-cable system assets. The four divisions that would have separate stocks were U.S. cable and telephony, programming, international operations and technology. When that didn't help, TCI cut back on its hardware purchases, squeezed its programmers and pared its employment rolls. In October, 1996, it halted all purchases of new equipment for the remainder of the year. In December it dropped half a dozen networks, which it labeled as underperforming, from its channel lineups. The next week it announced plans to cut its workforce by 2,500, or 6.5%.

Time Warner, laboring under a $15 billion debt load, restructured to break out its cable and telecommunications holdings into a separate unit that it would jointly own with partners U S West, Itochu and Toshiba.

The move did little to calm the nerves on Wall Street where analysts continued to trade rumors that Time Warner would have to sell off assets to meet its debt or perhaps sell the entire company. Jerry Levin, the soft-spoken, mild-mannered dreamer who had succeeded Steve Ross as CEO, was not viewed as having the cutthroat instincts needed to manage the far-flung empire whose many princes had enormous egos.

In August of 1995, as the rumors continued to swirl, Levin called his old friend Ted Turner and asked if he could come visit Turner and his wife, Jane Fonda, at their ranch in Montana. There Levin made a bold proposal: Time Warner would purchase Turner Broadcasting System Inc. for $7.5 billion in Time Warner stock. Ted Turner would become the vice chairman and largest shareholder of the combined companies.

Together, Levin told Turner, they would ensure themselves a place among the handful of giant media conglomerates that would surely dominate the entertainment industry in the 21st century. (ABC and Disney had just merged, creating a company with combined annual revenue of $20 billion, surpassing Time Warner's $16 billion a year to become the largest media company in the world. On the cable front it would have both ESPN and the Disney Channel as well as the ABC Television Network and the production facilities of the Disney studios).

Turner recognized that as a standalone he would never have the resources to compete with such entities as Disney-ABC and News Corp. His recent attempts to expand, by buying a broadcast network, had been frustrated sometimes by Time Warner or the other cable operators on his board and sometimes by the market. "After considering many options," Turner said when

Time Duo:
Time Warner Vice Chairman Ted Turner, BET Chairman Bob Johnson, Time Warner Chairman Gerald Levin and former FCC commissioner Andrew Barrett.

the deal was announced, "it became clear that a strategic partnership with Jerry Levin and Time Warner was the best way to ensure continued growth and expansion of Turner Broadcasting."

The deal needed the blessing of TCI, which owned a 21% stake in Turner and had a veto power over any major moves. After agreeing to pay TCI a premium for its shares in Turner and to long-term carriage agreements for TBS networks, Malone assented.

But the deal would not be an easy one to complete. Just hours after it was announced on Sept. 22, U S West, Time Warner's partner in its cable division, filed suit charging the deal (particularly the arrangement with TCI) violated Time Warner's non-compete agreement with U S West.

By the next week some 15 separate lawsuits had been filed challenging the deal. Other Turner shareholders, particularly Continental Cablevision and Comcast Corp., grumbled over the sweetheart deal for TCI. The Federal Trade Commission announced plans to review the merger. And shareholder eyebrows shot up when it was reported that Turner would receive a five-year compensation package worth $100 million and that his investment advisor, Michael Milken, would get a $50 million commission.

Time Warner's fortune continued to slide as it reported loses for the third quarter of 1995 that were triple the losses in the same period of the prior year. Michael Fuchs, who had built HBO into a programming powerhouse in the 1980s and 1990s, found he could not bring peace to the warring factions at Warner Music, where Levin had assigned him, and resigned.

Three months after the merger was announced, Malone offered the opinion that the deal had "zero chance" of winning government approval. Analysts figured that the government would view the deal as giving too much power to TCI, which would become a major shareholder of the new Time Warner. "The question you have to ask is: how much does the government want to get John Malone, the Teflon cable executive?" said Dennis McAlpine, a Josepthal, Lyon & Ross media analyst.

The deal was not completed until October, 1996, and only after TCI had agreed to remain a passive investor in the new company. But analysts continued to question its viability, expressing particular doubt that the volatile Turner would be happy in a number-two role. Under continued pressure, Levin began to talk about the possibility that Time Warner would have to sell or spin off its cable systems.

And the news from Orlando wasn't very heartening either, at least to those most attuned to the bottom line. The enormously elaborate Full Service Network worked, but its cost had proved so enormous and the results of its more exotic services so difficult to gauge that it was far from a financial success.

Orlando had started out as a demonstration of the power of the hybrid fiber-coax network, recalled Chiddix. "But it was spun by the dealmakers and the pitchmen until it got scarier and scarier. Pretty soon it was out of control. It became a rocket-sled ride."

Years later Chiddix and others at Time Warner and in the cable industry were able to look back at the FSN in Orlando and view it as a valuable experiment, much like Warner's 1970s QUBE systems. Although it did not work financially, it spawned a host of new ideas, technologies and services that would be applicable to other Time

Warner systems in the late 1990s. But by the end of 1996, when Time Warner announced it would not extend the experiment past the 4,000 original subscribers, the Orlando FSN had seemed nothing more than a costly experiment that had failed.

Even the drama of the Time Warner-Turner merger did not shake Wall Street's preoccupation with the weakened position of the cable companies in the wake of the FCC rate rollbacks and in the face of increased competition from the telcos and DBS.

The Paul Kagan Stock Index, which had stood at 1,506 in November of 1993, prior to the FCC rate rollback, dropped 38% to 937 by the end of 1996. The Dow Jones Industrial average, meanwhile rose to 6,346 from 3,664, an increase of 73%.

Individual stocks were hammered even worse. A share of TCI which had been selling for over $31 just after the sale to Bell Atlantic was announced, skidded to a low $11.31 in 1996.

Time Warner, which had traded above $46 a share in 1993, plunged to $29.75 share at its low in 1996. Cablevision Systems dropped from a high of $75 a share in 1993 to a low of $25 a share three years later. Century slid from $13.84 a share in 1993 to $5.56 in 1996. Adelphia plummeted from a high of $26.50 in 1993 to a low of $6 in 1996. Comcast dropped from $42.25 to $13.75.

Still Reed Hundt continued to demonstrate that he had the soul of an apparatchik. Told that many cable companies were increasing their investments overseas, where the regulatory framework was more favorable, Hundt basically said he didn't care.

"If the best way for a cable operator to maximize future earnings is to invest American consumers' dollars into cable in the United Kingdom, then that's what cable ought to do," he said. "There is no way our government here should keep you from making profitable investments there."

It evidently never occurred to Hundt that, as an employee of the United States government, he had a responsibility to create a regulatory environment that would encourage companies to invest in the U.S.

Nor did he particularly care that his iron hand was preventing investments by cable companies in future technologies and services. In fact, he expressed doubt that such a thing should even be done.

"I'm very concerned that consumers would be highly dissatisfied with this notion of being forced to pay more for today's enhanced basic in return for something hard to define in an uncertain future," he said in a speech in December, 1995.

(Relations between Hundt and the cable industry were not helped when Malone told *Wired* that the best thing that could happen to cable would be if somebody would "shoot Hundt.")

Hundt's comments demonstrated that the battles he was having with the cable industry did not revolve around tactical matters. This was not a dispute over how best to reach a common goal. Hundt, quite evidently, had nary a clue how the modern economic world worked. He completely missed the central fact of the business world: that the profits generated by a company today are essential to build the services and technologies of tomorrow. Without that process no progress would ever be made in any industry.

As the bitter year of 1996 came to a

close the cable industry faced the worst regulatory environment in its history, worse even than the rules that had nearly brought the business to its knees in the late 1960s and early '70s. Nor was there any relief in sight. The reelection of Bill Clinton in 1996 ensured that Reed Hundt, or someone else selected by Al Gore, would chair the FCC for at least four more years.

Twenty five years earlier, when government regulation had crippled cable, the industry had found its salvation in a new technology, one that allowed new, unregulated sources of revenue that could be used to create bigger, even more sophisticated cable systems. It was satellite-distributed programming.

In the late 1990s the same scenario would play out: a new technology would come to the rescue of an industry at the depths of a government-induced depression.

Ratings Gains Were One Bright Spot In The Mid 1990s

While the mid-1990s were a bleak period for most of the cable industry, the one bright spot continued to be ratings for basic cable services. CNN continued to increase its dominance over the broadcast networks when major news was breaking. The OJ Simpson verdict propelled CNN to a rating of 10.1 and also helped E! Entertainment Television and Court TV, both of which followed the trial closely. Cal Ripkin's quest to set the record for the most consecutive major league baseball games played generated a 7.3 rating for ESPN in the fall of 1995. A made-for-cable movie, *The Jessica Savitch Story*, garnered a rating of 7.98 on Lifetime.

The blizzard of January, 1996, which devastated the East Coast, provided an audience avalanche for The Weather Channel which racked up a rating of over 2.9, the network's highest ever.

USA Networks' huge investment in original programming paid off. It led the cable networks in prime time ratings for five consecutive quarters in 1995-96, posting a 2.2 average prime time rating in the first three months of 1996. Nickelodeon, benefiting from the increasing penetration of cable into multiple sets in the home, edged out other networks to rank first in total day in the first quarter of 1996, posting an average rating of 1.6.

And the quality of cable programming continued to be recognized by those who had once dismissed the medium as inconsequential. In 1996 cable programming walked off with a total of 26 prime time Emmy awards. HBO led the parade, copping 14 honors, more than ABC or CBS.

During the November ratings period in 1996 basic cable networks posted a combined rating of 19.2, up 10% from the same period in 1995. Nearly 1.8 million more cable homes tuned into cable programming in November, 1996, than in November, 1995. Most of those gains came at the expense of the big three broadcast networks which collectively lost 1 million viewers during the period.

Cable Stars:
CNN's non-stop coverage of the O.J. Simpson trial and ESPN's live coverage of Cal Ripken's record breaking consecutive games streak in MLB boosted basic cable ratings in the mid-1990's.

A Bright Future

It began as a joke, cocktail party banter between cable's youngest CEO and the world's richest man.

Brian Roberts had met Bill Gates before but had never really had a chance to spend time with the founder of Microsoft until the spring of 1997 when Roberts and a handful of other cable executives took a tour of the top computer companies on the West Coast.

The pilgrimage to Silicon Valley had its origins when the cable industry found itself on the defensive the previous summer at the annual conference for media moguls hosted by financier Herb Allen in Sun Valley, Idaho. There, in the ski lodges and adjacent buildings, the likes of Warren Buffett and Michael Eisner rub shoulders with Barry Diller and Jerry Levin. And every so often one or more is invited to give a presentation to the group.

In 1996 the presentation was by Intel founder and CEO Andy Grove.

Grove reviewed the various new initiatives his giant chip maker was launching and at one point came to the issue of cable modems, the devices that allowed consumers to access the Internet via cable systems rather than phone lines. Such modems, Grove, said, would not make an appreciable difference to Intel in the near future because so few cable systems had upgraded enough to handle Internet access. With cable stocks depressed and the industry under attack in Washington, he said, there was little reason to expect cable would be a viable delivery system for Internet access in the near future.

The news sent stock analysts at the Allen meeting running for the phones to dump their cable stocks, already near rock bottom because of reregulation and competition from DBS and other alternative delivery systems.

The cable guys at the meeting – Chuck Lillis of MediaOne, John Malone of TCI, Jerry Levin of Time Warner – organized a counter attack. To make cable's case they selected Brian Roberts, the 37-year-old son of Comcast founder and chairman Ralph Roberts.

Despite his youth, Brian Roberts was already an industry veteran. He had followed his father around as a young boy to meetings with investment bankers, to strategy sessions and briefings on technology, to the print shop to proofread the prospectus on Comcast's first

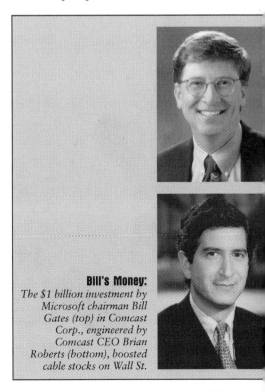

Bill's Money:
The $1 billion investment by Microsoft chairman Bill Gates (top) in Comcast Corp., engineered by Comcast CEO Brian Roberts (bottom), boosted cable stocks on Wall St.

public offering, and to meetings of Comcast's system managers. After graduating from the Wharton School, his father's alma mater, the younger Roberts had pleaded with his father for a job at Comcast.

The elder Roberts felt it would be better if his son started at another company.

"Now I had a problem with that," Brian later recalled. "My dad's 40 years older than I am. He was 61 years old at the time, and I thought 'Gee, how many good years are we gonna have together? Why do I have to prove myself? I mean, I'm either gonna be good or bad. If there's some good reason for me to work somewhere else, fine. But don't you love me?' That got him."

Roberts started at the bottom, selling cable door-to-door, climbing poles and doing installations. Those who worked with him found him to be an almost painfully earnest young man, eager to succeed and with none of the affectations one might expect in the CEO's son.

"He wanted so desperately to do well," recalled Bob Clasen who served as Comcast president during the young Roberts' apprenticeship. "Everybody was charmed by him. He endeared himself to everybody. There was never any suggestion he wanted the right to go directly to the top. Everybody knew Brian in those days, and we all felt better knowing he was going to be the heir. We felt it was an insurance the company wouldn't be sold."

So there was Brian Roberts, (his father had skipped the Allen meeting) preparing to take on Andy Grove in front of the most prestigious gathering of media executives in the world. The cable guys decided it might help to plant a question from the audience. They asked Bill Gates to pose it.

In the middle of his presentation Gates stood up and directed a question at Grove. "You didn't really mean to say that the cable modems won't work?" Gates asked. "You just mean they won't have a big impact on Intel's earnings."

And Grove backed off, conceding that the technology would work, but that Grove simply felt that not many of the modems would be rolled out in the near future.

To educate Grove and his colleagues about the progress cable was making on Internet access and other services, the cable CEOs organized, through CableLabs, a tour of Silicon Valley the following spring. They started with Netscape and followed with visits to Intel and Microsoft. At each stop they delivered the message that cable was ready to do what the telephone companies had promised but had failed to deliver: build a high-bandwidth network capable of delivering voice, video and data at high speeds and with great reliability.

"When we made the presentation to Gates and told him that two-thirds of the cable plant in the country would be rebuilt by the year 2000 you could almost hear him say 'wow'," Roberts recalled.

The evening after the Microsoft presentation Gates hosted the cable CEOs at dinner in a restaurant overlooking Seattle harbor. At Gates' table were Malone, Roberts and Cox Cable CEO Jim Robbins. The talk turned to cable's depressed stock prices and how badly the industry needed a shot in the arm to raise the money to finish rebuilding its plant.

"Gee, Bill," said Roberts, "why don't you just buy 10% of everybody in the room."

Gates laughed, and everybody else laughed and the laughter died down, and

Gates asked, "Well, how much would that cost? Because I have $10 billion in cash."

A quick calculation arrived at a price of about $5 billion. Everybody laughed again, and the talk returned to Gates' upcoming trip to the Amazon River with fellow Microsoft founder Paul Allen.

Then, a few minutes later, Gates got back to the subject again.

"Would there be any regulatory problems if I bought into the cable business?" he asked. Nobody could think of any.

The dinner went on, and the next morning Roberts took some ribbing from Robbins and Malone about his brash attempt to sell the cable business to Bill Gates.

The laughter had died down the next day when, back in Philadelphia, Roberts received a call from Microsoft's chief financial officer saying that Gates had asked him to follow up on the idea of buying a share of Comcast.

A month later the deal was done. For a price of $1 billion Gates bought an 11.5% share of Comcast.

It was the single most important deal in the history of the cable business. The world's best-known businessman, the guru of the computer world, had placed his stamp of approval on cable as the best road to the future of telecommunications. In a press conference announcing the deal, Gates said that despite the promises of telephone companies to build high-speed data delivery systems, cable "has a very strong advantage" in carrying high volumes of digital information.

Where Bill Gate led, hordes of investors soon followed. The Kagan Cable MSO average jumped 18% the week after the announcement. Shares of Cablevision Systems, widely viewed as the next likely takeover target, jumped 47% on the news. Century was up 23%, Adelphia 21%, TCI 23% and Jones 21%.

By the end of the year Adelphia's stock price would be up 191%, Cablevision Systems' 209%, TCI's 112% and Comcast's 82%. The dark days of reregulation were at an end.

Gates' investment in Comcast was not an act of philanthropy. Over the next 18 months, his investment would triple in value. But more than that, he would have a chance to help develop the new information infrastructure for the 21st century that would use cable's hybrid fiber-coax network to delivery a vast array of new services to TV sets and computers, many of them embedded with Microsoft software.

The deal, while universally welcomed by cable operators, also caused some to be cautious about Gates' intentions. Microsoft had built a reputation in the computer industry as a ruthless practitioner of monopolistic tactics. Cox CEO Jim Robbins warned that the industry would need to "keep an eye on our new friend's investment" to make sure that Microsoft would not monopolize the software links between the cable system and the TV and personal computer. The dilemma was solved when TCI decided in January of 1998 to use Sun Microsystems' Java programming language in many of its digital boxes, ensuring that the industry would have more than one source of the software needed to run the new services.

At the core of the Gates decision was the belief that cable would be able to deliver the vast streams of digital bits needed to run advanced computer programs.

The company that had blazed the trail to high-speed delivery of digitized data more than any other was Continental Cablevision.

The birth and growth of Continental was a classic cable story. The company was the brainchild of Amos Hostetter, a 1960 graduate of Harvard Business School who had first encountered the industry while working for a venture capital firm. One of the people who came to see him in search of financing was Bill Daniels, looking for an additional $50,000 to close the deal to sell a cable system in Keene, N.H. to Naragansett Capital.

In analyzing the opportunity, Hostetter sought out an old acquaintance, Ray Armstrong, head of the Starwood Office, a financial house in New York that had provided financing for several cable companies. "I went to New York and spent two weeks at the knee of Ray Armstrong who taught me everything he knew about cable," Hostetter later recalled. "A light bulb just went on, and I knew this is what I wanted to do. I thought the key skills the industry would need would be marketing and finance. For construction, engineering and operations, we could hire people."

Hostetter called his friend Irv Grousbeck, who had been a year ahead of him at Amherst College and Harvard Business School and had concentrated in marketing while Hostetter had focused on finance.

On the living room floor of Grousbeck's apartment outside Boston, the two poured over maps of the United States and the broadcast television signal coverage maps of *The Television & Cable Factbook*. They identified all the cities with broadcast stations and then drew circles around them on the map to outline the reach of those broadcast signals. They were looking for communities beyond the reach of broadcast signals. They then looked those communities up in the *Factbook* to see which ones did not yet have a cable system.

Pickings were slim. "We found that almost all of the markets with no broadcast station or only one station had been wired," Hostetter recalled. "So we decided to be pioneers in wiring communities with two stations." With an investment of $2,000 each, the partners formed Continental Cablevision.

The closest communities to Boston with no cable and only two broadcast stations were in central Ohio. Hostetter and Grousbeck bought a couple of plane tickets and paid the area a visit.

The first town they looked at was Mansfield. Grousbeck would knock on doors and explain that he was being transferred to the town, that his wife was an invalid and spent much of the day watching TV. Then he asked if he could come in and see how good the reception was.

In Mansfield, they found, the broadcast reception was pretty good. So they

Working The Streets:
Continental Cablevision Chairman Amos Hostetter welcomes the MSO's first subscribers in Tiffin, Ohio, to the 30th anniversary party marking Continental's entry into the cable business.

moved down the road to Findlay. There the signals were weaker. But a franchising team from Cox Cable, had already arrived. (James Cox, founder of Cox Enterprises, was a former governor of Ohio.) Cox offered them a deal; if they would leave Findlay to Cox, Cox would let them have a free hand in two other nearby towns: Tiffin and Fostoria. They went to Tiffin first.

"The head of the town council was a barber," Hostetter recalled. "Irv and I had our hair cut once a week whether we needed it or not."

At the franchise meeting the two had to answer many of the questions that citizens all across the country in those days asked about cable: Would cable suck all the signals out of the air and make it impossible for sets to pick up over-the-air broadcasts? Would radiation from the system make everyone in town sterile?

After answering these questions and paying for a couple of dozen haircuts, Continental won its first franchise. In neighboring Fostoria (a good German town where the names Hostetter and Grousbeck sounded local) they also seemed well on the way. All they needed was the money to build the system. That turned out to be not so easy.

"We went to 50 banks between Boston, New York, Cleveland and Chicago," Hostetter recalled. The only one who offered any encouragement was Arthur Snyder of New England Merchants National Bank, the bank that had provided loans to many of the high-tech, startup companies that lined Route 128 outside Boston. Snyder told Hostetter, "Find out what the other banks will do and then come back to see me. I'll see if I can do better."

Hostetter struck out everywhere else and went back to see Snyder. The banker told him, "Before you say anything else, here is the deal I will make: I will give you a seven-year term loan, with payments in the first two years of interest only." He offered to loan the company half the $600,000 they needed to build the system.

Now Continental needed investors to kick in the other half. Over an eight-month period, Hostetter got $50,000 from his father, another $25,000 from a Harvard classmate, $25,000 from a Harvard professor and $150,000 from Boston Capital, the venture capital firm. The final $50,000 came from a broadcast television station broker in Cincinnati.

The ordeal of raising the financing had been a long one. And the folks in Ohio weren't very patient. While Hostetter was putting together the financing, the Tiffin city council had awarded a competing franchise to a company headed by local radio broadcaster Malrite and Texas cable operator Fred Lieberman.

Boston Capital was not pleased. "They hauled us all into a room at the Boston Statler Hotel," Hostetter recalled, "and said, 'Okay, who is going to buy and who is going to sell?'" Finally Malrite was persuaded its main business was broadcasting and agreed to sell out for $85,000, what they said was the cost of the work they had already done to get ready for the construction.

Continental eventually charged that the price had been inflated by $22,500. The dispute raged back and forth until, Hostetter recalled, "I found out Lieberman was a tennis player." Hostetter had been a collegiate tennis player. "I offered to play him for it. I guess it was too obvious. He told me he would have a better chance if we just flipped coin."

Hostetter won the flip.

In January, 1965, Continental Cablevision turned on its system in Tiffin and Fostoria. It offered, in addition to better reception of the two stations consumers could already receive, signals from an ABC affiliate, from the Canadian Broadcasting Corp. and from the Public Broadcasting System that area residents could not get off-air.

Hostetter and Grousbeck set off across the country to seek new franchises, winning initial bouts in Quincy, Ill.; Stockton, Calif.; and Concord, N.H.

Hostetter took on the task of lobbying Washington, joining the NCTA board in 1968. With his blue-chip education, the tall, handsome, articulate Hostetter was a good addition to the rough-and-tumble crowd that dominated the NCTA. (But he remained a spectator at the poker games that took place every night in Bill Daniels' suite during the NCTA conventions. Daniels, Bob Magness, Martin Malarkey, Irving Kahn and others played for pots of thousands of dollars. "The numbers were just too intimidating for me," Hostetter recalled.)

As the industry waged its political battles in the late 1960s and early 1970s, Hostetter emerged as a leader of the industry. His devotion to the business and his genuine desire to be a good corporate citizen — paying attention to public service as well as the bottom line — stood him in good stead.

So did his tennis game. For several years he had teamed as a doubles partner with FCC commissioner James Quello. In the late 1960s when the FCC imposed a freeze on construction of new cable systems in the big cities, Continental's build in Stockton, Calif., was caught in the middle. As the freeze continued, "Our franchise was starting to grow whiskers," Hostetter recalled. Continental sought an FCC waiver of the freeze.

When the issue came to a vote the Commission rejected the request on a vote of 4-3, with Quello voting no.

Then the commissioner sitting next to Quello leaned over and said, "Jim, do you know you just voted against your tennis partner?" Quello changed his vote, and the Stockton build went forward.

Throughout his career, Hostetter was convinced Continental should remain a private company, a tenet that had been drilled into him at Harvard Business School. "We didn't want to have the personal intrusion of being public," he recalled, including having to file financial reports that would make public the salaries of the company's officers, its debt load and other details.

Staying private had its downside as well. When Grousbeck wanted to cash out in the late 1970s, there was no way for him to sell his share of the company. To give him a way out, Hostetter bought the shares himself for $300 million. He borrowed the money from a group of banks which insisted on so many covenants that they presented him a set of golden handcuffs as a symbol of the deal.

Staying private proved to be a liability in the franchising arena as well. The public companies spread rumors that Continental was skating on thin financial ice.

"We got to the point where we felt we had no choice but to go public, and in late 1981 we filed for a public offering. The very week we were to go public I got a call from Lehman Brothers (the investment banking house). They had a client who wanted to buy the entire offering." It was Dow Jones & Co.

Dow Jones CEO Warren Phillips came up to Boston a few days later and asked what the price would be. Hostetter told him that they had planned on a price between $22 and $25 per share, although the investment bankers had warned him the market might not go that high. "$25 dollars will be fine," Phillips told him.

"Out of nowhere, this angel had arrived," Hostetter recalled. Continental was able to remain private, could provide some liquidity for its investors, and could tell franchising authorities that it had the backing of the prestigious publisher of *The Wall Street Journal*. Dow Jones would exit the business at the end of the decade, having made about $600 million on its investment of $70 million.

Hostetter was deeply rooted in New England. He maintained a summer home on Nantucket Island, vacation spot for some of the most elite businessmen in the nation. And his office in the Pilot House on Boston Harbor had a picture window that made it seem that one of the yachts could, at any moment, tie up right at Hostetter's desk.

As Continental expanded it became the biggest cable operator in New England, outside of Boston, which had been won by Cablevision Systems. And perhaps more than any other operator, Continental prided itself on having state-of-the-art systems and superior customer service. (Asked in the mid-1980s who ran the best systems in the country, Daniels & Associates vice chairman John Saeman didn't have to think before answering "Continental.")

Because it was located in Boston, the company had the benefit of regular communication with the high tech, Route 128 firms and the area's prestigious universities.

In the late 1980s the vice president of engineering for Continental's New England systems, Kevin Casey, proposed interconnecting all the Continental systems in the region via fiberoptics, creating a central customer data center with sophisticated billing and information management. At the same time the company began to offer data services to other businesses using its fiber network to link the separate offices of such firms as Wang Labs.

Casey continued to look for new ways to use the network. "One day Kevin came to me and told me about this thing called the Internet," Hostetter recalled, "and told me we could do this faster and cheaper. I told him 'Let's get on this.'"

By August, 1993, Continental announced it had formed a partnership with Performance Systems International, a leading maker of equipment to link personal computers to the Internet.

The idea was to use cable systems to deliver digitized information to personal computers at speeds Casey said would approach 10 megabits per second, hundreds of times faster than the speeds available using phone lines. He predicted the service would quickly evolve to allow high-speed electronic mail and other services.

In March, 1994, Continental launched its Internet access service in Cambridge, Mass.

Reaction was mixed. The *Wall Street Journal* noted that the new service would allow consumers to download information from the Internet at speeds 200 times faster than phone lines and to access services such as full-motion video, which required transmission of high volumes of data.

The story also pointed out that the price was steep: $125 a month. "Even

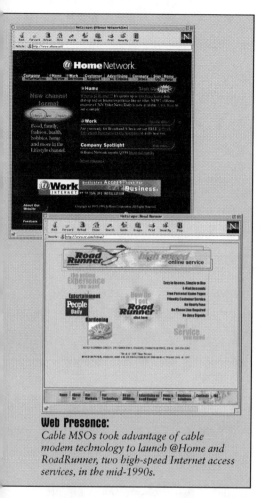

Web Presence:
Cable MSOs took advantage of cable modem technology to launch @Home and RoadRunner, two high-speed Internet access services, in the mid-1990s.

if cable subscribers could pay cheaper rates for Internet access," the story noted, "some observers doubt the service will catch on because the network is so difficult to navigate."

But other cable operators recognized the opportunity. Within a year most major MSOs had launched high-speed data projects and Internet access services.

Tele-Communications Inc. in 1995 teamed with a group of San Francisco entrepreneurs led by publishing heir William Randolph Hearst III to form an Internet service provider called @Home. The service charged $150 for a hookup and $39.95 a month for modem rental, unlimited Internet access and a package of local and national content developed by @Home. TCI brought in seven other MSOs as equity partners: Cox, Comcast, Shaw Communications, Rogers, Cablevision Systems, Intermedia Partners and Marcus.

By the middle of 1998 @Home counted more than 100,000 customers, and company CEO Tom Jermoluk predicted that number would hit 300,000 by the end of the year and one million by the end of 1999.

Time Inc. quickly followed with its own service, dubbed Road Runner. By the middle of 1998, Road Runner was serving some 90,000 customers nationwide and signing up 1,000 new subscribers per week.

Adelphia's Internet access service, PowerLink, counted 10,000 subscribers by the middle of 1998 and was adding 1,000 per month.

The business got a jump start when several MSOs, including TCI, decided to launch service using phone lines for the upstream channel and the hybrid fiber-coax system for the downstream channel until the cable systems were fully ready to handle two-way communications. CableLabs entered the picture early, issuing RFPs to ensure that the Internet access modems were interoperable and that they could therefore be sold at retail.

At the same time that the cable industry was launching its Internet access business, it was also beginning to see the first rollouts of long-delayed digital video service.

TCI launched its digital system, dubbed ALLTV, in Hartford, Connecticut, in October, 1996. The service offered a package of more than 150 channels, including 40 channels of pay-per-view, 25 premium networks and 18 special-interest channels. The company charged $15 for installation and between $34.99 and $69.99 a month, depending on how many channels were ordered.

By the end of April, 1998, TCI had more than 275,000 digital customers using about 345,000 set top boxes. By the end of July it had more than 600,000 digital subscribers. Company president Leo Hindery predicted TCI would have more than one million digital customers by the end of the year.

Cash-flow margins on the revenue from digital tiers approached 60%. Within a few months most other major MSOs began to roll out digital tiers.

Smaller cable companies were able to offer digital services as well. Buford Television, a 150,000-subscriber MSO, said the penetration rate for digital tiers had hit 13% in its Heath, Texas, system less than a year after it launched. "I think it's a 20% (penetration) business," said Buford executive vice president Ron Martin. He estimated that the company was producing a profit of just under $4 per month on each digital customer.

Translated over a national base of some 60 million basic cable homes, it was possible to imagine digital services creating some $1 billion a year in incremental cash flow for the cable industry.

Cable telephony also arrived in 1996 and expanded with a vengeance the following year. Cox got the jump, launching in the San Diego market a personal communications service in alliance with long distance company Sprint Communications, the day after Christmas, 1996. Cablevision Systems reported 12% of the first 3,900 homes offered telephony signed up in the first six months. Company CEO James Dolan, son of founder Charles Dolan, called the results better than he had expected and predicted that the ultimate take rate for cable telephony would be between 20% and 25% of the market. Cablevision planned, he said, to offer phone service to 60,000 homes by the end of 1998 and 200,000 by the end of 1999.

MediaOne, the cable operating company spun off by U S West, launched its cable telephony service in April of 1998. Company vice president Greg Braden predicted the business would reach $2 billion in revenues nationwide in 10 years.

As cable's first half-century drew to a close there were also signs the industry had turned a corner in Washington. Replacing Reed Hundt as chairman of the FCC was William Kennard. The new chairman continued to hammer away at what he called unacceptable cable rate increases (the average hike in 1998 was 8.5%). But he also conceded that imposing a blanket freeze on cable rates would be like "performing brain surgery with a meat cleaver."

The biggest issue before the Commission, and the media industry, in 1998 was how to treat digital transmissions

Digital Jump:
Cox Communications launched personal communications services in San Diego in 1996 in a joint venture with Sprint Corp.

by broadcasters. The FCC mandated that all broadcasting be converted to digital within 10 years. But the major question for cable operators was whether the additional channels broadcasters planned to provide via digital transmissions would be subject to must-carry and, if so, what standard would be used and how much space the new channels would therefore take up on a cable system.

While these debates continued, the industry had the advantage of a restructured and restored NCTA. Passage of the 1992 Cable Act and the subsequent FCC regulatory actions had shattered the cable industry's relations with Congress. It also created friction between the NCTA and various segments of the industry. NCTA president James Mooney tended to play his cards close to his vest, attempting to limit the number of players involved in the negotiations with Congress and other parties over legislation. This strategy may have made sense in terms of negotiations, but it left many segments of the industry feeling left out of the process and angry at the results.

Programmers, small operators and state associations were among those who felt that the NCTA had not included them in the debates and discussions that led to the 1992 Act. Small cable companies ended up forming their own organization, the Small Cable Business Association, to promote their interests. Programmers nearly did the same. Mooney, not a very social person, especially angered some of the programmers by not showing up at some of the premieres and other promotional events they staged in Washington and New York.

At the same time, members of Congress and the Administration felt the industry was not to be trusted. They recalled that cable had initially expressed a willingness to support rate regulation legislation and then changed its position. Some of that ire was directed at Mooney who was perceived in some quarters as arrogant and aloof because, for one reason, he was generally willing to deal only with members of Congress directly and not with their staff aides.

When Mooney left, he was replaced by a very different kind of leader, one who was much more of a people person. Decker Anstrom had grown up in the West, the son of a teacher who was also a labor organizer and would spend a few years in one town organizing a union and then head to another.

Anstrom had come to Washington and initially landed a job at the Department of Health, Education and Welfare in the Nixon Administration. When Jimmy Carter became President, Anstrom joined the Office of Management and Budget and later the White House personnel office.

When Carter lost, Anstrom formed a consulting firm before an old friend from the White House, Bert Carp, called to say he was leaving his job at the NCTA to head up Turner Broadcasting's D.C. office and asked Anstrom if he wanted to take over at NCTA. Anstrom joined NCTA as executive vice president in 1987.

When Mooney stepped down in 1993, Anstrom told the NCTA board that he would serve as acting president while they found a replacement, but that he did not want to be considered for the job.

"I just felt that the NCTA needed a clean start," he later recalled, "and I felt that hiring from within was not the best way to do that."

334

But after six months of searching the NCTA board asked Anstrom to take the job on a permanent basis, and he agreed.

He moved quickly to make major changes in the way the group worked. First he hired as executive vice president June Travis, a former ATC executive who was then president of Rifkin & Associates, the small MSO that former ATC chairman Monroe Rifkin had founded in the late 1980s.

The decision to ask Travis to move to Washington was a stroke of genius. She was the first person with experience as a cable operator to serve in a top position in the NCTA staff. With nearly 30 years in the industry, she provided an important bridge between the world of Washington and the world beyond the Beltway. She was able to explain to the NCTA staff how cable worked and to cable operators what the Washington developments would mean for their business. When she spoke to members of Congress or the FCC she could speak as a real cable operator, not some hired gun. The difference was enormous.

She had an interest in politics which had been honed while she served as head of CablePAC, and a devotion to such cable-sponsored public service organizations as C-SPAN and Cable in the Classroom.

Anstrom and Travis set out to restore the NCTA's reputation among the industry. They included more programmers on the board of directors, and created a place for programmers on the executive committee. (In 1996 Ted Turner became the first non-cable-operator to serve as NCTA chairman.)

Programmers played a key role in the effort to shape the "going forward rules" that provided, the industry with the first relief from the onerous rate rollbacks the

Cable Guy:
NCTA president Decker Anstrom donned a "Cable Guy" cap at an NCYA convention.

FCC had institute in the wake of the 1992 Act. Cable network executives such as Lee Masters of E! Entertainment Television, Dubby Wynne of Landmark Communications and John Hendricks of Discovery Networks testified and lobbied tirelessly on the issue. And members of Congress found it a lot tougher to say no to the backers of the Discovery Channel than to big, faceless MSOs.

NCTA instituted regular calls with the state organizations that had been largely ignored in the Mooney years. Anstrom spent countless hours at the state association meetings, not just making speeches but hanging around for the wine-and-cheese receptions where anybody could come up and talk with him.

They organized a "key contact" program where every member of Congress had one person in the industry whose job it was to keep contact with that member, attending fundraisers, meeting with him or her in the home district and visiting him or her in Congress.

Under the direction of NCTA chairman Larry Wangberg, president of Times Mirror Cable, the organization reached out to members of the "entrepreneurs'

Change Agent:
Leo Hindery led TCI's turnaround in 1997 by stemming subscriber losses, reaching out to political groups and entering into joint operating ventures with other MSOs.

club," a loose organization of cable company owners such as Alan Gerry, Glenn Jones and Jeffrey Marcus, some of whom had felt left out of the key decisions during the debate on the 1992 bill.

With respect to Congress, the aim of the NCTA, Anstrom recalled, was to "get in front of the public policy debate." A group of industry statesmen – Dick Roberts of TeleCable, Amos Hostetter of Continental, Brian Roberts of Comcast and Bob Miron of Newhouse – worked to establish a policy under which NCTA, rather than opposing prospective changes in policy that might threaten the business, would instead work to shape legislation in a way that might be positive. The aim was to become an instrument of change rather than a backer of the status quo.

The first major challenge came in 1994 when Congress considered legislation to reform the entire telecommunications industry. The big decision, Anstrom recalled, was to "embrace competition." By that he meant that the industry would put aside its opposition to telephone company entry into the cable business in return for permission for cable to offer telephony, Internet access and other services free of regulation. It also sought a date by which regulation of cable rates would end.

In the late summer of 1993, while still acting president, Anstrom laid out his proposed strategy to the board of directors at a meeting at the Blantyre mansion in western Massachusetts. "They took a deep breath, looked around the room and said 'Okay,'" he recalled.

In the end the legislation did all of those things that the industry had sought, and in the marketplace the cable operator proved to be much more adept at getting into the telephone business than the telcos proved in getting into cable.

The NCTA also took a pro-active approach to the issue of television violence. Under the leadership of Showtime CEO Tony Cox and senior vice president MacAdory Lipscomb, Jr., NCTA developed a proposal for ratings for all TV programs, getting the jump on the broadcasters who were reluctant to go along.

And Anstrom was able to teach the industry how to lighten up. When a new movie, "The Cable Guy," starring Jim Carrey, was slated to be released, many advance reviews predicted it would further damage cable's image. Some in the industry wanted a strategy to combat the film. But Anstrom simply donned a promotional hat for the film at the next NCTA convention and declared that he was proud to be a "Cable Guy." The film flopped at the box office.

Such tactics, big and small, helped the NCTA restore its relations with Congress and with its own constituents and enter the 21st century stronger than it had ever been.

On the cable equipment front, the industry at century's end was taking the first steps toward an entirely new way of dealing with the in-home equipment such as set-top boxes, digital converters and modems. From the time converters were first used in the mid-1960s, cable operators would purchase them and then lease them to consumers as part of the cable bill. By the mid-1990s the cost of such an effort had grown to enormous proportions. More than 50 million converters were in the field at a cost of about $100 each, for a total investment that exceeded $5 billion. Cable operators had to buy those converters, install them, maintain them and then retrieve them when service was canceled. It was a huge logistical headache.

For years some operators had dreamed of the day when they could get out of the businesses of owning and leasing in-home equipment. The launch of digital services gave them their chance.

In the mid-1990s both General Instrument Corp. and Scientific-Atlanta Inc. announced partnerships with major consumer electronics retailers. GI's deal was with Sony and S-A's with Pioneer and Toshiba. The alliances laid the groundwork for a system in which consumers would select, purchase and install their own in-home equipment. The change was akin to what happened in the telephone business in the 1970s when it became possible for consumers to own their own phones rather than leasing them from AT&T.

The same strategy was in play from the advent of the cable modem business. CableLabs served as the catalyst for the strategy. Its DOCSIS (Data Over Cable Service Interface Specifications) and OpenCable projects aimed to persuade manufacturers of all in-home cable devices to make their products interoperable so consumers could purchase any of a variety of set tops or modems and install them in any cable system in the country. The FCC, seeking more consumer choice, backed the movement with a series of rulings.

The years 1997 and 1998 saw several major MSOs solve some nasty internal problems as well.

At TCI the company and CEO John Malone settled a lawsuit that had been brought by the heirs to company founder Bob Magness, who died in November, 1996. Under the settlement, Magness' heirs agreed to put their shares in a trust that would vote in tandem with Malone, effectively ensuring that Malone would retain control of the company. The settlement resolved a major uncertainty about control of the company that had existed since Magness' the death. TCI also moved to purchase the supervoting shares of TCI shareholder Kearns Tribune Corp., further consolidating control in the hands of Malone.

But more important in the resurrection of TCI was the appointment of a new company president, Leo Hindery, who had been CEO of Intermedia Partners. (Malone had become TCI's chairman when Magness died.)

Hindery moved quickly to institute

Internet Access:
Cable operators took advantage of the public's interest in the World Wide Web to offer consumers high-speed access to the Internet via cable modems.

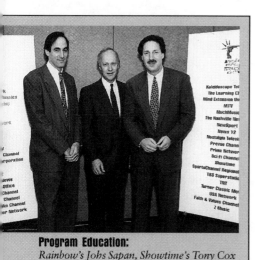

Program Education:
Rainbow's Johs Sapan, Showtime's Tony Cox and NBC Cable's Andy Friendly helped launch the cable industry's Voices Against Violence campaign.

sweeping changes in the beleaguered company. He cut corporate staff, forcing the exit of many of the young executives recruited from outside the cable business by Brendan Clouston, whom Hindery replaced as president.

Malone had already brought back to the company such veterans as JC Sparkman and Marvin Jones, ordered pay cuts for senior staff and sold off the company's corporate jets. Hindery scrapped the organizational structure Clouston had created – along technological lines – replacing it with more traditional regional operating divisions. He also decentralized operations, moving to the operating units many of the functions Clouston had maintained in-house.

"This is still a local business," Hindery said, "and it should be run as a local business. It's measured by only two things: the quality of the service you are providing your customer and the perception of the price-value relationship you're providing."

As Jones put it, "You need management autonomy and responsibility as close to the customer as you can get it. You have to make sure that the beast isn't so large that it takes too long for the message to get from the brain to the tail."

But most important, Hindery began to execute a series of deals with other MSOs. In a market where TCI and another company had systems, TCI would suggest combining operations under a partnership managed by the other company. Or it would offer to swap its systems in the area to the other company in return for shares of that company's stock. In either case the systems could then be operated by the other company off the TCI books, but with the TCI programming discounts and with the efficiencies that came with clustering.

The first such deal was with Hindery's old company, now run by former Jones Intercable president Bob Lewis. The two concerns agreed to merge their operations in Kentucky into a partnership that would be managed by Intermedia.

Similar deals followed with Cablevision Systems, Falcon, Adelphia, Comcast, TCA, Marcus, Bresnan, Lenfest and others. By the end of 1997 some 4 million of TCI's 14 million subs had been transferred to operation by other cable companies.

The deals took much of the debt off TCI's balance sheet and placed it on the shoulder of the partners. With the partnerships and with corporate cost-cutting, TCI was able to reduced its debt-to-equity ratio and regain the coveted investment grade bond status it had lost after reregulation.

Hindery also took on many of the public relations tasks that Malone had always disliked. Hindery was a frequent visitor to Washington, accessible at the NCTA conventions, and involved in such industry ventures as C-SPAN.

The changes had their impact. In the fourth quarter of 1997 TCI posted an increase in cash flow of 37% from the year before. It stemmed the loss of basic subscribers.

Wall Street noticed. When Hindery took office at TCI in February 1997, the company's stock had sold for $13.44 a share, just off its post- reregulation low of $11.31. By July it was up to $16.50 and in February of 1998, a year after Hindery's arrival, TCI's stock was back up to $29 a share, within striking distance of the $33 a share it had sold for just after the Bell Atlantic deal was announced.

Time Warner was also on the rebound. Its merger with Turner Broadcasting had been greeted with skepticism when it was announced in 1996. Time Warner partner U S West sued, charging the deal violated terms of the Time Warner-U S West deal. Analysts speculated that Ted Turner would never be happy as second in command. And rumors circulated that TCI's Malone would work to undermine Time Warner chairman Jerry Levin and perhaps launch a hostile takeover bid for the company.

These prophets of doom failed to understand the strong ties that already existed between these companies and their CEOs. Levin, Turner and Malone had known each other well for the better part of 20 years. They had in effect already been partners as shareholders in Turner Broadcasting Corp. where TCI and Time Warner had a deal to vote together on any major proposal.

Many of the top aides to all three principals had known each other for years. Bill Grumbles, head of distribution for Turner, for example, had worked for Levin at HBO. And most of the Turner operations remained under the control of the same team that had run it before the merger, led by Terry McGuirk.

It turned out Turner was happy to be relieved of some of the day-to-day duties of running a company so he could focus on more strategic objectives, which were his greatest strength. Malone held to his promise (needed to get government approval for the deal) to remain a passive investor in Time Warner. And even U S West retreated from the barricades, dropping its lawsuit to block the merger.

"I don't think there has ever been an acquisition that has gone better," Levin later recalled. "Maybe it's because we had known each other for such a long time. We had the same value systems. Terry McGuirk, for example, started in the industry at the same time I did. I know his family. There is a bond between us because we all went through the rough period when the vested interests were all against us. We have a shared sensibility, a special dimension of respect even if we have different opinions on things.

"Most businesses are very impersonal. In acquisitions the transactions are done by bankers. There are layoffs. But here it was a group of people coming together who had shared values."

Part of the success of the merger was attributable to Levin's style of management. Soft-spoken, clearly brilliant and with a deep belief in the cable industry, Levin had the ability to listen to a wide range of views and make shareholders, partners and subordinates — some of whom had very strong egos and had been used to running their own ships — feel part of a team.

It also helped that the company's bottom line began to improve in the late 1990s. For the first quarter 1997 Time

Warner's cash flow increased to $499 million from $338 million the year before. The cable division's cash flow improved 18%. Such improved results and the smooth integration of Turner boosted the company's stock from $37 a share in January 1997 to $88 a share by the middle of 1998.

On the competitive front, cable appeared by the end of 1998 to have escaped the worst fates predicted by those who had forecast that DBS would be the end of cable. DBS services continued to grow, adding some 700,000 new subscribers in the first four months of 1998 to bring the total to about 7 million nationwide, according to *DBS Investor* newsletter. But there were also signs that many of these DBS subs were keeping their basic cable connections and downgrading only premium services which they could get direct from satellite. This gave cable the chance to reclaim them as subscribers or, at the least, to offer them enhanced services such as Internet connections and telephony they could not get from DBS.

And some of the DBS providers hit bumps. By the middle of 1998 none had been able to find a way to use their technology to offer Internet access, telephony or other advanced services that cable systems were launching. A proposed merger between EchoStar and Rupert Murdoch's News Corp. fell apart when cable operators, led by TCI, gently reminded Murdoch that cable was the primary delivery mechanism for his Fox programming services. EchoStar struggled in its efforts to find a way to deliver local broadcast signals via satellite. And the company managed to go through three presidents in a six-month period. Primestar continued to find itself blocked in its effort to secure FCC permission to use a high-powered satellite to deliver its DBS service.

But the best news for cable on the competitive front came from Toms River, N.J. There Bell Atlantic had announced with much fanfare in 1995 that it planned to build a state-of-the-art system that would take away most of the customers from incumbent cable operator Adelphia Cable. It built a switched digital broadband network and leased it to a separate company that actually operated the system (a necessary division at the time when telcos were prohibited from operating cable systems in their service areas). The newcomer pledged that its prices in Toms River would always be 20% below what Adelphia charged.

Adelphia responded aggressively, rebuilding its plant to 750 MHz, offering high-speed Internet access, digital video, long distance and local telephony and other advanced services. The company also boosted its customer service and marketed like crazy.

At the same time the Bell Atlantic strategy changed. In the wake of the Telecommunications Act of 1996 it took over programming and marketing the system itself. It wasn't exactly Bell Atlantic's area of expertise. And the huge telco found that the switched network it had built wasn't the most cost-effective way to deliver cable service. Finally, it began to dabble in other areas, including a deal which had its cable system sell dishes for DirecTV.

By the middle of 1998 Bell Atlantic threw in the towel, announcing it would raise its rates to match Adelphia and would henceforth regard Toms River as an experimental system with an architecture that would not be widely deployed. At the time it made these announcements Bell Atlantic had signed up only 3,000 of the 110,000 subscribers in the town.

Adelphia executives held down the cheers. "It shows when faced with competitive pressures the cable operators are successful," said Jerry Clark, the Toms River government/community affairs manager who had been with the system for 20 years.

The battle of Toms River offered a deeper lesson, one that had been in play a decade earlier when ABC and Westinghouse attempted to take on Ted Turner in the cable news business. The incumbent fought desperately to hold on to his primary business, knowing that if he lost he would be out on the street. The challenger, larger and better financed, nevertheless had a wide array of businesses and was unwilling to risk years of losses needed to put a tough little competitor out of existence. It was like a Discovery Channel documentary about how a small animal fighting for its life can sometimes defeat a larger predator just looking for another snack.

As the industry began to get its house in order politically and as new services came on stream, the basic cable networks continued to increase their audiences. In the first quarter of 1998 basic cable posted a 14% gain in household ratings, to 22.4 average in prime time. The charge was led by USA Network, which saw its average prime time ratings jump to 2.6 from 2.0. The beat continued in the second quarter as basic cable posted a 12% gain in ratings. The broadcast networks continued to drop, with the figures suggesting that sometime early in the 21st century the combined ratings for cable networks would exceed the combined ratings of the major broadcast networks.

The advent of digital tiers changed the equation for ad-supported networks, working to the advantage of companies with one or more well-established networks and to the disadvantage of standalone, startup networks.

At MTV, for example, company president Mark Rosenthal said the rollout of digital tiers "really allows us to superserve our audience in specific genres.

"There is lots of music we can't play on MTV," he noted, either because there is not enough time in the day or because the music does not appeal directly to the MTV audience. To allow that music to be played, the programmer launched in mid-1998 a "suite" of digital services, all of them carried on the same satellite transponder as MTV's flagship services: MTV, Nickelodeon and VH1. The new networks included M2, MTV X (active rock), MTV S (Spanish language), VH1 Smooth (jazz and New Age), VH1 Country and VH1 Soul.

The economics of launching such a wide array of new networks, Rosenthal said, were "very attractive." Although he would not discuss the financial details of any of his new services, several

Going Shopping:
The MTV brass (l-r), Rich Cronin, Judy McGrath, Tom Freston, Mark Rosenthal, John Sykes and Geraldine Laybourne join model Cindy Crawford to debut MTV's electronic retailing test in the mid-1990s.

programming officials said that it was possible for an established network such as MTV to launch a spinoff network like M2 for as little as $100,000. That's because there were essentially no additional costs for transponders, ad sales teams, affiliate relations personnel or even programming assuming the flagship network's library could be used to program the new network.

That price for a new network compared to the $30 million to $40 million it cost to launch a new service a decade earlier.

Digital was to television what cold type had been to publishing. In the 1950s it cost so much to print and produce a publication using linotype machines that only those with the largest circulation could prosper. That was the era of *Life, Look* and *The Saturday Evening Post*. When cold type and desktop publishing were introduced in the 1970s the cost of printing dropped and it became possible to publish and make a profit with much smaller titles, serving narrow niche audiences. Newsstands became filled with such publications as *Bow Hunter, Doll World* and *Southwestern Art*.

But while the advent of digital favored such incumbent programmers as MTV, allowing them to leverage over a larger number of networks the investments they had already made to launch their flagship services, the reverse was true in the case of standalone networks.

For a standalone service attempting to launch in the late 1990s, being relegated to a digital tier meant giving up large chunks of the potential audience that could be found on the more widely distributed analog tiers of service. At the same time, new government regulations threatened to revive the very channel crunch that digital was supposed to eliminate.

When Michael Fleming took over as president of the Game Show Network, a venture backed by Sony Corp., he quickly learned that the game had changed considerably since he had been on the affiliate relations team at Turner Broadcasting in the late 1970s and early 1980s.

The FCC's going forward rules had allowed cable operators to raise prices for basic cable only if they launched new services. The ruling set off a scramble for networks to be among those launched. E! Entertainment Television led the way, offering itself to operators for free for several years in exchange for long-term carriage deals. It wasn't long before networks began to offer cash payments to operators in return for carriage, severely damaging the economic model for new services that was based on a dual revenue stream.

Fleming noted that this change alone forced GSN to redo its business plan, doubling the amount of time needed before the networks would break into the black.

Must-carry and retransmission consent, meanwhile, continued to give an advantage in the carriage wars to networks owned by broadcasters. Such net-

In Charge:
U S West Media Group CEO Chuch Lillis oversaw the company's Continental purchase and the MSO's move to Denver, which caused Continental chairman Amos Hostetter to resign.

works as ESPN2, MSNBC, CBS Eye on People and Fox News were all launched with the idea that cable operators would be willing to carry them in exchange for an agreement by the networks' parent companies not to charge for carriage of their broadcast stations.

Another government regulation hit GSN like a two by four. The FCC had ruled that all networks must deliver their programming with closed captions for the hearing impaired. For networks producing a lot of original programming this was not such an onerous provision. They simply added it to their production costs. But for GSN, which relied almost exclusively on a library of old game show programs, the cost was huge. Fleming estimated that the requirement would cost his network about $20 million.

While Sony had deep pockets and plowed ahead in the rough waters of the late 1990s it seemed almost impossible that an individual such as a Brian Lamb or John Hendricks with merely a good idea could get a network launched. "I feel bad for people who are trying to start a totally new programming service from scratch," MTV's Rosenthal said of the situation in 1998. "Operators won't pay you for the service and Madison Ave. won't even look at you until you have 20 million subscribers. And since most of the conceivable niches have been filled, you would almost certainly be competing with some existing network in some way."

The advent of the Internet changed cable programming as well. Many channels – led by ESPN and CNN – established their own web sites, offering viewers a chance to access more information than could be delivered on the network and to offer comments. Two new

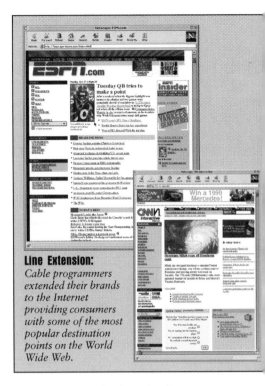

Line Extension: *Cable programmers extended their brands to the Internet providing consumers with some of the most popular destination points on the World Wide Web.*

networks launched in the late 1980s with joint online and video versions. One, MSNBC, was a joint venture of NBC and Microsoft. The other, ZDTV, was a subsidiary of computer industry publishing giant Ziff-Davis. Each would aggressively promote the Internet aspect of its network, effectively whetting the appetites of viewers for the very kinds of data and Internet access services cable was gearing up to offer.

But for the cable operator, it didn't matter much who produced the new programming. The bottom line was that there were dozens of new channels to fill the digital tiers, most of them spinoffs from existing analog channels that would promote the new tiers like crazy. MTV, for example, was full of spots touting its new digital sister networks.

With stocks up, new services coming on line, the regulatory situation under some degree of control and the financial

Free Ride:
Boxing promoter Don King wheels Showtime Entertainment Television executive vice President MacAdory Lipscomb Jr. through the halls of the MGM Grand Casino in Las Vegas in the spring of 1994. King and SET produced some of cable's most successful PPV, many featuring King's best-known client, Mike Tyson. (Lipscomb was recovering from a skiing injury.)

situation looking better and better all the time, it began to seem to some operators that the time had come to sell.

The first to do it was Amos Hostetter. "We had over 400 employees with substantial interest in the company and no liquidity window for anybody," he recalled. "Over the years lots of stock had been given away to institutions — Princeton and Stanford for example — and they had no way to turn that into cash." The lack of liquidity also complicated estate issues.

At the same time, he recalled, the company would have to find a way to raise $1 billion a year for the next five years (1995-2000) to upgrade its plant to offer data and telephony services and enough new digital networks to compete with DBS.

Hostetter began to look for a buyer. From 1994 to 1996 he "talked to virtually everybody." From the beginning, though, the logical buyer had been U S West. Most Continental systems were outside the U S West service area, so the federal ban on telcos owning cable systems in their own service areas would not be as much of a problem. U S West had already bought cable systems serving some 500,000 subscribers in the Atlanta market from Prime Cable and had operating experience from the investments it made in United Kingdom cable systems a decade before.

In March, 1996, Continental announced it had reached an agreement to sell its systems to U S West for a total of $10.8 billion, including the assumption

Top 10 All-Time PPV Events

Event	Date	PPV Homes (millions)	Buy Rate (percent)	TotalBuys (millions)	Gross Revenue (millions)
Holyfield-Tyson II	6/97	34.5	5.72	1.97	$99.2
Holyfield-Tyson I	11/96	32.4	4.91	1.59	$77.8
Tyson-McNeeley	8.95	28.1	5.52	1.55	$67.0
Tyson-Bruno	3/96	29.8	4.60	1.37	$57.5
Holyfield-Foreman	4/91	17.1	8.19	1.40	$53.2
Tyson-Ruddock II	6/91	17.9	7.00	1.25	$43.8
Tyson-Seldon	9/96	31.4	3.22	1.01	$43.7
Holyfield-Douglas	10/90	14.5	7.30	1.06	$38.7
Bowe-Holyfield II	11/93	21.2	4.32	0.92	$35.5
De La Hoya-Whitaker	4/97	33.8	2.54	0.86	$35.2

Source: Paul Kagan Associates

of $5.5 billion in debt. The price worked out to $2,190 per subscriber or 11.1 times cash flow. The deal left Hostetter with $1.3 billion in cash and stock. It was a pretty sweet return on the $2,000 initial investment he'd made to start Continental three decades earlier.

The companies announced that Hostetter would continue to manage the enterprise and that most of the Continental employees would remain in place. The agreement provided Continental not only with the cash it needed to give liquidity to its shareholders and rebuild its systems, but also a corporate parent with deep pockets and lots of experience in delivering telephone services.

Hostetter would report to Chuck Lillis, CEO of the U S West Media Group.

But it wasn't long before Lillis and Hostetter found they would not be able to coexist. The break came when Lillis decided in 1997 to transfer the corporate offices of the newly renamed MediaOne cable company from Boston to Denver. He argued the company would better be able to operate closer to its roots at U S West headquarters (the telco had taken the first steps toward a complete spinoff of MediaOne).

The decision outraged Hostetter who had promised many of his employees they would not be transferred and had assured city councils across the country that Continental management would remain intact. Hostetter was unwilling to make the move himself. He appealed to the board of directors of U S West but was rebuffed and resigned. Along with him went many of Continental's top employees. To replace Hostetter, Lillis named Jan Peters, president of U S West's wireless telephony division.

The ouster of Hostetter was the most

Big Deal:
AT&T's Michael Armstrong and TCI's John Malone seal the deal that would bring TCI under the ownership of long-distance telephone giant AT&T.

stunning change in cable company management since the removal of Bob Rosencrans by United Artists a decade earlier. Like Rosencrans, Hostetter was a universally respected figure in the industry.

While the bitterness among Hostetter and some of his top employees lingered, it was softened somewhat by that universal balm: money. When the headquarters was moved, the decision triggered the stock option clauses in many Continental executives' compensation packages, creating more than a hundred new millionaires overnight. And MediaOne stock prices rose along with the rest of the industry, jumping from $22 a share in August, 1997, to $42 a share by July of 1998.

But while the U S West purchase of Continental demonstrated the belief of a telephone company that cable and not telephone systems offered the best road to the future, such investments had been made before. It was Bill Gates' investment in Comcast that really opened the eyes on Wall Street.

Just a few months after Gates made his entry into cable operating companies, his Microsoft co-founder and fellow multi-billionaire, Paul Allen, followed suit, purchasing 94% of Marcus Cable Partners for $2.8 billion, or about $2,350 a sub. Allen quickly boosted his cable holdings by spending $4.5 billion to buy the 1.2 million subscriber Charter Communications. The price worked out to an eye-popping $3,700 per subscriber, one of the most expensive deals ever. The purchase, analysts said, was the first to fully factor in the revenues that would

Top 10 Highest Rated Programs
(1987-1998, excludes CNN news coverage of Gulf War, OJ Simpson trial)

Rank	Date	Program	Network	Rating/Share
1.	Nov. 9, 1993	NAFTA Debate	CNN	18.1/25.1
2.	Dec. 6, 1987	Chicago-Minnesota	ESPN	14.4/N/A
3.	Dec. 25, 1994	Detroit-Miami	ESPN	14.2/28.2
4.	Sept. 17, 1995	Dallas-Minnesota	TNT	14.0/21.5
5.	Dec. 4, 1994	Buffalo-Miami	ESPN	12.9/19.2
6.	Dec. 3, 1995	Buffalo-San Francisco	ESPN	12.8/19.1
7.	Dec. 30, 1990	Pittsburgh-Houston	ESPN	12.5/19.3
8.	Dec. 3, 1989	Chicago-Minnesota	ESPN	12.1/17.7
9.	Nov. 11, 1990	San Francisco-Dallas	ESPN	12.0/17.3
10.	June 6, 1994	OJ Simpson Chase	CNN	11.9/17.9

Top 15 Highest Rated Entertainment Programs
(excludes breaking news and sports)

Rank	Date	Program	Network	Rating/Share
1.	Nov. 9, 1993	NAFTA Debate	CNN	18.1/25.1
2.	March 22, 1987	National Geo/Titanic	TBS	11.4/N/A
3.	Sept. 6 1990	MTV Video Music Awards	MTV	8.5/15.1
4.	Jan. 31, 1990	China Lake Murders	USA	8.4/13.5
5.	Jan. 18, 1987	Rio Lobo	TBS	8.3/N/A
6.	Jan. 24, 1988	WWF Royal Rumble	USA	8.2/N/A
7.	March 15, 1998	MobyDick, part I	USA	8.1/11.8
8.	March 16, 1998	Moby Dick, part II	USA	8.1/11.6
9.	Nov. 9, 1993	NAFTA (post-show)	CNN	8.1/11.9
10.	Dec. 13, 1989	Gone With the Wind	TBS	8.0/13.6
11.	Sept. 4, 1995	Almost Golden	Lifetime	7.9/12.5
12.	March 6, 1991	Diamonds Are Forever	TBS	7.9/12.4
13.	Oct. 15, 1992	Presidential Debate	CNN	7.7/10.7
14.	April 23, 1995	Day of Mourning (Okla. City)	CNN	7.6/15.8
15.	Sept. 4, 1995	Intimate Portrait: Jessica Savitch	Lifetime	7.6/12.5

Source: Turner System Research from Nielsen Media Research

come from new services such as digital tiers, Internet access and telephony.

Like Hostetter, Marcus' new owner promised that Jeff Marcus and his team would continue to run the company from its headquarters in Dallas, where Marcus was as deeply entrenched as Hostetter had been in Boston.

But like Hostetter, Marcus saw the promise soon ignored. After the Charter deal, Allen announced that the combined companies would be run from the Charter headquarters in St. Louis.

Another long-time cable leader who exited the business in 1998 was Glenn Jones. His departure was less than graceful, albeit well-greased. In 1994 Jones had sold a third of his cable company, with an option to buy more, to Bell Canada. The agreement came on the heels of the Bell Atlantic deal to buy TCI and the Southwestern Bell offer to purchase Cox. Of the three, the Jones deal was the only one that lasted.

But relations between the two companies quickly went sour. In 1997, Bell Canada sued Jones, charging that he had inked sweetheart deals between Jones Intercable (in which Bell Canada had invested) and companies Glenn Jones owned entirely. In particular, the Canadians charged that Jones had awarded the Jones Intercable contract for Internet access to the Jones Internet Channel, 100% controlled by Glenn Jones, without seeking bids from other Internet access companies such as @Home or Road Runner. (Jones named nearly all his subsidiary companies after himself.)

In May, 1998, a U.S. District court ruled in favor of Bell Canada. The telco said it hoped to resume its relations with Jones. But within a month it had sold its shares in the company, together with its option to buy a controlling interest, to Comcast. Comcast, in turn, elected immediately to trigger its options to buy a controlling interest, buying out Jones' super voting shares for $69 a share.

Public shareholders, whose stock rose only to $28 a share after the deal was announced, were outraged. "It's extremely disappointing that in negotiating his own exit from the company (Glenn Jones) didn't insist that public shareholders be taken out as well," Gordon Crawford, executive vice president of Capital Research and Management, a Jones shareholder, told *Broadcasting & Cable* magazine. "He has basically left his public shareholders once again to dangle."

But the difference didn't bother Jones who had started in business in 1964 by convincing consumers in Georgetown, Colo., that the lousy reception they were getting was the fault of their TV sets, not his cable system.

As *Cable World* reported "Asked how he's justifying a deal whereby he will get a big premium and his common shareholders get nothing, (Jones) said 'Hey, it's America.'"

Later Jones said that the deal had been in the best interests of the shareholders, noting that the price of the common stock, which had traded at about $20 a share prior to the deal, had risen to more than $33 a share by the end of 1998.

"While some may have been miffed that they had to wait for their paydays," Jones said, "I know the way the transaction was designed was the right thing to do for our shareholders, customers and associates. When you are a CEO of a company and you have to consider the interests of a diverse group of people when you make decisions, it's an unfortunate fact of life that sometimes no matter what you do, some are not satisfied."

Cable In The Classroom: Cable Gives To Education

Along with C-SPAN, the most important public service from the cable industry is Cable in the Classroom, a joint effort of programmers and operators to wire the nation's schools and deliver a package of commercial-free video programming that can be used by teachers.

The effort came as the industry's response to a venture started in 1989 by Christopher Whittle. Whittle's plan was to create a 12-minute daily news program with commercials that would be sent free to schools around the country and used by current events teachers. Whittle planned to use the ad revenue to pay for the satellite dishes, monitors and other equipment schools would need to receive the transmissions.

Ted Turner, whose CNN had already developed its own news service for schools — Week in Review — saw the Whittle effort as a threat to CNN. The Whittle plan also raised objections from educators who opposed the introduction of commercials into classrooms.

When he heard of Whittle's plan, Turner called his Washington representative, Bert Carp, and told him that the cable industry would have to respond. Carp in turn called Robert Sachs, at Continental Cablevision, and Bob Thomson of TCI, each the head of his company's governmental relations office.

Turner, Carp, Sachs, Thomson, Continental chairman Amos Hostetter and TCI executive vice president J.C. Sparkman – with staff support from the NCTA vice president of public affairs Louise Rauscher – announced Cable in the Classroom in April 1989. Hostetter later recalled that the decision to have no commercials on Cable in the Classroom programming was made by Turner in the elevator on the way to the press conference announcing the project.

As its first executive director CIC hired Bobbi Kamil, an educator with a Ph.D. in instructional technology from Syracuse University who had worked on the Annenberg/CPB project to encourage the use of college level courses for credit via TV. In 1997 she was succeeded by Megan Hookey, a former Storer Cable official who had been Kamil's deputy.

In 1989 Cable in the Classroom began with some 6,165 schools. By the middle of 1998 cable operators had wired some 78,000 of the nation's 100,000 elementary and secondary schools (some were hooked up via the Primestar DBS service). Each day those schools receive via satellite a package of programming from 28 cable networks such as Discovery Channel, C-SPAN, CNN and A&E, that can be used by educators to help teach such diverse subjects as civics, current events, ecology and drama.

GOOD WORD:
The cable industry, under the guidance of Bobbi Kamil, developed the Cable in the Classroom initiative that provided noncommercial content to school teachers across the country.

The exits of Jones, Hostetter, and Marcus were dwarfed by the announcement in June, 1998, that AT&T would purchase TCI for $48 billion or roughly 14 times cash flow. The deal came within a few months of the 50th anniversary of the construction of the first cable system in the U.S. It also marked a full turn of the circle for TCI chairman Malone. The TCI chief had begun his career working for Bell Labs where he penned a long treatise on how AT&T could operate in an unregulated environment. The effort, he later recalled, earned him a bum's rush from the AT&T boardroom by the company chairman who told him on the way out that if he was able to make a single change in Ma Bell during his entire career he would be lucky.

The deal was the apex of an astounding rebound by TCI and the entire cable industry from the cellars they had hit following the 1992 Act and the FCC regulations that followed.

The deal was a classic Malone operation, enormously complex but ingenious as well. It was, of course, a stock deal so that Malone and the other TCI shareholders would not have to pay taxes on what they got from AT&T (in Malone's case the value was around $4.5 billion. He would also be the new AT&T's biggest shareholder.)

The agreement also provided that AT&T, after the merger, would be split into three divisions: one for business services, another (with Hindery as president/COO) to handle the long distance, wireless, Internet and cable holdings of the company. The third, which would have Malone as chairman and Liberty Media president Robert Bennett as president, would supervise all TCI's programming holdings, including its shares in Time Warner, USA Network, Discovery Channels, Encore and Fox Sports.

The Malone unit, to be called AT&T Liberty/Ventures, would trade as a separate "tracking" stock from the other units but would have the umbrella of the parent company's credit rating. The deal called for Liberty-owned cable networks to have a first crack at channels on AT&T/TCI cable systems.

The deal had something for everybody. AT&T got control of TCI's cable systems, allowing it to bypass local telephone companies in offering long distance service. It also would be able to compete with local phone companies in offering local telephone and Internet access services.

The deal was sweet for TCI shareholders, who would be given AT&T stock worth $51 for each share of their TCI stock, which had traded below $12 a share only two years before and $33 a share just a week before the deal was announced.

Malone himself got a huge payoff, including a premium for his Class B

Cablevision Dynasty:
Cablevision Systems Corp. chairman Chuck Dolan refused buyout offers, preferring to hand the business over to the next generation, son James Dolan, president and CEO.

supervoting TCI shares. As chairman of AT&T/ Liberty, he would have the chance to run a company that seemed small and entrepreneurial, but enormously well-funded, compared to the huge, highly leveraged cable company he had headed. (Malone once said that running TCI was like steering a giant oil supertanker: After the captain had turned the wheel, it seemed to take forever for the ship to respond and alter course.)

Recalling the collapse of the merger with Bell Atlantic, Malone included in the purchase agreement a clause that required AT&T to pay TCI almost $2 billion should the deal not be completed.

The TCI sale to AT&T, together with the exits of such figures as Alan Gerry, Glenn Jones, Amos Hostetter and Jeff Marcus from the operating side of cable, spelled the end of an era in the industry.

For its first 50 years, the cable industry had been like a small American town. It had its pioneers, who had first settled the community and remained active for decades. It had its chamber-of-commerce types and civic boosters. It had its conservative bankers and its wild west car dealers. The residents of the community competed with each other, for financing and franchises in the case of operators, for viewers and carriage in the case of programmers and for sales and new technology in the case

Looking Back And Looking Ahead

The National Cable Television Center and Museum grew out of a desire of some of the industry's pioneers — particularly George Barco and Ben Conroy — to provide a place to preserve the industry's history. Initially the organization was based at Pennsylvania State University under the direction of Penn State professor Marlowe Froke.

Landmark: *Under the guidance of Marlowe Froke, the Cable Center raised more than $50 million to build a permanent cable museum and education center in Denver.*

In 1996 the Cable Center board, under the leadership of Bill Bresnan, voted to move the organization to Denver where Denver University, headed by former Group W Cable CEO Dan Ritchie, had offered to make room for it.

Froke was named president of the Center and began a drive to raise enough money to build a permanent site for the Center and to fund a variety of programs.

By the middle of 1998 the Center had raised almost $50 million, including a gift of $10 million from Alan Gerry, to build a permanent home on the DU campus and launch a series of programs. These included projects to compile oral histories of the industry's founders, to develop a series of seminars and lectures, to create advisory and reference services, to launch a "demonstration academy" to educate the public officials and other community leaders about cable.

The Center was scheduled to open in the fall of the year 2000.

of manufacturers. But when they were attacked by somebody from outside the community, they banded together to defend their common home.

Now the big chain stores had come to town offering huge incentives for the original founders of the local businesses to sell out. Some would decline the offers, preferring to hand the businesses they had built on to the next generation (Brian Roberts, James Dolan and the Rigas sons for example). But many would see the chance to sell out as a great opportunity to watch their businesses become part of a much larger effort.

The open question was whether the new owners – most of them huge, multibillion dollar, publicly held companies – would be able to expand the home-grown, closely-held businesses while at the same time retaining at least some of the entrepreneurial spirit responsible for the industry's dramatic rise over the previous half century.

Whatever the future held in terms of the business world, it was clear as the 20th century drew to a close that the cable industry had changed the world of communications in a fundamental way, constructing an enormously powerful network of fiber and coaxial cable capable of offering some 60 million U.S. homes a huge array of video services, advanced voice communications and Internet access that would shape the world in the 21st century and well beyond.

Cable's New Revenue Streams

Cable Revenue by Segment
(figures in billions)

Catagory	1998	2008
Basic	21.3	35.4
Pay*	5.7	9.5
Adv. Analog/digital	0.44	7.4
Data	.13	4.9
Phone	.34	16.4
Other**	5.7	13.4

*Includes PPV, NVOD, VOD, gaming.
**Includes local advertising, home shopping, rental fees, installs

Cable Digital Revenue
(figures in billions)

Catagory	1998	2008	CAGR
Basic	21.5	35.4	5.1%
Other*	11.4	22.8	7.2%
Data	0.1	4.9	44.6%
Advanced Analog/digital	0.4	7.2	32.8%
Cablephone	0.3	16.4	47.2%

CAGR-Compound annual growth rate
*Includes pay/minipay, PPV/NVOD/VOD, gaming, local advertising, home shopping, equipment and install fees

Cable Telephony Revenue
(figures in billions)

Catagory	1998	2003	2008
Residential	.04	3.0	12.7
Bussiness	.31	1.6	3.7
Other Cable	33.4	51.9	70.0
Total	**$33.8**	**$56.4**	**$86.4**

Source: Paul Kagan Associates

Cable Digital/Data Subscribers
(figures in millions)

Catagory	1998	2003	2008
Total Subs	65.9	70.2	72.8
Digital Subs	1.1	23.2	40.7
Rev/sub/month	$11.01	$12.17	$13.44
Data subs	0.5	8.9	15.2
Rev/sub/month	$42.00	$32.00	$27.00

Final Thoughts

"In the beginning was the word." St. John, the apostle and writer of the New Testament, understood the power of communications. Had he lived 2000 years later he might have noted that nobody ever went broke overestimating the desire of humans to communicate.

That fact is most clearly demonstrated by the history of communications devices. When a new method of communicating arrived, the widespread assumption was that it would kill off the older ones just as tractors eliminated oxen in farming and indoor plumbing eliminated the outhouse. But radio did not kill newspapers. Television, in turn, did not kill radio. Cable didn't put the broadcasters under and DBS didn't kill cable.

At the end of the 20th century even the most ancient forms of communication, live theater and handwritten letters, were still very much alive, coexisting comfortably with the Internet and satellite television.

Humans had an almost insatiable appetite for communications. That hunger drove the growth of cable television in the last half of the 20th century. But the industry would not have grown as fast as it did, and might not have survived at all, had it not been for an extraordinary confluence of events and individuals.

The first, and most important, was the system of entrepreneurial capitalism practiced in the United States. That system made it possible for an individual, even with little or no resources, to transform a dream into reality.

The cable industry was not built by giant corporations. It was built by private individuals and small businessmen – Bill Daniels, John Rigas, Amos Hostetter, Alan Gerry, Bob Magness, Glenn Jones and others. In fact, when the big companies did get into the business, as often as not they stubbed their toes. American Express, CBS and Westinghouse all found cable too difficult a business and exited.

Part of the reason was that cable was a different animal financially from the traditional American business. Cable made no profits. To find the financial resources it needed to build a huge new communications infrastructure, cable first had to persuade bankers, stock analysts and others in the financial establishment to abandon a decades-long belief that profit was the best measure of a company's health.

They had to be educated that profits were not the only, or even necessarily the best, measure of a company's strength. That task of education was largely shouldered by John Malone.

Reporting profits, according the Malone doctrine, was a mistake. Profits in America were taxed twice, once when the company made them and

Thomas P. Southwick

again when the individual investor received them in the form of dividends. Malone argued that it was better to plow profits back into the company, generating future growth. The best measure of success, he said, was the growth of the company's cash flow, the difference between revenue and expenses before deducting such items as depreciation. That cash flow, in turn, could be used to borrow more money to grow the company even more, all the while avoiding posting taxable profits and avoiding having to share even a nickel with Uncle Sam.

To pursue such a strategy in which profits were reinvested rather than distributed, it was necessary for cable companies to remain private, or at least closely held. Most cable companies that went public – TCI, Jones and Comcast, for example, — created supervoting classes of stock held by a small group or an individual. That way they could continue to reinvest their cash flow in more acquisitions and construction without having to worry that the lack of earnings would hurt the stock price and would open the door for a hostile takeover by an outside party.

Malone and other innovative financiers such as Julian Brodsky made use of a wide array of financial instruments – from Eurobonds to industrial bonds to limited partnerships – to raise the money needed to build cable systems across the country and overseas.

In addition to the doubts on Wall Street, cable operators had to overcome the adverse rulings by the federal government, which at several junctures seemed bent on crushing the industry. But government proved to be an extraordinarily inefficient way to regulate an industry that was growing as fast and changing as quickly as cable television. It was, after all, a government designed by people who wore powdered wigs attempting to regulate an industry that was building the information pipeline for the 21st century.

It was hardly surprising then when an action by the government produced a result that had been unintended or unforeseen or was exactly opposite of what the regulators had sought.

Sometimes government regulations had the unintended consequence of helping cable, as with the freeze on new broadcast stations in the late 1940s which helped cable get started.

At other times the government unintentionally clobbered the industry, as with the highly leveraged transaction rules which were intended to shore up the banking industry but at the same time nearly drove cable under.

Even when the government wanted to achieve a specific goal with respect to cable, its regulations sometimes produced results exactly opposite of what had been intended.

Such was the case in the late 1960s when the Federal Communications Commission set out to promote diversity and universality of television service. To do this, it reasoned, it had to protect broadcast television stations from competition. To do this, it had to limit the growth of cable. But it soon became clear that cable and not broadcast would be the medium that offered the most programming diversity.

The reason was the fundamental economics of the two businesses. Each broadcast station had only one source of revenue: advertising. That forced broadcasters to air shows with the widest possible appeal to the lowest-common-denominator viewer.

Cable, on the other hand, had two sources of revenue: advertising and subscription fees. That allowed cable networks to program to niche audiences. The cable audiences were smaller, but far more loyal and devoted than the broadcast audiences. Cable subscribers were willing to pay directly for the programming. Cable also gave advertisers a way to reach specific segments of the audience most likely to buy their products.

The most dramatic example of government's ineptitude with regard to regulation of cable came in 1992 when Congress passed a bill designed to hold down cable rates. But the rules adopted by the FCC to implement the law had exactly the opposite effect for millions of subscribers, raising their rates.

In the end, the regulators did prevail. But they did so only with the most heavy-handed tactics. Rate rollbacks drove several cable companies into bankruptcy and ushered in a recession for all the others. It forced many cable companies to cut back on expansion plans, to lay off workers and to swap programming with high affiliate fees for networks such as home shopping channels that paid for carriage. The regulations cost millions of dollars and millions of people-hours to implement and decipher. All this to save consumers a dollar or two per month on their basic cable bill.

Rates were always a bone of contention between cable operators and the government. The government itself actually was responsible for a good portion of the monthly cable bills. City-mandated requirements for such things as government access channels, local access studios, franchise fees, and other items not directly related to basic cable service accounted for as much as one-fifth of the cost of cable service. And the HLT rules adopted in 1990 forced many cable systems to increase rates in order to maintain the debt-to-equity ratios mandated by the new federal banking rules.

Cable operators were at times capable of the most crass kind of greed. Many of the rate increases that so incensed Congress and the public in the late 1980s came because a speculator had purchased a cable system for a huge amount and then needed to pay back the loans that had financed the deal. And even the big MSOs, in business for the long term, proved extraordinarily inept in executing even modest rate hikes.

Despite the undeniable and seemingly widespread instances of unjustified rate hikes, the fact remained that the biggest cause of rate increases in the 1980s and beyond was the increased cost of programming.

Yet few cable subscribers and few lawmakers understood that cable systems actually paid programmers and that increased rates were needed to purchase better programming.

The cable industry was never able to communicate to its customers this basic element of the business. People understood that the price of gasoline at the pump was related to the price of oil at the well. They knew that the price of a T-bone steak at the restaurant was related to the price of corn off the farm. But they did not understand the link between basic cable rates and programming.

This was not the only thing about cable that consumers did not understand. Fifteen years after the launch of C-SPAN, for example, millions of cable customers were unaware that the network was supported by the cable industry as a public service. Many consumers, and a fair number of public officials, continued to believe C-SPAN was funded by the government.

Cable operators, who built the biggest and most powerful communications network in the world, were largely unable to use it themselves to explain to their own customers how the industry worked and what it was doing to make the world a better place.

Despite its inability to project a positive public image for itself, the cable industry managed to escape every time it seemed to be trapped by either government or by competitors. Each time the rescuer was new technology.

In 1974, when limits on distant signal carriage and other regulations had nearly halted the growth of cable, satellite-distributed programming came along to give the business new life.

Twenty years later, when rate regulation had delivered a hammer blow, digital transmission and fiberoptics again launched a series of new businesses for cable that the government had not even imagined when it passed its legislation.

But each new technological advance threatened, or appeared to threaten, the existing business of some powerful other sector of the media business. Theater owners, movie studios, telephone companies, video rental stores, television manufacturers and others at various times felt threatened by cable and attempted to get the government to intervene on their behalf.

The most brazen attempt came in the early 1960s when theater owners in California persuaded voters to pass a ballot measure banning pay television in the state.

But the industry most threatened by cable was broadcasting. Initially the broadcasters were content to watch cable grow, thinking of it as a means to bring signals to those areas outside the reach of broadcast stations. But as cable and broadcast expanded, they found themselves battling for viewers in the same markets.

It was the broadcasters who demanded the regulations that strangled cable in the late 1960s. And it was the broadcasters who persuaded Congress to resurrect must-carry in the 1992 Act after the Supreme Court had killed it.

Must-carry is an important issue both because of the way it distorts the marketplace and because of the damage it does to a cornerstone of the U.S. Constitution.

By decreeing that broadcasters had an inherent right to be carried on cable, Congress gave that industry a status above all other media. Broadcasters could not only demand to be carried, they could also demand that cable systems pay them for carriage. The impact of that was not to strengthen local broadcasting, as Congress intended. After all, Congress never came to the point of forcing people to watch the broadcast stations. But must-carry and retransmission consent did give the big broadcast networks and group station owners a leg up on creation of new cable programming networks.

Services such as MSNBC, Fox News, ESPN 2, and CBS Eye on People might well have made it on their own. But they all had a head start because Congress in effect forced cable operators to carry them just because they were owned by broadcasters.

The First Amendment to the Constitution states that "Congress shall pass no law infringing on freedom of the press." It does not state "no law except where necessary to protect local broadcasters." It says no law, period.

Must-carry and retransmission consent were the most egregious infringement on the First Amendment in the history of the nation outside of wartime censorship. One can imagine the reaction of newspapers if Congress attempted to dictate what they should carry on their front pages or gave a wire service, say, the right to automatic space in every newspaper under the theory that the wire service was in the public interest. Yet that is exactly what must-carry did in the world of electronic publishing: dictated what cable should carry on its lowest tier of service and gave broadcasters and their sister cable networks the right to carriage on cable before other networks.

The success of the broadcasters in persuading Congress to reinstate must-carry was a tribute to that industry's ability to position itself as somehow more important than any other in serving the public needs.

Yet the cable industry itself was all to willing to allow the erosion of its First Amendment rights.

Part of the reason was that most cable operators were businessmen, with no roots in the world of journalism that had spawned many broadcasters.

Broadcasters understood far better than cable operators how to influence the political process. They made sure each member of Congress knew the station managers in his district. And members of Congress understood that these managers could transmit the images of politicians to millions of voters.

Cable was even better positioned to achieve local political power than the broadcasters. Cable had grown out of a political process: franchising. And cable was even more local than broadcasting. Thousands of communities that had no local broadcast station had a local cable system.

Yet cable systems were slow to add local news shows to their channel lineups. Many viewed local news as a little-watched revenue loser. Broadcasters, on the other hand, were able to make their local news shows sources of both profit and political clout.

Systems were also reluctant to get involved in local political frays for fear of upsetting their franchising authorities.

Cable was also largely silent when the First Amendment rights of other media were challenged. The industry did not get involved when the broadcasters fought the fairness doctrine and equal-time provisions. And while cable did defend the video version of *Playboy*, it was largely silent when the print version came under attack.

So it is not surprising that cable received little sympathy from other sectors of the media when cable's First Amendment rights were challenged.

But the biggest reason cable was unable to assert its status as an electronic publisher, with all the First Amendment rights of a newspaper, was that it was willing to sell those rights in return for a better business deal.

Rights that are not defended are soon lost. Cable operators, even when they were backed by the courts, were willing to make deals with broadcasters and the Congress on must-carry in return for concessions in other areas. They always appeared more concerned with defending their bottom line than their Constitutional rights.

Despite its own shortcomings and

despite the opposition of powerful competitors and the government, cable television in just 50 years was able to revolutionize communications. And along the way, cable television changed society, largely for the better.

In the 1950s most viewers had a choice of, at most, three television stations. In effect, the decision about what people could watch on TV was made by a group of three network executives in offices within a stone's throw of each other in New York City. Their decisions were based on the same set of numbers, brought to them each morning by the Nielsen ratings company.

By the end of the century Americans were accustomed to 50 or more channels of programming, with dozens more in the pipeline or on the drawing boards. The decision about what to watch each night was made by the viewers, not by the network executives.

Americans don't always choose to watch the most uplifting programming. But at least they were deciding. And among the choices cable offered them, and increasingly among the shows they watched, was programming that clearly would make the society a better place. And these shows – for children or minorities, for women or opera fans, for public affairs junkies or classic movie buffs – would never have been aired in the world of just three commercial broadcast networks.

The most widely cited cable network in such discussions is C-SPAN. But while C-SPAN is an extraordinary story, even more remarkable in some respects are those networks that managed to serve the public and generate profits at the same time. Nickelodeon, BET, Discovery, A&E, CNN, among others were all started by entrepreneurs with little or no background in television and without the backing of huge, well-heeled corporations. They all managed to prosper in the world of hard-headed businessmen and women closely attuned to the bottom line.

Their existence offers proof that in business it is possible to achieve both social good and financial success.

The impact cable programming has had on society is beyond calculation. Cable brought the Joffrey Ballet to rural Montana, the debates of the British parliament to the suburbs of Chicago, the O.J. Simpson trial — live and unedited — to every living room in America, the Gulf War to every part of the world, and fourth grade lessons to a little girl in Strongsville, Ohio, too sick to go to school.

One example of the power of cable programming came during the NCTA convention in Atlanta in 1990. It was just after the fall of the Berlin Wall and the collapse of the communist regimes in Eastern Europe.

The final panel session at the convention included Ted Turner and was moderated by NBC TV correspondent Ken Bode.

Bode opened the session by recounting how he had just returned from Romania where he had covered the first free elections in that country's history. In a small rural town he interviewed people as they waited in line to vote.

He asked one elderly woman how Romanians were able to overthrow a brutal dictator. She told him she had watched what happened in East Germany on television and figured that if the East Germans could topple their regime, so could the Romanians.

Bode asked her how she had learned of what had happened in East Germany

357

when the state-controlled television station had blacked out such news.

"Oh," she answered. "I watched it on CNN."

Bode told her he would soon be in Atlanta and would be meeting with the man who had founded CNN, Ted Turner. "Please tell Mr. Turner thank you," she told Bode. "Without him the revolution in Romania would never have happened."

Bode invited Turner to the podium.

The founder of CNN recalled that when CNN had started he didn't know if it would succeed, and that there were days when it seemed bankruptcy was just around the next corner.

"But I knew that with the support of cable operators we could make it," he told the crowd, "and when we did CNN would shine a light into the darkest corners of the world. And in that light, no dictatorship, no tyranny can survive."